Numerical Radiative Transfer

Numerical Radiative Transfer

Edited by

WOLFGANG KALKOFEN
Center for Astrophysics
Harvard University

CAMBRIDGE UNIVERSITY PRESS

Cambridge
New York New Rochelle
Melbourne Sydney

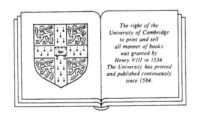

CAMBRIDGE UNIVERSITY PRESS
Cambridge, New York, Melbourne, Madrid, Cape Town, Singapore, São Paulo, Delhi

Cambridge University Press
The Edinburgh Building, Cambridge CB2 8RU, UK

Published in the United States of America by Cambridge University Press, New York

www.cambridge.org
Information on this title: www.cambridge.org/9780521115179

First published 1987
This digitally printed version 2009

A catalogue record for this publication is available from the British Library

ISBN 978-0-521-34100-4 hardback
ISBN 978-0-521-11517-9 paperback

CONTENTS

PREFACE

This volume treats the numerical solution of radiative transfer problems in optically thick, radiating atmospheres. The volume addresses two topics: the solution of line and continuum transfer problems by fast and efficient methods, and the transfer of polarized radiation.

In 1973 Cannon proposed a novel method for solving radiative transfer equations subject to constraints such as radiative or statistical equilibrium. His method was based on the perturbation of the exact equations with an approximate but numerically simpler and more convenient operator. Even linear equations that could be solved directly were instead solved iteratively, but with the potential of a significant advantage in computation time. It was not appreciated then that this was the beginning of a new era in computational radiative transfer. Powerful general methods for solving transfer problems had been developed during the 1960's and 70's. The most widely employed among them was Auer and Mihalas' complete linearization method for the solution of non-linear equations. But cost in computer time and memory limited its use, particularly in time-dependent applications. It seemed only a matter of time, however, until a new generation of computers would permit an attack on problems still beyond reach. Cannon's approach revealed a new vista, but it took nearly a decade to realize its promise. This was accomplished by Scharmer who formulated a line transfer problem in terms of integral equations, writing the exact equation as a perturbation series with an approximate integral operator. His equations were efficient, saving time both in the construction of the operators and in the solution of the resulting system of equations. And the solution was obtained in much less time than with conventional direct differential or integral methods.

Cannon's and Scharmer's method opened a new approach and, what may be even more important, a new way of looking at radiative transfer. It has spawned a host of methods that are built on the same general principles. This development is still in progress, as is well documented by the present book which contains several new numerical methods, on operator perturbation as well as on polarized radiative transfer, that are described here for the first time. This volume will achieve its purpose if it stimulates further research on these topics.

The book is conceived as a manual of modern numerical methods for solving radiative transfer problems. Both sections, on operator perturbation and on polarized radiation, open with introductory articles, and the introduction to the entire book give synopses and cross references for all articles. While their focus is on astrophysical plasmas, the methods are easily adapted to applications involving other media where self-absorption of radiation is important. Stratification in plane-parallel layers is generally assumed but most methods can be extended to other geometries. Although the book is intended primarily for graduate students and workers in the field of radiative transfer, the level of presentation is designed to make it accessible also to advanced undergraduates.

A book such as this needs the diligent work and willingness of all its authors to conform to an overall plan. I take this opportunity to express my gratitude to my collaborators for their efforts. In particular I extend my warmest thanks to Bengt Gustafsson, president of the Commission of the International Astronomical Union on Stellar Atmospheres at the time of its meeting in 1985 in New Delhi, India, for having invited me to organize a session on radiative transfer, where reports on some of the research included here were first given.

Cambridge, Massachusetts W.K.
July 1987

Wolfgang Kalkofen
Harvard-Smithsonian Center for Astrophysics, Cambridge, USA

ABSTRACT: The papers in the two sections of this book address two topics: the efficient solution of radiative transfer problems by means of operator perturbation and related numerical methods, and the transfer equation for polarized radiation. This introduction provides synopses of the papers, assessing their specific importance and relevance within the broader context. The first section begins with a survey of numerical methods contained in this volume or forming their background; then the topics covered concern the use of diagonal operators, the acceleration of the convergence of the resulting equations, line transfer for a time-dependent two-level atom, the formulation of the transfer and statistical equilibrium equations for multi-level atoms in terms of equivalent two-level atoms, the construction of stellar atmospheres with non-LTE line blanketing, and the derivation of operator perturbation equations of low order from high-order equations. The second section, on polarized radiative transfer, also has its own introduction; the topics concern the use of the Feautrier equation, the derivation and solution of the transfer equation using real matrices, or complex matrices. The last two papers describe a discrete space technique and a generalization of the formal integral of the transfer equation to the case of an absorption *matrix* — methods that can be used also for line transfer with *partial redistribution*; apart from these two papers and a critical analysis in a paper in the first section of operator perturbation methods for partial redistribution, only *complete redistribution* is treated.

1. OPERATOR PERTURBATION

The task considered in the first section of this book is the solution of the equation of radiative transfer subject to constraints such as statistical equilibrium or radiative equilibrium. The mathematical problem can be stated in the form of a system of coupled differential equations in which the unknown functions are either the source function, or the intensity at all frequencies, angles, and depths, or, as in the construction of model atmospheres, the fundamental variables of the medium, such as temperature, pressure, and particle densities. Three numerical difficulties characterize typical transfer problems: The equations are non-linear, the system of equations has large order, and important characteristic length scales differ by many orders of magnitude — factors as large as 10^4 or 10^6 are not uncommon. The first section of this book deals with ways of overcoming two of these hurdles: it addresses the efficient solution of systems of stiff equations having very large order. Non-linearities may occur, but they are only incidental.

The aim of the numerical methods described here is to solve the equations fast, yet accurately. The tool used is operator perturbation. The procedure is to separate a problem into two parts, the approximate calculation of corrections to a provisional solution, and the accurate calculation of the error made by the provisional solution in a conservation equation. The error becomes the driving term in the next cycle of correction calculations. Thus, this approach uses iterations to arrive at an accurate solution, even when the equations are linear.

In the error calculation, individual (*i.e.*, uncoupled) transfer equations are solved for known source function. This calculation treats only the short-range interactions of the radiation field, where the region of influence of the gas extends over a distance of the order of the mean-free-path of the respective photons. The long-range influence of scattering and the diffusion of photons in frequency must be contained in the correction calculation, where an approximate operator describes the radiative transfer. This places great emphasis on the construction of the approximate operator to ensure that it model the overall structure of the solution. It is therefore interesting to inquire into the design principles of operator perturbation methods with a view of learning how to construct approximate operators and how to modify or improve existing methods. This is the task considered in **Chapter 1** where I survey the basic ideas on which operator perturbation and related methods are built.

In **Chapter 2** Hamann describes a solution method for multi-level line transfer problems in extended atmospheres for the Sobolev limit of high differential velocity and applies it to a Wolf-Rayet atmosphere. To construct the approximate transfer operator he assumes that the photo-absorption rate is equal to the photo-emission rate except for photons that can escape in a single free flight. Thus the local intensity is equal to the local source function multiplied by a factor describing the fractional contribution by the line core to the absorption. The equations amount to a restatement of Rybicki's (1972) core saturation method; in the limiting case of a two-level atom without induced emission they are identical with the core saturation equations (*cf.* Rybicki 1985). The essential contribution of Hamann (and of Werner & Husfeld, 1985; *cf.* Chapter 3) is to have embedded the equations in the operator perturbation formalism. Since the resulting approximate transfer operator is diagonal, the solution of a line transfer problem is similar to Λ-iteration; it would be identical to Λ-iteration were it not for the leakage of photons in the wings of a line. This is the decisive difference. Hamann's equations converge whereas Λ-iteration fails, except in trivial cases. Of course, convergence is not spectacular. It is, in fact, no better than that of the core saturation method. But that does not seriously limit the usefulness of this method whose real strength is rooted in the fact that it separates the transfer equations from the equations of statistical equilibrium, which are now entirely local. The iterative solution process therefore alternates between the scalar transfer equations for the individual lines and the equations of statistical equilibrium. The method can be viewed as an accelerated Λ-iteration in which the amplification factor varies from being approximately equal to the inverse of the wing fraction of the absorption profile near the surface of the atmosphere, to the inverse of the scattering parameter ϵ at large depths. For strong lines this amplification can have enormous values far from the surface: a factor of several million is reached in one example.

The separation of the problem into the solution of the transfer equation and the equations of statistical equilibrium has two major advantages: It makes the order of the matrices occurring in this formulation for a given problem very small, thus allowing more ambitious problems to be attacked; and it makes programming the equations very simple. The drawback of the method is that it requires a relatively large number of iterations to solve a problem (*cf.* Chapter 4 for an acceleration method applied to these or similar equations). For small atomic systems, where the computer time is dominated by the solution of the transfer equations in the error calculation, this method is therefore not competitive with other operator perturbation methods.

Among interesting numerical questions discussed are: the calculation of the boundary between line core and line wings for a moving medium in the comoving frame formulation; the free parameter that defines the boundary between line core and wings and its effect on convergence; the formal solution in the error calculation treats individual frequencies but the equations are coupled over angle (or impact parameter) by the Thomson scattering term, a complication that is handled by means of the variable Eddington factor technique.

The formulation of the statistical equilibrium equations makes use of net radiative bracket expressions. A full discussion of these is given in Chapter 6. See Chapters 3 and 4 for similar operator perturbation methods.

Werner's approach to radiative transfer in **Chapter 3** is similar to that described in Chapter 2, showing the same provenance from the group under Hunger in Kiel. The operator perturbation method is based on approximate Λ-operators that separate the transfer equation from the equations of statistical equilibrium, which then contain no explicit depth-coupling. In addition to the purely diagonal matrix used by Hamann, Werner also employs a triangular operator first suggested by Scharmer (1981, 1984); and for lines with an underlying continuum he investigates a different, but again diagonal, operator, with a larger probability for photon escape from the medium (*cf.* also Chapter 2); the free parameter defining this probability is chosen so as to improve the stability of the method.

In many instances the triangular operator is almost as convenient to use as the diagonal operator since at any given depth in the atmosphere the contribution of the lower layers to the radiation field is known; its speed advantage in the number of iterations over the diagonal operator can be as large as a factor of two, but more typically it is only slight. A possible disadvantage is that the equations of statistical equilibrium must be solved sequentially, and from the lower boundary in the outward direction, opposite from the direction in which the hydrostatic equilibrium equation is integrated. On parallel processors the purely diagonal operator is to be preferred since the calculations at the various depths are then independent of one another and can be carried out in parallel. This is also true in the construction of model atmospheres if the pressure is updated outside the Newton-Raphson iteration cycle.

Two types of problem are solved here with this method: multi-level line formation in atmospheres with given gross structure, and the construction of hydrogen line-blanketed model atmospheres in plane-parallel geometry, for effective temperatures between 30,000 and 100,000K. A problem is solved by the linearization of the conservation equations (*viz* radiative, hydrostatic, and statistical equilibrium; particle and charge conservation). The procedure is analogous to the complete linearization method of Auer & Mihalas (1969) except that the transfer operator is a diagonal or triangular matrix. In line transfer calculations for a given atmosphere the problem can often be simplified further by cancelling the core frequencies analytically; the statistical equilibrium equations resulting with the diagonal operator are then linear in the particle densities and no Newton-Raphson perturbation is necessary for their solution.

The operator perturbation method is more convenient than the complete linearization method because of the smaller size of the matrices and because of ease of programming, and it allows much larger atomic systems to be attacked (*cf.* however Chapter 6 for expressing large problems in terms of equivalent two-level atoms). The author estimates that up to 100 non-LTE levels can be included in a calculation — the number of lines follows from the number of levels — whereas the complete linearization method is restricted to approximately 20 levels, the limits being set by the size of the largest matrices occurring in the formulations. The demonstration problems in the paper are for a hydrogen atom with five bound levels, and both line and continuum transitions are treated. Typical iteration numbers are 10 to 30 in the line formation calculations for given atmospheric structure, and 15 to 40 in model atmosphere constructions, the higher iteration numbers required at the lower temperatures; a maximum of 85 iterations is needed with the diagonal operator for a 30,000K model atmosphere. Werner presents detailed documentation of the properties of the method, including the tuning of the free parameter defining the separation of the line into core and wings and the effect on the rate of convergence and the stability. His experience with the two approximate operators suggests that the transfer in the wings, as implied by the triangular operator for example, may not have to be treated in line transfer problems but is important in the construction of model atmospheres, where wing transfer has a major impact on the rate of convergence.

While this method requires too many iterations to be competitive in the solution of small atomic systems with more efficient operator perturbation methods, such as Scharmer's (1981, 1984) or the one described in Chapter 5, the ease of formulation and programming and the small size of the matrices occurring in it make it ideal for many-level atoms.

In **Chapter 4** Auer offers a remedy for the relatively slow convergence of operator perturbation methods based on diagonal operators, reducing by a factor of at least two the number of iterations required to solve a problem. Any linearly convergent iteration scheme can be accelerated by this approach.

Unlike the Newton-Raphson equations, arising in complete linearization methods for example, operator perturbation equations have only linear convergence. The convergence rate is rapid in spite of that when the characteristics of the approximate operator match those of the exact operator, *i.e.*, when the maximal eigenvalue of the matrix in the perturbation expansion of the solution vector (*cf.* Kalkofen 1984, eq. 2.15) is *small* compared to unity. But diagonal approximate operators typically give eigenvalues *near* unity (*cf.* Olson, Auer & Buchler 1986, figs. 1-5), resulting in slow convergence when the new solution is based exclusively on the preceding solution (or in divergence for eigenvalues exceeding unity). To speed up convergence, information from earlier solutions must be used as well. Auer shows how this can be accomplished. He constrains the new solution by requiring that it be as close as possible in the least squares sense to an estimate of the converged solution of the equations. This requirement yields the coefficients for the extrapolation.

The method does not change the nature of the iterations, which remains linear. But it does drastically reduce the number of iterations necessary to satisfy a convergence criterion. Auer graphically depicts the improvement in the performance of solution methods that rely on diagonal approximate operators. He varies the number of preceding iterates he uses in the extrapolation and compares weighted and unweighted accelerations for approximate diagonal operators that are either obtained from estimates of the diagonal elements of the exact operator or constructed with the core saturation method. The improvement in his test cases is large. In one instance the number of iterations drops from more than 100 to about 20; weighting of the corrections by the inverse of the source function saves a few iterations. It is worth noting that the speed-up of convergence is obtained at almost no expense.

An important parameter in a line transfer problem is the thermalization length, the maximal distance over which features in the gross structure of an atmosphere are communicated by the line radiation to the gas. In a semi-infinite atmosphere this is also the depth to which the boundary influences the gas, and where the source function of a two-level atom saturates to the Planck function. Other parameters important for the general character of the solution are frequencies at which the atmosphere becomes transparent at points closer to the surface than the thermalization depth. The respective frequencies are located between line center and the frequency for which the monochromatic photon mean free path is of the order of the thermalization length. Frequencies further out in the wings are unimportant for the overall structure of the solution since the amount of energy escaping is too low to have an effect. Thus the broad features of the solution of a line transfer problem are defined by a small set of parameters. In **Chapter 5** I describe an operator perturbation method that makes use of this fact. The method employs two angle-frequency sets, a coarse set, typically consisting of a single angle point and a few well-chosen frequency points, and a dense set with many angle-frequency points. The coarse set is used to construct the approximate operator of the problem and the dense set to construct the exact operator.

The basic formulation is that of the integral equation method of Scharmer (*cf.* Scharmer & Nordlund 1982, Scharmer 1984), in which the driving term of the equation for computing corrections to a provisional solution is the error made by that solution in the equations of statistical equilibrium. But the integral equations are used merely as a device to derive a perturbation series. The actual numerical solution involves no integral equations. Instead, both the exact and the approximate integral equations are expressed in terms of their differential equation equivalents. This procedure of deriving the differential equations via intermediary integral equations insures that the perturbation series is equivalent to the exact equation and does not contain any extraneous terms (*cf.* Chapter 8). Thus, only systems of Feautrier equations are solved. In the case of the correction calculation the system is coupled but has very low order, which is given by the number of angle-frequency points in the sparse set; and in the case of the error calculation the system has high order, but the equations are uncoupled. As generally in these operator perturbation methods, the accuracy of the solution is given by the accuracy of the error calculation, which is second-order for the ordinary Feautrier equation and fourth-order with Auer's (1976, 1984) Hermite form.

Time-dependent problems in a moving medium are solved efficiently by using the converged solution at any time step as the starting solution at the subsequent time step and by perturbing the exact operator about the approximate operator that corresponds to the static medium. As a consequence, the profile function for the approximate operator is symmetric, halving the already low order of the system of equations for the correction calculation; in addition, the matrices need to be constructed and inverted only at the beginning of the computation, leading to substantial savings in computer effort. The drawback of this procedure is that the method is suitable only for problems in which the macroscopic flow velocity does not exceed a few times the speed of sound, a restriction that is typical, however, of formulations of transfer problems in the observer frame (*cf.* Chapter 7 for a similarly restricted method, and Chapter 2 for the treatment of high differential velocities). This method assumes that a multi-level transfer problem is expressed as a sequence of equivalent two-level problems (*cf.* Chapter 6).

Two numerical examples are given in the paper. For line transfer in a model chromosphere with a velocity amplitude of three Doppler widths the exact problem is described with 4 angle points per hemisphere and 30 frequency points in the half profile, and the approximate operator is computed for a symmetric profile with a single angle point and 5 frequency points. This amounts to a reduction by a factor of about 50 of the order of the system of equations. Convergence is rapid. It is further accelerated by assuming that consecutive corrections at a given depth point define a geometric series, reducing the number of iterations required for the maximal error of the solution to drop below 1% to typically three or four, except for the first time step, where five or six iterations are necessary.

The typical line transfer problem in a stellar atmosphere consists of the equations of statistical equilibrium for an atomic model with several levels and the equations of radiative transfer for the corresponding lines and continua. One of the standard methods of solving these coupled equations is the complete linearization method of Auer & Mihalas (1969). This method accounts to first order for all the interactions of all the levels of the atom and at all optical depths. If the linearization is done fully consistently the resulting Newton-Raphson equations converge quadratically. This is a very satisfying property; but to achieve it one pays a high price in the order of the system of equations unless the number of lines is very small. A limit on the number of lines, and hence levels, that can be treated is soon reached. In the integral equation method the limit concerns primarily the number of levels rather than lines, but the result is the same. One way to increase the practical limit on the size of the atomic model has been described in Chapters 2 and 3, where the diagonal operator of the transfer equation permitted an exact solution of the coupling of all the interactions taking place within the model atom at any given optical depth, but at the expense of approximating the depth dependence of the transfer. Another way is to do just the opposite, to treat the depth dependence within each line exactly but to approximate the interactions among the various levels of the atom; since this approach treats the radiative transfer of a single line at any one time, the model is referred to as the *equivalent two-level atom.* In both cases, the numerical problem separates into one in which the order of the system of equations is given by the parameters of a single line, either the number of depth points or the number of angle-frequency points, and by the parameters of the model atom, such as the number of bound levels. How to set up the equations of the equivalent two-level atom is described in **Chapter 6** by Avrett & Loeser.

The paper addresses two main questions: how to write the equations of statistical equilibrium, whether in terms of individual photoexcitation and de-excitation rates or in terms of net radiative rates; and which level equations to single out of the equations of statistical equilibrium for deriving the source function of the equivalent two-level atom, whether the equations of only the upper and lower levels for the transition in question, or of all the levels.

When the statistical equations are written in terms of *single* upward and downward rates, the numerical results can be meaningless at large depths. This is especially true for the multiplet problem, discussed in the paper, in which the lower level is common and the upper levels are strongly coupled collisionally. On the other hand, when the equations are expressed in terms of *net* radiative rates, it can happen that the populations near the surface are negative, a result from which an iteration scheme might not recover. The best approach is to mix the two formulations, with the single and net rate expressions contributing each one half near the surface, and the net rates being used exclusively at large depths. All strong lines are treated the same way, the transition from one to the other formulation occurring in a particular layer in the atmosphere. The weak lines are frequently solved by a different procedure; since their source function is determined largely by the transfer in the strong lines, their radiation field can be obtained from a Λ-iteration. Avrett & Loeser describe both procedures for deriving the source function equation of an arbitrary model atom, *i.e.*, using either just the two combining levels, or all levels. No general prescription is given as to which procedure might be better, whether the former, which contains all the other unknown population ratios, or the latter, which contains all the other unknown radiation fields; in their numerical code the authors allow both procedures since they have found that different applications have different requirements. The two sets of source function equations are identical in the example given in the paper, that of a three-level atom.

Note that in spite of the symmetrical relation of the procedure described in this Chapter with the operator perturbation methods of Chapters 2 and 3 concerning the separation of the equations into two blocks, one for the individual lines and the other for the atomic interactions, the equivalent two-level atom does not constitute an operator perturbation method. It may be used in conjunction with any approach to the solution of non-LTE problems, including operator perturbation methods (*cf.* Chapter 5), for which it extends the scope by removing the limits based on matrix size, leaving only the practical limit from managing large numbers of interactions. In this respect it resembles the methods based on diagonal operators. Note also that in any iteration the source function interaction coefficients (*i.e.*, ϵ and $\mathcal{B}$) in the transfer equation of a particular line depend only on information from the preceding iteration; they do not (in this formulation) depend on the order in which the transfer equations are solved. Thus the transfer equations can be solved in parallel, an operation that could be carried out efficiently on parallel processors.

In **Chapter 7** Anderson describes a method for constructing line-blanketed model atmospheres in statistical equilibrium. In an ingenious way he uses the related, critical observations that the overall structure of an atmosphere depends mainly on the gross properties of the opacity and on the long-range communication between distant parts of the atmosphere. To exploit these features he groups the photons of the fine-grained frequency set needed for a full accounting of the opacity in a line-blanketed medium into coarse-grained frequency blocks in which all photons experience approximately the same interactions, *i.e.*, photons in a given frequency block have all nearly the same mean free path and probability of scattering into other layers. Thus, a block might consist of frequency points in the cores of the lowest resonance lines of some atom, or in the near or far wings, or of frequency points in subordinate lines of another atom. It follows that the frequencies in a block may be drawn from different, non-contiguous parts of the spectrum.

The transfer and conservation equations written for frequency blocks are treated in much the same way as the equations in the complete linearization method of Auer & Mihalas (1969). The main difference is that explicit reference to particle densities is eliminated. Therefore the order of the system of equations is independent of the number of energy levels in the equations of statistical equilibrium and is given essentially by the number of coarse-grained frequency blocks. In the example of this paper the number of block equations is on the order of 100, reduced from the 2000 equations at the fine-grained frequency points necessary for the treatment of line-blanketing. The Newton-Raphson equations resulting from the complete linearization of the equations would converge quadratically were it not for the interaction coefficients in the block transfer equations. These contain frequency form factors, *i.e.*, sums over the fine-grained frequency set, which play the same role as the variable Eddington factors in the treatment of the transfer equation in differential form. The form factors are updated between the iteration cycles for correcting the variables of the problem; these variables are the kinetic temperature, the hydrostatic pressure, and the block energy densities of the radiation field.

Anderson compares his method with Cannon's (1973a, b; 1984), citing corresponding equations in his and Cannon's paper. Yet his is not an operator perturbation method in the sense of Cannon's or Scharmer's (1981, 1984). The basic approach of Anderson's method is that of the variable Eddington factor technique. In either case the aim is to reduce the order of the system of equations, which Anderson achieves by using form factors that are updated outside the correction calculation. This reduction is possible here since the fundamental parameters of the problem are temperature and pressure, from which the source terms of both the fine and the coarse-grained intensities can be constructed (perhaps with the help of form factors and frequency averages). As a consequence Anderson's approach does not suffer from the difficulties of Cannon's, where the calculation of the intensity in the fine-grained frequency grid from the solution of the coarse-grained equations leads to inconsistent equations (*cf.* Chapter 8). Thus, if the number of frequency blocks is adequate and the assignment of frequency points to the blocks is well chosen, the block equations may be expected to converge to the solution demanded by the fine-grained equations.

The limits on the size of the problem that can be attacked by means of this method are not set by the order of the matrices occurring in the correction calculation, which is moderate, but by the time taken to evaluate form factors, partition functions, and their derivatives with respect to the block intensities. The work load thus is similar to that of other methods of constructing models of line-blanketed atmospheres.

In two important papers Cannon (1973a, 1973b) showed how radiative transfer problems may be solved efficiently by perturbing an exact operator about a simpler, approximate operator. One of these papers (Cannon 1973b), in which the exact integral equation of a line transfer problem with complete frequency redistribution is written as a perturbation series in integral equations, forms the basis of Scharmer's (1981, 1984) method and of several others that build on Scharmer's (*cf.* Chapters 2, 3, 4, 5). The other paper (Cannon 1973a) dealt with the perturbation solution of systems of differential equations and was intended for line transfer problems with partial frequency redistribution. The method of that paper is flawed. In **Chapter 8** I examine Cannon's approach and show that the perturbation series contains extra terms that are not present in the original equation; the extra terms, which arise from formal solutions of the transfer equation in which the intensity for the high-order system is generated from the source function for the low-order system, are found also in the simpler case of complete redistribution. Since Cannon's method for partial redistribution has been offered in recent publications (Cannon 1984, 1985; Auer 1986) it is important to point out the error in this perturbation approach. This Chapter shows how the inconsistency arises, taking as the example the simplest transfer problem that exhibits it, that of isotropic scattering of monochromatic radiation.

The task that was considered by Cannon was how to write a system of differential equations of high order as a perturbation series in systems of low order. Chapter 8 shows how this can be accomplished, but the solution given applies only to complete redistribution or to other, analogous problems in which the critical function to be determined depends only on depth and not also on angle or frequency. The thrust of the method by which this reduction in the order of the system of equations is obtained is similar to that of Chapter 5: the exact system of differential equations is expressed as an integral equation; this equation is perturbed according to Cannon (1973b) and Scharmer (Scharmer & Nordlund 1982, Scharmer 1984) and is written as a series of integral equations for an approximate integral operator that has a simpler representation as a differential operator than the exact integral operator does. The basic equations of the approach are therefore integral equations, but the equations solved numerically are systems of differential equations. The transformations between integral and differential equations are obtained from a formulation that separates the radiation field into diffuse internal and and transmitted external fields. The approach and the numerical properties of the method are shown in the example of a grey atmosphere in radiative equilibrium with prescribed net flux.

2. POLARIZED RADIATIVE TRANSFER

A pencil of radiation is characterized by three properties: intensity, frequency, and polarization. The second part of this volume is concerned with polarization. This part opens in **Chapter 9** with *A Gentle Introduction to Polarized Radiative Transfer*, in which Rees eases the transition from the customary scalar transfer equation for the intensity to the vector equation for the Stokes parameters, sparing the reader most of the complexities of the matrix algebra ordinarily burdening the subject. In the process, with careful attention to sign conventions and handedness, he defines the Stokes vector in a series of intensity measurements, conceived as a Gedankenexperiment, using a quarter-wave plate and a polarizing filter. This establishes the relation between the Stokes parameters and the electric field vector of a plane monochromatic wave in classical electromagnetic theory. A similar definition is given by van Ballegooijen in Chapter 12, and the Stokes vector is defined in terms of the second quantization of operators in quantum electrodynamics by Landi Degl'Innocenti in Chapter 11.

Continuing with his *Gentle Introduction*, Rees shows with very simple examples how circular polarization arises by optically thin emission in a magnetic field from the longitudinal Zeeman effect, and how the Mueller matrix for Faraday rotation follows from the transfer equation. He then shows in a beautiful and simple derivation how the transfer equation can be integrated analytically in a plane-parallel atmosphere when the source function is a linear function of optical depth. For a taste of some of the fearsome mathematical acrobatics in matrix algebra that Rees has shielded us from, we are invited to look at the elegant solution of a more general problem shown in Landi Degl'Innocenti's paper.

Rees gives an extensive list of references to non-LTE, polarized transfer and to physical situations where the transfer equation has been solved numerically, and he assesses the importance of the other papers on polarized radiative transfer contained in this volume. For his summary the reader is referred to his section 6; not included there are the papers by Schmidt & Wehrse and by Peraiah describing methods that *can* be used to attack polarized radiative transfer but do not deal with that topic or treat it only peripherally. These papers are summarized only in this Introduction.

In **Chapter 10** Rees & Murphy discuss two problems: the Zeeman effect of a non-LTE line split by a strong magnetic field, and the polarization due to scattering of a resonance line in an atmosphere without a magnetic field.

In the Zeeman effect the authors treat the case where the magnetic field affects the intensity only in the degree of polarization, without changing the mean integrated intensity or the line source function (a self-consistent solution method is described by van Ballegooijen in Chapter 12). Hence they separate the problem into two parts, the calculation of the line source function ignoring the magnetic field, and the formal solution of the vector transfer equation for known source function and opacity in a magnetic field, which may have arbitrary strength and orientation. They discuss only the formal solution of the vector transfer equation for which they offer two different methods; one is based on differential equations, the other on integral equations.

In the differential equation method, the first-order differential equation for the Stokes vector is transformed into the Feautrier equation: The transfer equation is written separately for the outward directed beam at a frequency in one line wing, and for the inward directed beam at the corresponding frequency in the other line wing; in the latter equation the Stokes vector is modified by changing the sign of its U-component. Both equations then have identical opacity matrices. Since the emission function is symmetric in this problem, the two transfer equations can be combined into first-order differential equations for the even and odd parts of this modified intensity, and further combined into a second-order differential equation for the even part alone. The resulting equation is the generalization of the Feautrier equation for the polarized case. It is differenced in the usual manner, leading to a block tridiagonal system of algebraic equations whose solution yields the Feautrier variable and the emergent Stokes vector. In addition, a recursion relation gives simple approximate expressions for the evolution operator of the formal solution and for the contribution vector, *i.e.*, the generalization of the contribution function.

The integral equation approach skillfully exploits the observation that the diagonal elements of the absorption matrix in the differential equation for the Stokes vector are identical. It is therefore convenient to separate the absorption matrix into two matrices, with only diagonal or only off-diagonal non-zero elements, respectively. When the *diagonal* matrix is taken to the left-hand side of the transfer equation, the operator acting on the Stokes vector is the same as in the scalar transfer equation with the same opacity. Thus the transfer equation for the Stokes vector can be integrated formally with the exponential integral kernel, familiar from the scalar transfer equation for the specific intensity. This formal integral now contains the unknown Stokes vector also in the integrand; the equation thus is an integral equation for the Stokes vector. A linear interpolation scheme for the matrix expression in the integrand transforms the equation into a block bidiagonal system of algebraic equations whose solution yields the Stokes vector in the outward direction. This approach, too, results in approximate expressions for the evolution operator and the contribution vector.

The second problem the authors discuss is the linear polarization due to scattering by a resonance line in the absence of a magnetic field. The symmetry of the physics allows this problem to be described by a simplified Stokes vector of only two components. Instead of the opacity *matrix*, the transfer equation contains a *scalar* absorption coefficient; the two components of the equation are coupled via the source function. Correlations in frequency, angle, and polarization are accounted for by a generalized redistribution matrix. The transfer equation is solved in an approximation in which part of the angle dependence of the redistribution operator is integrated over analytically, resulting in two relatively simple coupled differential equations of transfer. Following standard procedure in deriving the integral form of the transfer equation from the differential form, the differential equations are integrated formally (with the same Λ-operator) and the formal solutions for the Stokes vector are inserted into the equations defining the two source functions, resulting in two coupled integral equations. They are solved iteratively by assuming that polarization is weak so that the polarization term may be lagged. This scheme then requires (1) the solution of only one integral equation for a function that corresponds to the angle and frequency-dependent line source function in partial redistribution when polarization is neglected, and (2) the evaluation of an integral for the polarization component. This is a new method by the authors for which this paper describes in detail only the version that assumes complete redistribution in the integral equation.

In **Chapter 11** Landi Degl'Innocenti discusses the symmetries of the absorption matrix for polarized radiation, a crucial property if the transfer equation is to be expressed in the form of the Feautrier equation, for example. He names physical effects described by the various components of the absorption matrix: Faraday rotation (in a magnetic field), rotation of the plane of linear polarization, transformation of linear into circular polarization (Faraday pulsation). Landi draws a distinction between non-LTE effects of the first kind, where the magnetic sublevels of the atom are populated according to their statistical weights and phase relations between the sublevels are negligible (unpolarized atomic system), and the converse situation, called non-LTE of the second kind. In the former case, which is the one usually meant by the term non-LTE, the source function vector has only one non-zero component. Note that except for a discussion of resonance polarization in Chapter 10 the polarization problems discussed in this volume concern only non-LTE of the first kind, which ignores phenomena like the Hanle effect that depend on phase relations between the magnetic sublevels.

Landi formally solves the differential equation for the evolution matrix, an operator that is used to express the formal solution of the transfer equation, employing the chronological ordering operator; for the case of an absorption matrix that is constant (except for a depth-dependent scale factor) he obtains the exponential of the absorption matrix, which is analogous to the well-known scalar damping factor; this matrix exponential can also be obtained by diagonalizing the transfer equation (*cf.* Chapter 14). (Note that the solution for the evolution matrix does not require any special treatment to avoid growing exponential terms since without coupling of oppositely directed radiation beams these terms do not arise; they would if the source function contained a scattering term. For the more difficult case of this coupling, see the paper by Schmidt & Wehrse) Landi decomposes the matrix exponential using a linear combination of complex 4×4 matrices with commutation properties similar to those of Pauli matrices. He obtains a decomposition in terms of real 4×4 matrices, which allows easy numerical integration of the transfer equation for a depth-dependent absorption matrix on an arbitrary spatial grid, using standard numerical methods, but it requires the source function to be known. The approach is most advantageous when the largest and smallest eigenvalues of the absorption matrix differ by a large factor since then the frequently-used fourth-order Runge-Kutta method becomes inefficient because of the deep optical depth space needed for the problem. This approach can of course be employed in the solution of LTE as well as non-LTE transfer equations. For non-LTE problems Landi suggests decoupling the calculation of the non-LTE source function from that of the Stokes parameters and solving by iteration. This is analogous to Cannon's (1973a) perturbation method for partial redistribution problems (for the distinct risk in that approach to converge to the wrong solution, *cf.* Chapter 8). A direct, self-consistent solution method for that problem is discussed in Chapter 12.

One of the difficulties of polarized radiative transfer is that the familiar, simple scalar differential equation is replaced by a matrix equation for the four-component Stokes vector. If the vectors and matrices of the theory had only two components, much of the work in determining eigenvalues, eigenvectors, matrix inversions, and the solution of differential matrix equations could be done by hand. In **Chapter 12** van Ballegooijen describes just such an approach in which he formulates the transfer equation in terms of complex 2×2 matrices. He first reviews some of the history of observations of solar and stellar magnetic fields via the Zeeman effect, and then presents a thorough discussion of the physical approximations that make his line transfer problem with polarization numerically tractable: a magnetic field that is strong enough for the Zeeman splitting to be larger than the natural line width in order to exclude the interferences between magnetic substates that lead to the Hanle effect, and collisional coupling between the magnetic states that is stronger than the radiative coupling for the substate populations to be given by their statistical weights. The departures from LTE of the atom are then fully described by the familiar non-LTE source function for the combining energy levels. Apart from these conditions the case he treats is completely general: the Doppler width may vary with depth, as may the strength and orientation of the magnetic field; and the medium may have systematic motions. Of course, the paper assumes plane-parallel stratification of the atmosphere, although the method is in principle not tied to any geometry since the depth integration is carried over rays.

Whereas Rees (in Chapter 9) pays much attention to handedness and sign conventions for *monochromatic* beams only, van Ballegooijen considers them with an eye to the frequency-coupled equations he needs to solve numerically. Thus he makes unconventional choices for signs in order to arrive at symmetry properties of the profile functions that are similar to those of unpolarized radiation. Recall that the profile function for unpolarized light in the observer frame is invariant under a simultaneous change of the sign of the direction μ and of the frequency relative to line center $\Delta\nu$. For polarized radiation the Zeeman components have to be interchanged as well (*cf.* also Chapter 10). The transformation of the transfer equation for polarized radiation from the form in terms of real 4×4 matrices for the Stokes parameters into the form with complex 2×2 matrices results in a simple expression for the evolution matrix, an operator that is used in the formal solution of the transfer equation. (For another justification of the differential equation satisfied by the evolution matrix and a different approach to its solution *cf.* Chapter 11, equations 12-24). To construct the Λ-operator, a fair number of products of matrices, their inverses, and their Hermitian adjoints has to be evaluated. But the result of all this work, which makes ample use of the theory of linear equations and matrix algebra, is an integral equation for the line source function that is in the familiar form of a two-level atom. This method thus presupposes a decomposition of a multi-level problem into a series of equivalent two-level atoms as described in Chapter 6.

In **Chapter 13** Peraiah describes a method he calls an integral operator technique. The first problem he considers is the solution of the monochromatic transfer equation for an atmosphere with spherical symmetry. His basic formulation is in terms of partial differential equations with derivatives relative to radius and angle. These equations are solved along logarithmic spirals, *i.e.*, along characteristic lines of constant angle with respect to the outward normal. The formal solution is cast into the form of the so-called interaction principle (for a detailed description see Peraiah's paper in *Methods in Radiative Transfer*), which relates emergent intensities from a given cell in the radius-angle grid to the corresponding incident intensities. This relation between the intensities is written in terms of transmission and absorption matrices, giving the equations the appearance of integral equations. In the second problem Peraiah considers the solution of the transfer equation for a Doppler-broadened line in a spherical medium, using the comoving frame formulation; and in the third problem he investigates first-order relativistic effects in continuum transfer for conservative, isotropic scattering. The basic approach is the same as the one used for the monochromatic transfer equation. The paper is concerned with numerical tests of such basic questions as the invariance of the intensity *in vacuo* and the continuity and uniqueness of the solution. This requires non-negative intensities, a condition that places an upper limit on the allowed basic cell size for constructing the elementary matrices. In the relativistic flow problem Peraiah investigates aberration and advection effects in high-speed flows. For velocities on the order of a percent of the speed of light he finds that the mean intensity can be changed significantly by the motion, the magnitude of the effect depending on the optical thickness of the atmosphere.

Since the transfer equation in Peraiah's approach is a first-order differential equation, or a system of such equations, the associated difference equations are stable and have second-order accuracy only if the grid size for the interaction matrices is small compared to the photon mean-free-path. In order to satisfy this condition the interval on the radius grid is subdivided until the elementary path length is sufficiently small compared to the photon mean-free-path. The matrices for the prescribed grid are then obtained by taking the appropriate products of elementary matrices. This can be done fairly rapidly since the opacity and the radius are assumed constant within each grid interval, making the elementary matrices identical. (For a different remedy of this problem, *cf.* Chapter 14.)

The derivation of the equations is different from what is usually done in radiative transfer (a more complete description of the method is found in Peraiah & Varghese, 1985, for the somewhat simpler problem of continuum transfer in a spherical atmosphere); therefore the procedure looks fairly complicated. A different derivation of the transmission and reflection matrices along lines that resembles more the traditional approach to radiative transfer is presented in Chapter 14. Note that Peraiah uses a sign convention for designating inward and outward directed radiation pencils that is the opposite of the usual one.

In **Chapter 14** Schmidt & Wehrse treat the solution of the systems of differential equations that occur in the transfer of polarized radiation or in line transfer with partial frequency redistribution. For spherical geometry the equations form a system of first-order *partial* differential equations; they are solved as a set of *ordinary* differential equations for an optical depth variable along logarithmic spirals defined by the (constant) polar angle. In this formulation the curvature term merely provides additional coupling of the intensity components. The equations of the transfer problem are discussed separately for depth-independent and depth-dependent absorption matrices.

When the absorption matrix is depth-independent except for a scale factor that can be absorbed in the definition of the optical depth, the equations are solved by the so-called matrix exponential method, which is the matrix analogue of the formal integral of the scalar transfer equation. A novel feature of this method is that the solution of the transfer equation for the emergent intensities is changed from a form that involves the absorption matrix in the evolution operator *directly* into a form in terms of transmission and absorption matrices; these matrices are manipulated so as to eliminate exponentially growing terms, leaving only decaying contributions. The equations can therefore be used without special precautions for semi-infinite media. The particular formulation shown in this Chapter assumes that the distribution of thermal sources is given in analytic form.

For Gaussian division points of the angle discretization in a spherical medium, negative intensity components are found when the curvature is large and the opacity is low. The authors provide criteria for the choices of angles and integration weights that ensure positive emergent intensities.

When the absorption matrix has arbitrary depth-dependence the formal solution, written in terms of an evolution operator, is more complicated than the analogous equation found in Landi's paper, for example, because of the coupling of the inward and outward directed intensities by the scattering term in the transfer equation. Then the formal expression for the emergent intensity as a function of the incident intensity and the internal sources contains again growing exponentials. As before, these are eliminated and the solution is written in terms of transmission and reflection matrices whose order is half the order of the system of equations in the transfer problem. But now, because of the depth-dependence of the opacity, these matrices are determined from the solution of first-order differential equations.

The methods given in this paper are applied to a grey, plane-parallel atmosphere of finite optical depth and to a polarization problem.

The treatment of the equations is interesting in the context of polarized radiative transfer since it shows how matrices containing exponentially growing functions of optical depth can be rendered numerically well-behaved. For polarization, as also for conservative scattering for example, the problem must be solved by the general method even for a depth-independent absorption matrix since the requirement of distinct eigenvalues of the matrix exponential method is not satisfied.

ACKNOWLEDGMENTS

I am grateful to my colleagues for critical reading of the synopses of their papers, and I thank Rudy Loeser for editorial comments.

REFERENCES

Auer, L. H., 1976, *J. Quant. Spectrosc. Rad. Transf.*, **16**, 931.

__________ 1984. *Methods in Radiative Transfer*, W. Kalkofen ed., Cambridge University Press, Cambridge, 79.

__________ 1986 *J. Quant. Spectrosc. Rad. Transfer*, submitted.

Auer, L. H. & Mihalas, D. 1969, *Astrophys. J.*, **158**, 641.

Cannon, C. J. 1973a, *J. Quant. Spectrosc. Rad. Transfer*, **13**, 627.

__________ 1973b, *Astrophys. J.*, **185**, 621.

__________ 1984, *Methods in Radiative Transfer*, W. Kalkofen ed., Cambridge University Press, Cambridge, 157.

__________ 1985, *The Transfer of Spectral Line Radiation*, Cambridge University Press, Cambridge.

Kalkofen, W., 1984, *Methods in Radiative Transfer*, W. Kalkofen ed., Cambridge University Press, Cambridge, 427.

Olson, G. L., Auer, L. H. & Buchler, 1986, *J. Quant. Spectrosc. Rad. Transf.*, **35**, 431.

Peraiah, A. & Varghese, B. A. 1985, *Astrophys. J.*, **290**, 411.

Rybicki, G. B. 1972, *Line Formation in the Presence of Magnetic Fields*, R. G. Athay, L. L. House & G. Newkirk, Jr. eds., Boulder: High Altitude Observatory, p145.

__________ 1985, private communication.

Scharmer, R. 1981, *Astrophys. J.*, **249**, 720.

__________ 1984, *Methods in Radiative Transfer*, W. Kalkofen ed., Cambridge University Press, Cambridge, 173.

Scharmer, G. & Nordlund, Å. 1982, *Stockholm Obs. Rep.*, **19**.

Werner, K., & Husfeld, D. 1985, *Astron. Astrophys.*, **148**, 417.

SURVEY OF OPERATOR PERTURBATION METHODS

W. Kalkofen
Harvard-Smithsonian Center for Astrophysics, Cambridge, USA

ABSTRACT: This introductory paper surveys operator perturbation methods without confining itself to papers in this volume. It discusses the salient features of methods employing integral as well as differential equations, and it emphasizes the physical principles on which the approximate operators are based.

1. INTRODUCTION

Radiative transfer plays a fundamental role in many astrophysical media, where energy transport by radiation may determine the structure of an atmosphere or control the energy loss from a shocked gas. The mathematical description of the radiation usually involves a system of coupled equations, often of high order. This can be the case even for a single spectral line, as when the medium is in motion and has very low density.

By the late 1960's and early 70's, such complex transfer problems were solved routinely to great accuracy. Significant advances in the field had been achieved with the complete linearization of differential equations by Auer and Mihalas (1969). The versatility of that approach for problems in which only a small number of depth-dependent functions had to be determined was further broadened by Rybicki (1971) with a formulation in which the structure of the differential equations became identical with that of integral equations. Kalkofen's (1974) complete linearization of the integral equations was an alternative to Rybicki's approach; starting from the formal integral of the transfer equation, it allowed the treatment of transfer problems with anisotropic scattering function. These Newton-Raphson linearization methods were highly efficient in terms of the number of iterations required to solve non-linear transfer problems, but inefficient in the use of computer time and storage, taxing computer resources. This provided a strong incentive to find faster ways of solving the transfer equation.

A new and significant departure in numerical radiative transfer was introduced by Cannon (1973a,b) who reduced the order of the system of equations without sacrificing accuracy, thereby gaining a distinct advantage in the use of computer storage and time, but at a somewhat slower convergence rate. The essence of the operator perturbation approach is to divide a transfer problem into two parts: in the first, corrections to a solution are calculated with the aid of an approximate (differential or integral) operator, and in the second, the error made by a solution in satisfying the conservation equation is determined. In this approach, even linear problems require iterations; but this is offset by a significant reduction in the needed computer time.

Cannon's method achieved its promise in the hands of Scharmer (1981) who turned an inspired idea into an efficient tool, the main ingredient of which is a vastly simplified description of radiative transfer. This simplification in the description of radiative transfer found its culmination with the use of local, *i.e.*, diagonal operators (Werner and Husfeld 1985, Werner 1986, this volume; Hamann 1985, 1986, this volume; Olson, Auer and Buchler 1986; Auer, this volume; also Rybicki 1985). To appreciate the thrust of the approach, consider the application of an operator perturbation method with diagonal operators to a large atomic system with K levels in an atmosphere with N discrete depth points: in the conventional formulation with integral equations, this problem is described by a matrix equation of order $K \times N$ (the matrices occurring in the differential equation formulation tend to have even larger order); diagonal operators reduce this problem to two sets of equations, one of order K, the other of order N. No further reduction is possible.

The aim of this paper is to inquire into the logic of operator perturbation methods. We begin the survey of the various approaches with Scharmer's, describing first its salient features and how the other integral equation approaches differ from it. The integral equation methods of Scharmer, Hamann, Werner & Husfeld, and Olson *et al.* are discussed in section 2. In section 3 we collect other methods, namely, Anderson's, which concerns differential equations directly, Kalkofen's, in which the perturbation equations in differential form are derived via integral equations, and Rybicki's core saturation method, a general approach not tied to any particular formulation. The final section summarizes the basic ideas.

2. INTEGRAL EQUATION METHODS

Following Scharmer's (1981) lead, a spate of operator perturbation methods were developed, based both on the integral and the differential equation forms of the transfer equation and applied to line transfer and to the construction of model atmospheres. The approach implies that even linear problems, which could be solved directly in a single step, are solved more efficiently by means of a perturbation approach that separates a problem into two parts, the calculation of an error, in which individual (*i.e.*, scalar) equations are solved very fast and to great accuracy, and the calculation of corrections to a solution, in which typically a low-order system of coupled equations is solved.

Scharmer (1981, 1984; Scharmer & Nordlund 1982; Scharmer & Carlsson 1985) writes the integral equations for complete redistribution problems in conservation form — unlike his equations for partial redistribution (Scharmer 1983), for which he uses an expansion similar to Cannon's (1973a, 1984, 1985). The order of his system of equations is equal to the number of depth points (times the number of unknown functions in the case of a multi-level line transfer problem). This is the same as in the standard integral equation formulation; thus there is no reduction of the order of the system of equations. In spite of that, the solution time for his perturbation equations is much shorter than that for the conventional direct solution; it is controlled by the calculation of the error term, at least for problems involving only a small number of atomic levels. The overall solution time therefore scales linearly with all grid sizes, *i.e.*, as the product of the numbers of depth, angle, and frequency points, and the number of iterations. The extraordinary speed of the method is achieved by omitting many of the steps required in conventional treatments. This is true both for the construction of the Λ-operator, which ordinarily couples all depth points of the medium, as well as for the solution of the system of simultaneous equations.

The construction of the Λ-operator is based on the fact that a photon travels on the average one mean-free-path between its points of creation and absorption or scattering. The result that the intensity is equal to the source function at a distance of one photon mean-free-path in the upstream direction is valid only far from boundaries and for source functions that depend linearly on optical depth. But in many media this relation is sufficiently well satisfied to provide an adequate approximation to the transfer of radiation. This permits the dependence of the intensity on emission throughout the atmosphere to be replaced by a dependence on emission at a few points. Two additional assumptions concern the surface layers: the outward intensity is assumed to be constant and given by a modified Eddington-Barbier relation, in which the emergent intensity is equal to the source function at some optical depth, a free parameter of order unity; and the inward intensity due to internal sources is assumed to be zero. With these approximations, the Λ-matrix has nearly triangular shape and can be constructed very rapidly.

The resulting system of equations can also be solved very rapidly if the sparseness of the matrix is taken into account; the solution time scales as the square of the order rather than the cube. In the course of the iterations, the matrix equation must be solved repeatedly, with the same matrix but different inhomogeneous vectors. The first, and time-consuming, step in this solution process needs to be performed only once if the equation is solved with an $L \times U$ decomposition of the matrix (*cf.* Scharmer & Nordlund 1982) and the respective elements are stored for later use.

If the number of simultaneous lines treated is sufficiently small, as it is in a two-level atom, the solution time of Scharmer's equations is dominated by the error calculation and hence it is proportional to the number of iterations. It is imperative therefore that this number be kept small. For Scharmer's equations this is the case, with the number of iterations to reach an accuracy of better than 1% in an isothermal, static problem, for example, as small as 3 (*cf.* Scharmer 1984).

A significant reduction in the system of equations for multi-level atoms is achieved by Hamann (1985, 1986, this volume) and Werner & Husfeld (1985; Werner 1986, this volume) with the use of local, *i.e.*, purely diagonal transfer operators inspired by the work of Scharmer (1981, 1984) and Rybicki (1972). The solution process is similar to Λ-iteration from which it differs by a depth-dependent amplification of the difference between consecutive source function iterates. The scale factor amplifying the correction of the solution is based on an estimate of the fraction of photons escaping from the atmosphere in a single flight. For thick media this scale factor can reach enormous values; in a case reported by Hamann (this volume) the amplification amounted to a factor exceeding 10^6.

The simplification in the description of the transfer based on diagonal operators comes at a price: it takes about ten times as many iterations as with a more detailed description (*cf.* Scharmer 1984; Kalkofen this volume); typically $20 - 50$ iterations are needed to reach an accuracy of 1%. But compared to straight Λ-iteration, which would require on the order of $1/\epsilon$ iterations (*cf.* Avrett & Hummer 1965), this is fast, especially if the scattering parameter is very small.

The diagonal operator used by both Werner and Hamann is based on the monochromatic Λ-operator

$$\Lambda_\nu(\tau_\nu, \tau'_\nu) = \begin{cases} \delta(\tau_\nu, \tau'_\nu) , & \tau_\nu > \gamma , \\ 0 , & \text{otherwise} , \end{cases} \tag{1}$$

where γ marks the optical depth τ_ν, or more typically the monochromatic optical distance $\tau_{\nu\mu}$ from the boundary along an inclined ray, at which the "core" for the frequency ν of the line in question begins (*cf.* Rybicki 1984, Fig. 1); typical values of the parameter γ lie in the range 2-5.

With the core saturation approximation of Rybicki (1972) the non-zero diagonal elements of the integrated Λ-operator become

$$\Lambda(\tau, \tau) = \int_{\nu_1}^{\nu_2} \varphi_\nu(\tau) d\nu \quad , \tag{2}$$

where φ_ν is the absorption profile of the line and where the limits of the integral are the frequencies beyond which the medium becomes transparent, (*i.e.*, where τ_ν or $\tau_{\nu\mu} < \gamma$). All off-diagonal elements are zero. Thus the approximate operator is diagonal, and the value of its elements is given by the core fraction of the line (in the case of a line transfer problem).

A second, related operator, used by Werner, is often almost as convenient and has the additional advantage of reducing the number of iterations required for convergence by up to about a third. It is given by

$$\Lambda_\nu(\tau_\nu, \tau'_\nu) = \begin{cases} \delta(\tau_\nu, \tau'_\nu) , & \tau_\nu > \gamma , \\ \frac{1}{2}\delta(\tau'_\nu, \gamma') , & \tau_\nu \leq \gamma ; \end{cases} \tag{3}$$

it contains a second free parameter, γ', which gives greater freedom than a single parameter would; typical choices are $\gamma' = 1$ (Scharmer 1984, eq. II-5) and $\gamma' = \gamma$ (Werner, this volume, eq. 4). The integral operator corresponds to the equations investigated by Scharmer (1981, equations 7-9; 1984, equations II−4 and II−5) in his derivation of the so-called probabilistic transfer equation (Athay 1972, Frisch & Frisch 1975, Canfield *et al.* 1984). If the parameters γ and γ' satisfy $\gamma \leq \gamma'$ then all elements of the Λ-matrix (2) below the principal diagonal, $\tau > \tau'$, vanish. Thus, if the solution of a problem is started at the lower boundary, the off-diagonal terms in a particular iteration give rise to known terms. This procedure can easily be used in a line transfer problem. But in the construction of model atmospheres the hydrostatic equilibrium equation must satisfy a boundary condition at the outer boundary. In this case, Werner separates the pressure calculation from the correction calculation and integrates the pressure between the normal cycles of the iteration. This allows him to integrate the transfer and energy equations from the inner boundary and thus to use the triangular matrix there as well.

The method proposed by Hamann and by Werner & Husfeld, using either the purely diagonal or the triangular operator, is highly effective for very large atomic systems, for which otherwise the high order of the matrix equations to be solved would be overwhelming. But for small atomic systems their method is expensive since it requires a relatively large number of iterations for convergence. Since the computer time in such problems is dominated by the formal solution it is much more efficient to choose a method that requires few iterations such as Scharmer's (1984), Kalkofen's (this volume), or Anderson's (this volume). Perhaps the greatest advantage of diagonal operators is the ease of formulation of problems in complicated geometries. While the method may not be as efficient in the use of computer time as some other methods, it can significantly reduce the time spent at writing and testing a program.

A similar method, but requiring fewer iterations and thus representing an improvement of the method of Hamann and Werner & Husfeld, has been described by Olson *et al.* (1986) and Auer (this volume); this method also employs a diagonal approximate operator, but instead of the core fraction (*cf.* equation 2), Olson *et al.* use a truncated form of the monochromatic Feautrier equation, followed by a frequency integration, to determine the non-zero elements of the diagonal operator. Their approach is more complicated and requires a larger effort, which seems unnecessary, at least for typical transfer problems in plain, static atmospheres. It also contains an arbitrary feature since it introduces a dependence of the values of the elements of the approximate matrix on the arbitrary grid spacing.

Since the operators of Olson *et al.* and of Hamann and Werner & Husfeld are very similar in concept, one might expect only trivial differences in their numerical properties. However, it is found that the method of Olson *et al.* is significantly faster. The improvement in the convergence properties is achieved by means of the acceleration scheme of Ng (1974), by which a new solution is estimated on the basis of several consecutive iterates. In their demonstration for an optically thick scattering problem with $\epsilon = 10^{-3}$, Olson *et al.* advanced the solution after the first three iterations with Ng's acceleration, reaching within 2%

of the exact result. Thus, two cycles of four iterations each would suffice for the error to drop well below 1%. This method thus requires only a fraction of the number of iterations of the method of Hamann and Werner.

The optimal approach in these methods based on diagonal or triangular operators is to construct them in the manner of Hamann or Werner & Husfeld and to use Ng's acceleration scheme as described by Olson *et al.* and Auer. This should be the method of choice for transfer in plane-parallel media with atomic systems of many energy levels and for transfer in multi-dimensional geometries.

3. OTHER METHODS

The basic design of the operator perturbation method of Kalkofen (*cf.* separate papers in this volume on line transfer problems and on model atmospheres) is that of integral equations. But these are used merely to derive the perturbation equations in differential form. The actual solution employs exclusively differential equations.

The derivation of the perturbed differential equations proceeds as follows: The transfer and the constraint equations are written as a set of coupled differential equations. These "exact" differential equations are transformed into the equivalent integral equation in conservation form (*cf.* Scharmer 1981, 1984; Scharmer & Nordlund 1982; Scharmer & Carlsson 1985). The integral equation is then written as a perturbation series in integral equations (Cannon 1973b, 1984, 1985), which are the approximate equations; and these, in turn are again expressed as systems of coupled differential equations. Thus, the perturbation theory applied to integral equations is used as a device to derive the perturbation equations in differential form from the exact differential equations. — The transformations between integral and differential operators are facilitated by constructing the integral operators from inverse finite-differenced differential operators.

The *exact* differential equations are solved as uncoupled equations for known source function (formal solutions) in the *error* calculation, and the *approximate* differential equations are solved as a coupled system for *corrections* to the solution.

Two sets of angle and frequency points are employed: (1) A coarse set, in which the frequency points are chosen so as to describe the overall structure of the solution. For this the important length scales are the thermalization length and the depth from which typical photons escape. These lengths must be represented by the monochromatic mean-free-paths that correspond to the frequencies in the coarse set. A single angle point is generally adequate. And (2) a dense set of mesh points, which has a sufficiently fine grid of angle and frequency points to give high accuracy in the description of the microscopic interactions between gas and radiation field. — The coarse set leads to a set of coupled correction equations that is of low order and can therefore be solved very rapidly. The high-resolution set, used in individual (scalar) transfer equations for the computation of the error in the conservation equation, insures the accuracy of the solution, as does the Feautrier equation, which is used both in the error and in the correction calculation.

As with other efficient integral equation methods, the solution time (for problems involving only a few lines) depends mainly on the time taken by the formal solutions in the error calculation. This is basically a very fast calculation which becomes time-consuming only because a large number of angle and frequency points may be needed in problems involving macroscopic motion in the medium. To conserve time it is important to keep the number of iterations low. In problems with moderate flow velocities, of the order of the velocity of sound, this is achieved by this method, two or three iterations often being sufficient to reach an accuracy better than 1%.

In time-dependent applications, the method is highly efficient by using the converged solution of the preceding time step as the starting solution at the current time and by basing the approximate operator on a time-independent atmosphere so that the operator need not be updated at every time step.

Rybicki's (1972) core saturation method is a general approach to transfer problems, relying on physical insight. It is not an operator perturbation method in itself, but it has inspired and influenced many of them. This is true of Scharmer's (1981, 1984) derivation of the probabilistic transfer equation, of Kalkofen & Ulmschneider's (1984) core saturation method for time-dependent problems in radiation gas dynamics, and of Kalkofen's (1985) suggested choice of frequency points for approximate Λ-operators.

For two-level atoms without induced emissions, Rybicki's core saturation equations amount to a diagonal approximate operator (*cf.* Rybicki 1985). And when his equation, modified to converge to the exact solution (Rybicki 1984, equation 5.12), is written as an operator equation, it implies the diagonal approximate operator (1) employed by Werner and by Hamann. Their contribution is to have cast Rybicki's equation in operator form and to have embedded the resulting diagonal operator in Cannon's (1973b) perturbation equations.

Anderson's differential equation method (this volume) for the solution of line transfer problems or for the construction of line-blanketed model atmospheres has many of the attributes of an operator perturbation method, such as a large reduction in the order of the system of equations and corrections based on an accurately computed error in a conserved quantity. As a consequence of the reduction in the order it is very much faster than a direct solution method. But the derivation of the equations is heuristic; thus it is not an operator perturbation method in the sense of Cannon's (1973a,b) equations.

Anderson achieves the reduction in the order of the system of differential equations by integrating the transfer equation over frequency intervals to form frequency blocks, resulting in a system of transfer equations whose order is given by the number of such blocks. These frequency blocks are defined by the requirement that photons within a block experience approximately the same physics, *i.e.*, have nearly the same mean-free-path and scattering albedo. In general, frequency intervals contributing to a block are not contiguous.

The block transfer equations and the constraint equations (statistical, hydrostatic, and radiative equilibrium) are completely linearized relative to the block intensities and the state variables of the gas and are solved in a Newton-Raphson iteration. During an iteration, frequency averages of the opacity are held constant. These frequency averages play the role of form factors, analogous to the variable Eddington factors in the angle distribution; they are updated between the iterations on the corrections to the solution. The driving term in the correction equations is computed for the high-resolution angle and frequency set (and not for the low-resolution block frequencies).

Anderson's method combines features of Lucy's (1964) approach, using angle and frequency form factors, with the Auer-Mihalas complete linearization method. It achieves the same aims as Kalkofen's operator perturbation method and arrives at virtually the same equations, albeit by a very different route (this implies that Anderson's equations, or similar ones, can be derived via bona fide perturbation theory). The success of the method depends on the skillful grouping of frequencies of photons with similar opacities and therefore similar fates. The resulting savings in computer time are substantial; the method allows the most concentrated attack thus far on the knotty problem of line-blanketing in early-type stars without assuming the gas to be in LTE (for another approach to this problem, *cf.* Werner & Husfeld and Werner).

4. SUMMARY

The operator perturbation methods owe their success to the exploitation of essential physical principles of the radiative transfer. We summarize the approaches discussed in this paper:

Scharmer's method uses a much simplified description of the relation between the intensity and its sources. Thus, in the interior of an atmosphere, radiation is due to emission from a distance of one photon mean-free-path; in the surface layers, the outward intensity arises at unit optical depth; and the inward intensity due to sources located in the medium is zero.

The diagonal operators used by Werner & Husfeld, Hamann, Olson *et al.* , and Auer put the intensity equal to the local source function, but with a reduction that depends on the fraction of photons escaping from the atmosphere in a single step. And the triangular operator of Werner & Husfeld and Werner contains an additional contribution to the outward directed radiation, but from deeper layers only.

The approximate operator of Kalkofen's method embodies the idea that the structure of an atmosphere depends only on the distance between photon creation and true absorption (*i.e.*, thermalization) or escape from the medium.

Anderson's method uses typical frequencies, frequency averages, and form factors that are based on the notion that the basic features of an atmosphere depend only on certain average values of the opacity and on typical photons. Like Kalkofen's method, it recognizes that the structure of the atmosphere is characterized by fewer parameters than is an accurate description of the microscopic physics.

In addition to these operator perturbation methods there are important supporting procedures and ideas that are very useful. Thus, the separation of a problem into a series of effective two-level atoms (*cf.* Avrett & Loeser, this volume) is convenient for methods that structure a problem in terms of such synthetic atoms. And the numerical solution of the statistical equilibrium equations is greatly facilitated if cancellations between upward and downward radiative rates are effected analytically (*cf.* Scharmer 1984). Finally, the core saturation approach has been extremely fruitful in isolating features important in the construction of approximate operators that are easy to build and yet are close (*i.e.*, in the sense of having similar eigenvalues) to the exact operators.

ACKNOWLEDGMENTS

I am indebted to Dimitri Mihalas for valuable comments on the manuscript and to Larry Auer and Gordon Olson for discussions of their paper.

REFERENCES

Athay, R. G. 1972, *Astrophys. J.*, **176**, 659.

Auer, L. H. 1984, *Methods in Radiative Transfer*, W. Kalkofen ed., Cambridge University Press, Cambridge, 237.

———— 1986 *Methods of Comp. Phys.*, submitted.

Auer, L. H. & Mihalas, D. 1969, *Astrophys. J.*, **158**, 641.

Avrett, E. H. & Hummer, D. G. 1965, *Monthly Notices Roy. Astron. Soc.*, **130**, 295.

Canfield, R. C., McClymont, A. N., & Puetter, R. C. 1984, *Methods in Radiative Transfer*, W. Kalkofen ed., Cambridge University Press, Cambridge, 101.

Cannon, C. J. 1973a, *J. Quant. Spectrosc. Rad. Transfer*, **13**, 627.

———— 1973b, *Astrophys. J.*, **185**, 621.

———— 1984, *Methods in Radiative Transfer*, W. Kalkofen ed., Cambridge University Press, Cambridge, 157.

———— 1985, *The Transfer of Spectral Line Radiation*, Cambridge University Press, Cambridge.

Cuny, Y. 1967, *Ann. d'Astroph.*, **30**, 143.

Feautrier, P. 1964, *Compt. Rend. Acad. Sci. Paris*, **258**, 3189.

Frisch, U. & Frisch, H. 1975, *Mon. Not. Roy. Astron. Soc.*, **173**, 167.

Hamann, W.-R. 1985, *Astron. Astroph.*, **148**, 364.

———— 1986, *Astron. Astroph.*, **160**, 347.

Kalkofen, W. 1974, *Astrophys. J.*, **188**, 105.

———— 1984, *Methods in Radiative Transfer*, W. Kalkofen ed., Cambridge University Press, Cambridge, 427.

———— 1985, *Progress in Stellar Spectral Line Formation Theory*, J. E. Beckman & L. Crivellari eds., 153.

Kalkofen, W. & Ulmschneider, P. 1984, *Methods in Radiative Transfer*, W. Kalkofen ed., Cambridge University Press, Cambridge, 131.

Lucy, L. B. 1964, *Proc. First Harvard-Smithsonian Conf. Stell. Atmos.*, Smithsonian Astrophys. Obs., Spec. Rep. No. **167**, 93.

Mihalas, D. 1978, *Stellar Atmospheres*, Second Edition, W. H. Freeman & Co., San Francisco.

Ng, K. C. 1974, *J. Chem. Phys.*, **61**, 2680.

Olson, G. L., Auer, L. H. & Buchler, J, R. 1986 *J. Quant. Spectrosc. Rad. Transfer*, **35**, 431.

Rybicki, G. B. 1971, *J. Quant. Spectrosc. Rad. Transfer*, **11**, 589.

———— 1972. in *Line Formation in the Presence of Magnetic Fields*, ed. R. G. Athay, L. L. House, & G. Newkirk, Jr. (Boulder: High Altitude Observatory), p 145.

———— 1984, *Methods in Radiative Transfer*, W. Kalkofen ed., Cambridge University Press, Cambridge, 21.

———— 1985, private communication.

Scharmer, R. 1981, *Astrophys. J.*, **249**, 720.

———— 1983, *Astron. Astroph.*, **117**, 83.

———— 1984, *Methods in Radiative Transfer*, W. Kalkofen ed., Cambridge University Press, Cambridge, 173.

Scharmer, G & Carlsson, M. 1985, *J. Comp. Phys.*, **59**, 56.

Scharmer, G. & Nordlund, Å. 1982, *Stockholm Obs. Rep.*, **19**.

Werner, K. 1986, *Astron. Astrophys.*, **161**, 177.

Werner, K., & Husfeld, D. 1985, *Astron. Astrophys.*, **148**, 417.

LINE FORMATION IN EXPANDING ATMOSPHERES: MULTI-LEVEL CALCULATIONS USING APPROXIMATE LAMBDA OPERATORS

W.-R. Hamann
Institut für Theoretische Physik und Sternwarte
Leibnizstraße 15, D-2300 Kiel, Fed. Rep. of Germany

Abstract. We study multi-level non-LTE spectral formation in spherically expanding atmospheres, using a perturbation technique. Our method employs approximate lambda operators which act only locally. Adequate operators are defined for continua and lines, the latter being constructed from considerations in the comoving frame. The method is tested for a two-level atom and then generalized to the multi-level case. The equations of statistical equilibrium become non-linear, and are solved by means of a linearization. In our method, the exact radiation transfer occurs only in the "formal solution", which is performed by means of the comoving-frame technique.
Test calculations are presented for a typical Wolf-Rayet star atmosphere consisting of pure helium, with the ground state of He I, ten bound levels and 45 lines of He II, and He III. Convergence is found to be satisfactory. The method is capable of handling He I and CNO additionally as complex atomic systems. It is superior to previous solution techniques for the problems under consideration.

1. INTRODUCTION

The main difficulties in non-LTE calculations arise because two sets of equations must be satisfied simultaneously: the radiation transfer and the statistical equilibrium. The "classical" method of solving this non-linear problem is the "complete linearization" technique (cf. Mihalas, 1978). This method fully accounts for the mutual coupling of all variables. Consequently, it is restricted in practice to simple geometries and a limited complexity of the model atoms.

However, there are some important problems in the field of stellar atmospheres which demand a more powerful solution algorithm. One of these problems is the construction of line-blanketed (plane-parallel and static) non-LTE model atmospheres. Another is the computation of synthetic spectra from expanding atmospheres, e.g. in order to analyze Wolf-Rayet stars.

A new approach to non-LTE problems was proposed by Scharmer (1981, 1984). He combined the operator perturbation technique (Cannon, 1973) with suitable ideas for constructing approximate lambda operators. The method was generalized to multi-level problems (Scharmer & Carlsson, 1984) by linearizing both the transfer equation and the statistical equations.

The full advantage of Scharmer's idea is gained if the approximate lambda operator is defined as an entirely local operator without spatial coupling. By means of such a "diagonal" operator the non-LTE problem separates into two independent steps which must be performed iteratively in turn: the formal solution of the radiative transfer problem (spatially coupled), and the solution of the statistical equilibrium equations (coupled in frequency) which are entirely local, but non-linear, due to the incorporation of the approximate radiation transfer. The application of this approach to multi-level non-LTE problems has been demonstrated by Werner & Husfeld (1985) and Werner (1986) for the case of plane-parallel, static atmospheres of hot stars (cf. also Werner, this book).

The same concept can be applied to spherically expanding atmospheres (Hamann, 1985b, 1986). In the present paper we describe how the operator perturbation technique can be adapted to that case. First we study the two-level atom as a test case (Sect. 2). In Sect. 3 the method is extended to the multi-level problem.

2. THE TWO-LEVEL ATOM TEST CASE

For resonance lines where the two-level atom approximation applies, the line-formation problem in expanding atmospheres can be readily solved by differencing the equation of radiation transfer in the comoving frame (Mihalas et al., 1975, 1976). Because the problem is linear in the radiation field, the equations can be solved for the intensities by means of elimination techniques. If the line is not strongly saturated, a lambda iteration, accelerated by an Aitken extrapolation, may be sufficient and even more economical (Hamann, 1980).

Hence, the two-level case serves as an exercise before we proceed to the multi-level problem in which we are actually interested. First, the operator perturbation technique is revisited under the restriction that the approximate lambda operator is entirely local (Sect. 2.1), giving some insight into the nature of this iteration procedure. In Sect. 2.2 we then give an appropriate definition for approximate lamda operators in spherically expanding geometry, while test calculations are presented in Sect. 2.3.

2.1. Iteration with approximate lambda operators

When the radiation field $\mathbf{J}_\nu$ (angle-averaged intensity at frequency ν) is calculated from a given source function, this procedure is called a "formal solution" of the radiation transfer problem and denoted symbolically by the linear operator Λ_ν :

$$\mathbf{J}_\nu = \Lambda_\nu \ \mathbf{S}_\nu \quad . \tag{1}$$

In the case of line scattering the averaged intensity $\mathbf{J}$ is defined by the "scattering integral"

$$\mathbf{J} = \int_{-\infty}^{+\infty} \mathbf{J}_\nu \varphi_\nu \ d\nu \quad . \tag{2}$$

φ_ν denotes the normalized profile function. Correspondingly, we define Λ by

$$\mathbf{J} = \Lambda \ \mathbf{S} \quad . \tag{3}$$

The main complications in non-LTE calculations arise because the radiative transfer has to be solved consistently with the equations of statistical equilibrium. In the case of a two-level atom, these "rate equations" degenerate to the well-known formula

$$\mathbf{S} = (1-\varepsilon)\,\mathbf{J} + \varepsilon\mathbf{B} \quad . \tag{4}$$

The first term describes the scattering process, while $\varepsilon\mathbf{B}$ denotes the thermal sources.

Hence, **J** must satisfy the equation

$$\mathbf{J} = \Lambda\,[(1-\varepsilon)\,\mathbf{J} + \varepsilon\mathbf{B}] \quad . \tag{5}$$

The solution may be found iteratively by a so-called lambda iteration:

$$\mathbf{J}_{new} = \Lambda\,[(1-\varepsilon)\,\mathbf{J}_{old} + \varepsilon\mathbf{B}] \quad . \tag{6}$$

This procedure requires only the performance of formal solutions, but fails if large optical depths occur. Alternatively, this linear Eq. (5) may be solved for the unknown variable **J** by a suitable elimination technique (cf. Mihalas, 1978).

Another idea is the construction of an iteration scheme which retains the simple structure of the lambda iteration, but avoids the inherent convergence problems. Following Cannon (1973), this is done by introducing a new, "approximate" lambda operator Λ^*. An iteration scheme is derived from the identity

$$\mathbf{J} = (\Lambda - \Lambda^*)\,\mathbf{S} + \Lambda^*\,\mathbf{S} \quad . \tag{7}$$

The iteration now may proceed as

$$\mathbf{J}_{new} - \Lambda^*\,\mathbf{S}_{new} = (\Lambda - \Lambda^*)\,\mathbf{S}_{old} \tag{8}$$

(c.f. Cannon, 1973; Scharmer, 1981). This makes sense, if Λ^* has the following properties:

a) Λ^* can be inverted easily, so that $\mathbf{J}_{new}$ and $\mathbf{S}_{new}$ can be

consistently determined with little effort;
b) $\Lambda - \Lambda^*$ does not yield contributions from optically thick line cores, so that the convergence problems of the conventional lambda iteration are avoided although this term acts on the old source function.

Inserting the "rate equation" (Eq. 4) and denoting the formal solution obtained from the old source function by J_{FS} leads to

$$J_{new} - \Lambda^* (1-\varepsilon) J_{new} = J_{FS} - \Lambda^* (1-\varepsilon) J_{old} \qquad (9)$$

or

$$J_{new} = \left[1 - \Lambda^* (1-\varepsilon)\right]^{-1} \left[J_{FS} - \Lambda^* (1-\varepsilon) J_{old}\right] \quad . \qquad (10)$$

A very simple operator Λ^* which satisfies the above requirements may be defined by picking up the idea of "core saturation" (Rybicki, 1972):

$$\Lambda^*_\nu S = \begin{cases} S & \text{in optically thick line cores} \\ 0 & \text{else} \end{cases} \quad . \qquad (11)$$

The suitable definition of frequencies ν_{red} and ν_{blue} (depth-dependent) which separate the optically thick line core from the wings will be discussed in Sect. 2.2. With these frequencies determined, we define the (again depth-dependent) "core fraction" f_c as that part of the scattering integral which falls into the optically thick domain:

$$f_c = \int_{\nu_{red}}^{\nu_{blue}} \varphi_\nu \, d\nu \quad . \qquad (12)$$

Hence the application of the approximate operator Λ^* becomes a simple multiplication (assuming complete redistribution, i.e. a frequency-independent line source function):

$$\Lambda^{*} S = f_c S \quad . \tag{13}$$

As this operator acts only locally, it can be easily inverted, so that Eq. (10) becomes

$$J_{new} = \left[1 - (1-\varepsilon)\, f_c\right]^{-1} \left[J_{FS} - (1-\varepsilon)\, f_c\, J_{old}\right] \quad . \tag{14}$$

The structure of this iteration process may be clarified by writing it in the form

$$J_{new} - J_{old} = \left[1 - (1-\varepsilon)\, f_c\right]^{-1} \left(J_{FS} - J_{old} \right) \quad . \tag{15}$$

This can be interpreted as an "accelerated lambda iteration": The iteration is driven by the difference between the old radiation field J_{old} and the new radiation field J_{FS} which is obtained by means of a formal solution. Compared with a conventional lambda iteration, however, the corrections are amplified by the factor $1/(1-(1-\varepsilon)f_c)$. The larger the core fraction, the larger the amplification factor.

2.2. Definition of the line cores

Up to here, there was no need to specify the geometry of the problem. In order to apply the iteration process of Eq. (14), the core fraction f_c must be estimated now from physical considerations.

As defined in Sect. 2.1, the core fraction denotes that part of the scattering integral which falls within the optically thick line core. For a static, plane-parallel atmosphere the procedure may be straightforward (cf. Scharmer, 1981): at the depth point considered, the line core covers those frequencies which have a monochromatic optical depth τ_ν - as seen from the boundary - exceeding a specified value γ of the order of unity. However, for practical reasons, this concept does not apply to rapidly expanding spherical atmospheres: the computation of the optical depths for all radius points, angles, and frequencies by integration in the observer's frame would not only be time-consuming but also very difficult, due to the necessary refinement of integration steps in the vicinity of the line resonance. Therefore, we prefer a different approach and estimate the line core from local quantities alone, using ideas which resemble the Sobolev approximation.

As throughout this paper, a spherical atmosphere, expanding with an accelerated velocity $v(r)$ ($dv/dr > 0$) and without continuum opacities, is under consideration. The line should be purely Doppler-broadened with mean velocity v_D. A frequency variable x (in Doppler units relative to the line center ν_o) is defined in the comoving frame of reference (CMF):

$$x = (\nu/\nu_o - 1) c / v_D \quad . \tag{16}$$

Note that any angle-averaging in moving media must be performed in the fluid frame, i.e. for a fixed CMF frequency. In the CMF the profile function φ_x is isotropic. The task now consists of determining at each depth point the frequency range (x_{red}, x_{blue}) of the optically thick line core. The core fraction f_c of the scattering integral then results in

$$f_c = \int_{x_{red}}^{x_{blue}} \varphi_x \, dx \Big/ \int_{-x_{max}}^{x_{max}} \varphi_x \, dx \quad . \tag{17}$$

The denominator is introduced to retain the exact normalization with respect to the finite frequency band $\pm x_{max}$ (typically 3 or 4 Doppler units) considered. For later use, we introduce a shorter notation by defining the integral of the profile function :

$$\Phi(x) = \int_{-x_{max}}^{x} \varphi_{x'} \, dx' \quad . \tag{18}$$

(For Doppler profiles, $\Phi(x)$ is related to the error function.) Eq. (17) then reads

$$f_c = [\Phi(x_{blue}) - \Phi(x_{red})] / \Phi(x_{max}) \quad . \tag{19}$$

We now choose a parameter γ and define a given frequency not to belong to the line core if a ray exists which reaches the (inner or outer) boundary via an optical path shorter than γ. The strict application of this condition would imply that optical depths should be calculated in all spatial directions (Fig. 1). Note that a photon which starts from a point at radius r with a CMF frequency x alters its CMF frequency whilst travelling through space, according to the differential motion between different points in the atmosphere. Hence, the exact determination of all optical path lengths may be extremely cumbersome.

For convenience one may pose simplifying assumptions in calculating the core fraction f_c. The appropriate requirements may depend on the specific problem under study. A poor estimate of f_c may result in slow convergence (if f_c is under-estimated) or

Fig. 1. Schematic representation of the expanding atmosphere. Photons from a point at radius r may escape in all spatial directions to a boundary. For estimating the core fraction f_c, we only consider the radial and transversal directions (heavy lines).

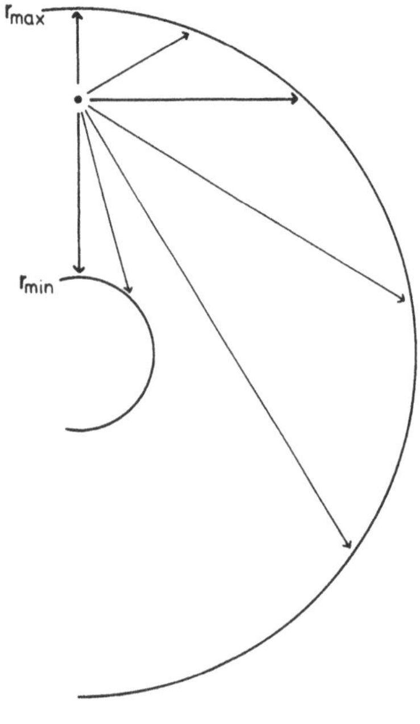

divergence (if over-estimated). In the following we describe a simplified estimate which has proved to be adequate for the test calculations presented in Sect. 2.3. We stress that, although important for practical applications, the way we estimate f_c is not essential to this method, but can be easily modified if necessary.

Our simplifications in determining f_c are:

a) Only two spatial directions are considered, namely the radial and the transversal rays;

b) Velocity gradient and opacity are assumed to be constant in the vicinity of the point under consideration.

The core-confining frequencies x_{red} and x_{blue} described in the following must be estimated separately for the two spatial directions considered. Different considerations apply to interior points and to points lying close to a boundary.

2.2.1. Interior points

A photon travelling along a ray with a given impact parameter p across the spherically expanding atmosphere encounters exact resonance with the line-absorbing atoms only at one spatial point at radius r. Actually, the interaction may take place within a finite spatial range around this exact resonance, due to the width of the absorption profile in the fluid frame. In this section we consider the case of "interior points" where this "scattering zone" lies entirely inside the boundaries of the atmosphere, and we estimate the optical depth across that zone.

The optical depth is defined as the opacity, integrated along the ray. In the situation under consideration (i.e. a spherically expanding atmosphere with $dv(r)/dr > 0$), any two volume elements recede from each other. Hence, a photon of a given (constant!) frequency in the observer's frame is continuously red-shifted in its comoving-frame frequency when travelling. This holds for all spatial directions, irrespective whether they are directed inward or outward. Hence, for the evaluation of the optical depth the integration along the spatial coordinate can be transformed into an integration over the CMF frequency, x, thus involving the velocity gradient (projected on the ray considered), v'(r,p).

Under the approximation that the line opacity and the velocity gradient remain constant over the scattering zone, its optical diameter finally becomes (cf. Castor, 1970)

$$\tau(r,p) = \kappa(r) \,/\, v'(r,p) \quad . \tag{20}$$

Note that the expansion velocity and its gradients are measured in Doppler units as used in Eq. (16) to define the dimensionless CMF frequency, x. The line opacity, $\kappa(r)$, refers to the corresponding unit frequency interval.

For the two spatial directions considered, the projected velocity gradient is

$$v'(r,p=0) = dv(r)/dr \quad \text{or} \quad v'(r,p=r) = v(r)/r \tag{21}$$

Fig. 2. Characteristics of the transport equation in the comoving frame (CMF) for inward and outward rays (schematic representation). When travelling in space, any photon is redshifted in its CMF frequency x due to the differential expansion. Within the finite frequency band $\pm x_{max}$ the line scattering is calculated.

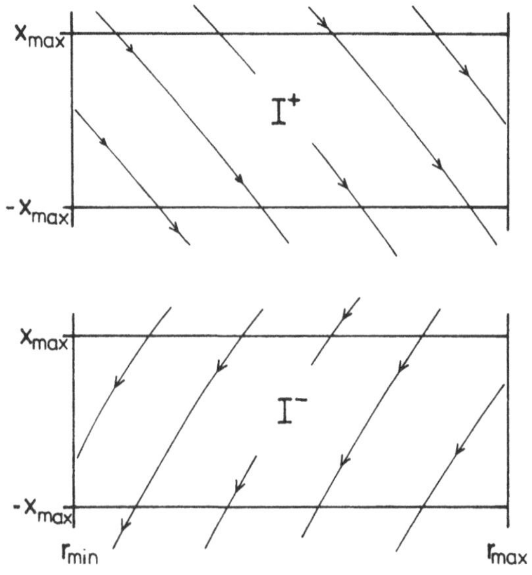

for radial or transversal rays, respectively.

If $\tau(r) < \gamma$ the profile is optically thin, i.e. we set $f_c(r) = 0$. Otherwise, an optically thick line core exists.

We recall that any two volume elements recede from each other, i.e. any photon is red-shifted in its CMF frequency when travelling (Fig. 2). Hence, a photon which starts at r with a CMF frequency in the blue wing always has to penetrate the whole line core before it can escape. Consequently, if $\tau(r) > \gamma$ the whole blue side of the line belongs to the line core in the sense that photons cannot escape, but are trapped more or less locally! Hence,

$$\text{if} \quad \tau(r) > \gamma : \quad x_{blue} = x_{max} \quad . \tag{22}$$

From the red wing, however, photons may leave the scattering zone. The confinement between wing and core, x_{red}, is derived again by transforming the optical depth integral into an integral over the CMF frequency x. One obtains

$$\Phi(x_{red}) - \Phi(-x_{max}) = \gamma / \tau \quad . \tag{23}$$

2.2.2. Boundary zones

The above considerations only hold for radius points which are well apart from the boundaries, i.e. from the red wing confinement x_{red} the boundaries cannot be seen:

$$x_{red} - (-x_{max}) < \Delta v \quad . \tag{24}$$

Δv is the relative velocity between the radius point considered and the boundary. In radial direction, the photon may escape via the outer or the inner boundary, i.e.

$$\Delta v = \max\left(v(r_{max}) - v(r) \;,\; v(r) - v(r_{min}) \right) \quad . \tag{25}$$

On a transversal ray, the velocity relative to its intersection point with the outer boundary (cf. Fig. 1) is

$$\Delta v = v(r_{max})\ (1-r^2/r^2_{max})^{1/2} \quad . \tag{26}$$

If condition (Eq. 24) is not true, we are considering a point in the "boundary zone": the "characteristic" (Fig. 2) which meets (x_{red}, r) crosses the spatial boundary before leaving the CMF frequency band. However, we demand that there is an optical depth of γ between (x_{red}, r) and the boundary. Hence, in this case the appropriate x_{red} must satisfy the non-linear equation

$$\Phi(x_{red}) - \Phi(x_{red} - \Delta v) = \gamma/\tau \quad . \tag{27}$$

(For verification, transform again the optical depth integral into an integration over frequency!) A discussion of this equation may be restricted to profile functions which are symmetric about the center and monotonically decreasing towards the wings. Then Eq. (27) has exactly two solutions if

$$\Phi(\Delta v/2) - \Phi(-\Delta v/2) = 2\ \Phi(\Delta v/2) - 1 > \gamma/\tau \quad . \tag{28}$$

Otherwise, the core is optically thin ($f_c = 0$). The smaller of the solutions, x_{red}, lies between the x_{red} for interior points (Eq. 23) and $\Delta v/2$. The other solution is x_{blue}, lies between $\Delta v/2$ and x_{max} and can be obtained from x_{red} because both solutions lie symmetrical to $\Delta v/2$:

$$x_{blue} = \Delta v - x_{red} \quad . \tag{29}$$

The core-confining frequencies have now been determined for the boundary zones as well as for interior points. This whole procedure is performed for both spatial directions under consideration; more directions could be treated if necessary in a similar manner.

Finally, at each radius point the minimum extension of the core, i.e. the largest x_{red} and the smallest x_{blue} of all directions, is inserted to estimate the core fraction $f_c(r)$ (Eq. 17).

2.3. Test calculations

The iteration procedure of Sect. 2.1, together with the core fraction estimate of Sect. 2.2, is now applied in a series of test calculations in order to study the convergence properties. The formal solutions are performed with the usual difference equation technique in the comoving frame (Mihalas et al., 1975; Hamann, 1981). However, we emphasize that the way in which the formal solution is obtained is optional, as this is a completely independent step within our iteration procedure.

2.3.1. Model assumptions

Basic assumptions of the calculations are: spherical symmetry, pure scattering ($\varepsilon=0$) by two-level atoms, pure Doppler broadening and complete redistribution. No continuum opacity is accounted for. The intensity which enters the atmosphere from the stellar core via the inner boundary is assumed to be constant in frequency and angle, while the outer boundary condition is given by $I^- = 0$.

The atmosphere extends from the inner boundary at $r_{min} = 1$ to the outer boundary at $r_{max} = 50$. The velocity field is of the usual form

$$v(r) = v_\infty \; (1 - b/r) \quad . \tag{30}$$

We set $v(r_{max}) = 2100$ km/s and $v(r_{min}) = 21$ km/s, which fixes the constants in Eq. (30). The Doppler-broadening velocity is $v_D = 100$ km/s. These values are typical for stellar winds from early-type stars. (Note that only the ratios of the quoted velocities enter the calculation.)

The opacity is set

$$\kappa(r) = \frac{\kappa_0}{r^2 \; v(r)/v_\infty} \tag{31}$$

according to the geometrical dilution of a spherically expanding flow. The parameter κ_0 is varied in order to study cases of different optical depths.

2.3.2. Results

Test calculations are now discussed for three different opacities (κ_0 = 10, 10^3 or 10^5, respectively), and the convergence behaviour is studied according to its dependency on the parameter γ. The iteration starts from the Sobolev approximation, and the maximum relative correction

$$\log \left(\max \frac{|S_{new} - S_{old}|}{S_{old}} \right) \tag{32}$$

Fig. 3. Convergence test of the example with κ_0 = 10 and different values of the parameter γ (labels). Maximum relative corrections (Eq. 32) are plotted logarithmically versus the iteration number.

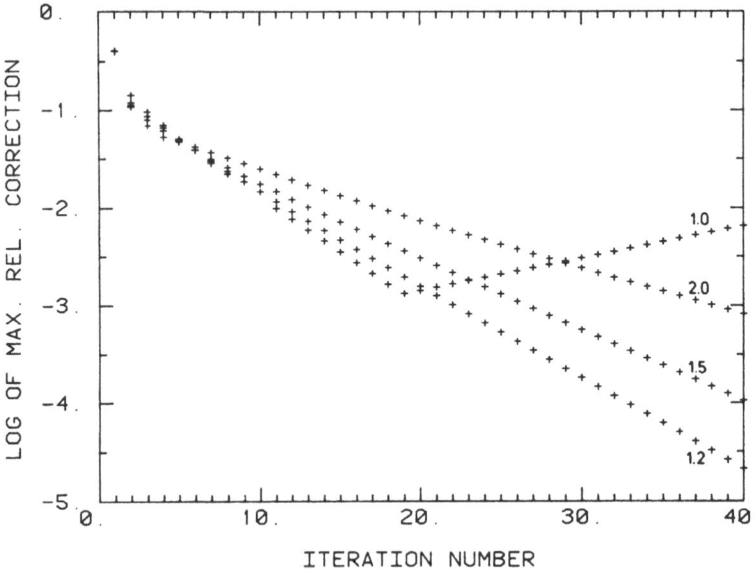

has been plotted logarithmically versus the iteration number (Figs. 3, 5, 6) for different values of γ.

An inspection of Fig. 3 reveals that for $\kappa_o = 10$ convergence is achieved if $\gamma \geq 1.2$. Higher values of γ yield slower convergence. This is expected because with higher γ, lower core fractions result, i.e. more work is left for the conventional lambda iteration which implicitly treats the wing contributions. For $\gamma = 1.0$ the solution tends to alternate and finally diverges, i.e. core fractions that are too large cause overcorrections that are too large.

The convergence, if any, is obviously linear, i.e. the corrections decrease in geometrical series. This demonstrates the progress against conventional lambda iteration, where the corrections tend to stabilize far away from the solution in optically thick cases.

Fig. 4. The extension of the "optically thick line core", as defined in Sect. 2.2, for the test case $\kappa_o = 10$, $\gamma = 1.2$. The CMF frequencies x_{red} and x_{blue} which confine the core against the wings are plotted versus the local velocity v(r) which serves here as depth coordinate.

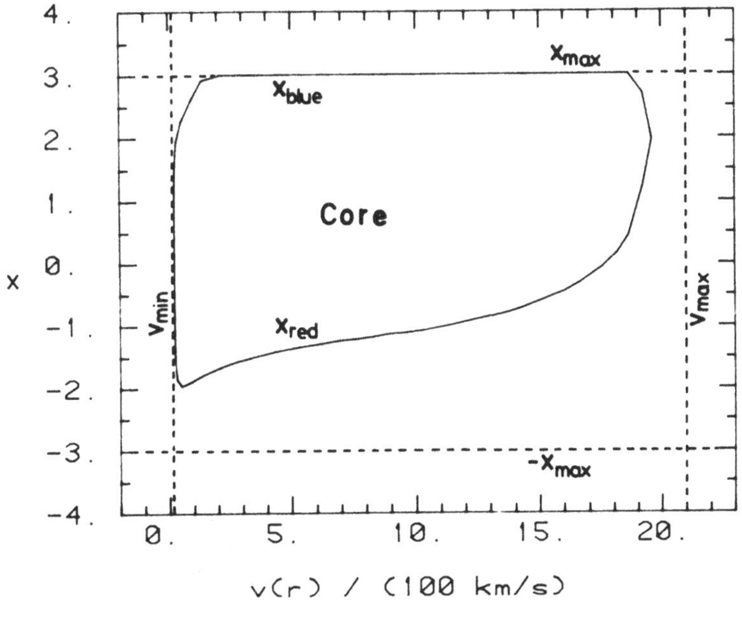

The definition of the line core (Sect. 2.2) is illustrated in Fig. 4 for the present example ($\gamma = 1.2$). Within the considered CMF frequency band (± 3 Doppler widths), the line core is confined

Fig. 5. Same as Fig. 3, but for $\kappa_o = 10^3$

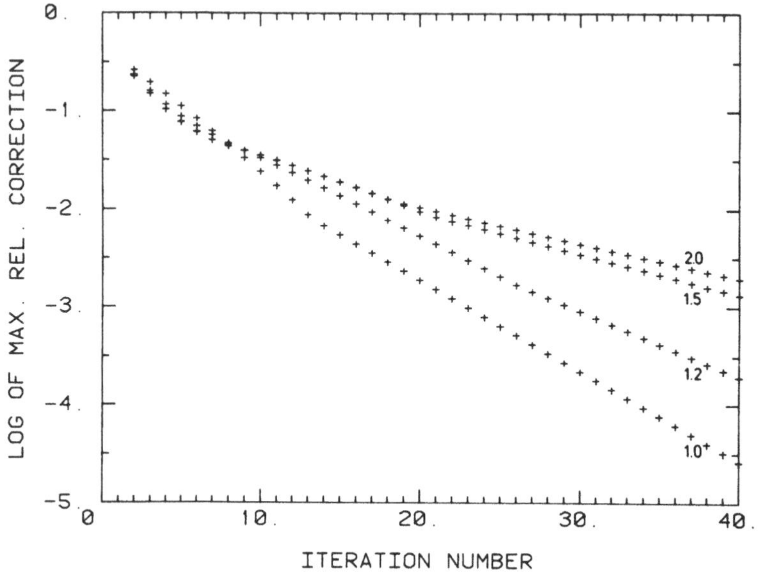

Fig. 6. Same as Fig. 3, but for $\kappa_o = 10^5$

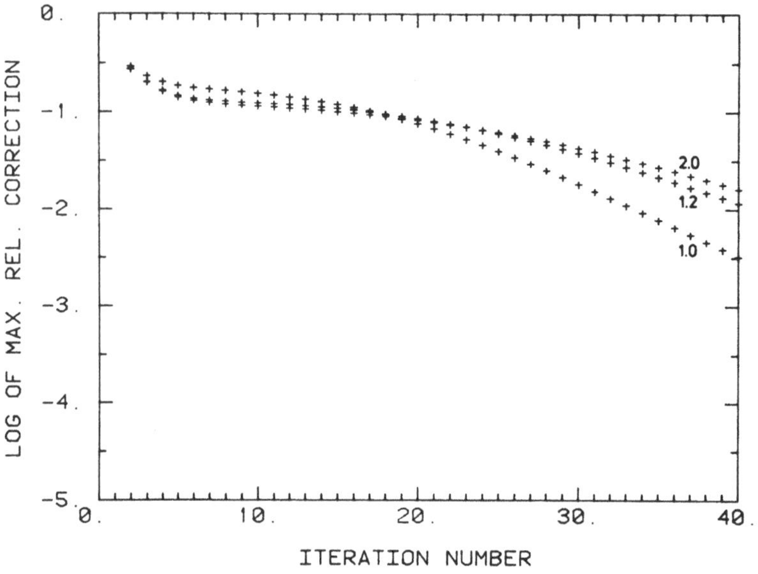

by x_{red} and x_{blue} against the wings. The velocity v(r) serves as depth coordinate in Fig. 4. In this example, the core fraction reaches a maximum of 0.9948, i.e. the corrections are amplified by a factor up to 191.3, compared to a conventional lambda iteration.

The next example (Fig. 5) deals with a larger optical depth ($\kappa_o = 10^3$) and behaves similar to the first case, although the convergence is somewhat slower. The maximum correction amplifier occurring in this example amounts to 3.1 10^4 (γ= 1.0).

The last test with $\kappa_o = 10^5$ (Fig. 6) handles an extreme optical depth which is typical for the He II resonance lines in Wolf-Rayet star atmospheres (for illustration: the radial integral over $\kappa(r)$ amounts to 10^7). A bandwidth of $x_{max} = 4.0$ Doppler units is now required. Even in this case, convergence is obtained, although at slower rate than in the previous examples. The maximum correction amplifier reaches 3.5 10^6.

All converged solutions agree with solutions obtained in the "classical" way, i.e. by elimination from the moment equations and iteration with respect to the Eddington factors (Mihalas et al., 1976).

3. THE MULTI-LEVEL ATOM CASE

After these encouraging results in the two-level test case, we now focus on the multilevel problem, motivated mainly by the so far unsolved task of Wolf-Rayet star spectrum synthesis. A solution of the comoving-frame radiation transfer in spherically expanding geometry, with simultaneous solution of the multi-level rate equations, can be achieved by the "equivalent two-level atom approach" (Mihalas & Kunasz, 1978). This technique was applied with some success for calculating He II spectra of Wolf-Rayet stars (Hamann, 1985a), but the computational effort is high, and convergence problems caused by complex level coupling are severe and sometimes even insurmountable.

The present approach allows the entire comoving-frame formalism for the formal solution of the radiation transfer to be retained. Only the statistical equations become modified, as they now incorporate the approximate radiation transfer (Sect. 3.1). The definition of approximate lambda operators (Sect. 3.2) profits from the two-level study. Test calculations are described in Sect. 3.3, and the convergence properties are studied in Sect. 3.4.

3.1. Iteration with approximate lambda operators

Denote the set of population numbers which is reached in a certain stage of an iteration procedure by $\mathbf{n}_{\mathrm{old}}$. The (line or continuum) source function calculated from these "old" population numbers is called $\mathbf{S}_{\mathrm{old}}$. As in Sect. 2, the radiation transfer equation is symbolized by the linear operator Λ and yields the radiation field $\mathbf{J}_{\mathrm{FS}}$ as "formal solution":

$$\mathbf{J}_{\mathrm{FS}} = \Lambda\ \mathbf{S}_{\mathrm{old}} \quad . \tag{33}$$

In the multi-level non-LTE the radiation field enters the radiative rate coefficients of the statistical equilibrium equations and, hence, the population numbers depend (non-linearly!) on the radiation field, symbolically:

$$\mathbf{n} = \mathbf{n}(\mathbf{J}) \quad . \tag{34}$$

The main complications in non-LTE calculations arise because the radiation transfer (Eq. 33) must be solved consistently with the equations of statistical equilibrium (Eq. 34). For practical application, efficient iterative algorithms are required for solving this high-dimensional non-linear problem.

As in the two-level case, a simple procedure is the "lambda iteration": new population numbers are calculated from the formal-solution radiation field,

$$\mathbf{n}_{\mathrm{new}} = \mathbf{n}(\mathbf{J}_{\mathrm{FS}}) \quad , \tag{35}$$

and the steps denoted by Eqs. (33) and (35) are repeated in turn.

Still in complete analogy to the two-level case (Eq. 8), we again follow Cannon (1973) by introducing an approximate lambda operator Λ^* and define the "new" radiation field by

$$\mathbf{J}_{new} = \mathbf{J}_{FS} + \Lambda^* (\mathbf{S}_{new} - \mathbf{S}_{old}) \quad . \tag{36}$$

This radiation field (Eq. 36) is now inserted into the rate equations (Eq. 35):

$$\mathbf{n}_{new} = \mathbf{n}(\mathbf{J}_{new}) \quad . \tag{37}$$

However, $\mathbf{J}_{new}$ is not explicitly known, as the right-hand side of Eq. (36) contains the new source function $\mathbf{S}_{new}$, i.e. implicitly the new population numbers which are about to be calculated. Hence, the rate equations are no longer linear in $\mathbf{n}_{new}$.

The exact lambda operator determines the radiation field at one spatial point from the source function at all points throughout the atmosphere. The decisive simplification now arises from our definition of approximate lambda operators which are entirely local. Thus, one avoids that the modified rate equations (Eq. 37) become spatially coupled. Let $\mathbf{n} = (n_1, \ldots, n_N)$ denote the row vector of population numbers of levels 1 to N at a certain depth point. The rate equations at this depth point then read as

$$\mathbf{n}_{new} \, \mathbf{P}(\mathbf{n}_{new}, \mathbf{n}_{old}, \mathbf{J}_{FS}) = \mathbf{b} \tag{38}$$

Row vector $\mathbf{b}$ contains essentially zeros except those columns representing charge- or number conservation. $\mathbf{P}$ is the usual rate coefficient matrix, composed of collisional and radiative terms. The latter depend on the new radiation field (Eq. 36), hence involving the new population numbers at the same spatial point. For given old population numbers $\mathbf{n}_{old}$ and formal solution radiation field $\mathbf{J}_{FS}$, both at the considered depth point, the non-linear Eq. (38) can now be solved with respect to $\mathbf{n}_{new}$ by linearization (Newton-Raphson iteration). Some effort, i.e. the use of "net radiative brackets", is necessary to overcome numerical problems arising in certain situations because the system of equations is poorly conditioned. Details of the formulation are given in the Appendix.

Radiation transfer (Eq. 35) and modified rate equations (Eq. 38) are repeatedly solved in turn, similar to the normal lambda iteration procedure. The interpretation in terms of an "accelerated lambda iteration" given for the two-level case (Sect. 2) also applies here: The iteration is driven by the inconsistency between (old) population numbers and the radiation field obtained by "formal solution". Compared to a simple lambda iteration, however, the corrections are drastically amplified, the more Λ^* approaches unity. Contrary to the two-level case, the amplification acts multidimensionally, so that the correction vector in the vector space of population numbers is not only lengthened, but also turned in direction.

3.2. Definition of approximate lambda operators

For the estimate of the "line cores" we adopt the same definition as the one we tested successfully in the two-level case (Sect. 2). Thus the underlying continuum opacities are further neglected for that estimate. Contrary to Sect. 2.2, the "core fractions" are no longer useful in the presence of underlying continua, as the total (line plus continuum) source function in the scattering integral is frequency-dependent. Hence, the integration over the line core must be performed numerically:

$$\Lambda^* S = \int_{x_{red}}^{x_{blue}} S_x \; \varphi_x \; dx \qquad (39)$$

In addition to the lines we now also need an approximate lambda operator for the continua. In the continua the Doppler shifts can be neglected. Hence, a radial optical depth $\tau_\nu(r)$ may be defined in the usual manner. We denote the maximum optical depth which is reached at the bottom of the atmosphere by τ_ν^{max}, and replace τ_ν afterwards by

$$\tau_\nu(r) := \min(\ \tau_\nu(r),\ \ \tau_\nu^{max} - \tau_\nu(r)\ ) \tag{40}$$

The approximate lambda operator may be defined by

$$\Lambda_\nu^*(r) = 1 - \exp(\ -\tau_\nu(r)\ /\ \gamma_c\ ) \tag{41}$$

with some adjustable parameter γ_c of order unity. This definition reflects a physically reasonable estimate of that fraction of photons which cannot escape from the atmosphere, i.e. which is trapped more or less locally.

With the alternative formulation

$$\Lambda_\nu^*(r) = \begin{cases} 1 \text{ if } & \tau_\nu(r) > \gamma_c \\ 0 \text{ else} & \end{cases} \tag{42}$$

we experienced convergence instabilities. The reason is that Eq. (42) neglects the (small) probability for spatial escape of photons from optically thick regions, although this effect can dominate over local photon destruction rates.

3.3. Test calculations

The method is tested for spherically expanding atmospheres of pure helium. The assumptions are typical for Wolf-Rayet stars. Model parametrization and the formal solution of the radiation transfer are the same as used by Hamann (1985a) for our small model grid calculated by the "equivalent two level atom approach", and the reader is referred to that paper for more details.

3.3.1. Radiation transfer

The formal solution in the continuum is calculated in spherical symmetry by the usual Feautrier technique. Each frequency point can be treated separately, while the equations are coupled over all impact parameters via the Thomson scattering term. Due to the large number of impact parameter points, this angle-dependent radiative transfer is computationally expensive. Therefore we use it only for the calculation of Eddington factors. The formal solution of the continuum transfer is actually obtained with much less computational effort from the moment equations, keeping the Eddington factors fixed over several iterations. In the test calculations presented here the Eddington factors are recalculated every three iteration cycles, but a less frequent update would also suffice.

Formal solutions of the line radiation transfer are calculated by means of the comoving-frame technique (Mihalas et al., 1975). Pure Doppler broadening with mean velocity v_D and complete redistribution are assumed.

3.3.2. Model assumptions

The stratifications of temperature and density must be specified in advance. The assumptions which allow the model to be constructed from a few basic parameters are only briefly summarized here.

With the mass loss rate $\dot{M}$ given, the density stratification and the velocity field are related by the equation of continuity. In the high-velocity region we specify the velocity by the usual law (Eq. 30). In the low-velocity part of the atmosphere we assume a hydrostatic density stratification for a specified effective gravity. The connection point of both domains is defined by the condition that the transition must be smooth. In practice it lies close to the sonic point.

A parameter T_* defines the temperature stratification, which is calculated by assuming a plane-parallel, grey LTE-atmosphere with $T_{eff} = T_*$. (In a plane-parallel, grey atmosphere the temperature approaches a constant boundary value, while in spherical geometry it would decrease continuously with radius.)

3.3.3. The atomic model

Pure helium atmospheres are considered only. He III, 10 bound levels of He II and the He I ground state are taken into account. All 45 line transitions are fully included.

3.4. Results

The new method has been successfully applied for many parameters in a grid of models (Hamann and Schmutz, in preparation). In the present paper we demonstrate the applicability and the convergence properties of the new technique. To this purpose we concentrate on one typical model having the following parameters:

$\dot{M} = 4\ 10^{-5}\ M_{\odot}/yr$

$T_* = 35$ kK

$R_* = 12.5\ R_{\odot}$

$\log g_{eff} = 3.5$ (cgs units)

$v_{max} = 2500$ km/s at $R_{max} = 60\ R_*$

$v_D = 100$ km/s

In the definition of the approximate lambda operators the parameters γ_C (for continua) and γ_L (for lines) were introduced. These parameters determine the convergence of the iteration. In the following we take the same value γ for continua and lines, $\gamma = \gamma_C = \gamma_L$, in order to simplify the representation. In practice, the convergence can be further improved in many cases by means of individual γ's, distinguishing not only between continua and lines, but also between resonance lines and subordinate lines.

The limit $\gamma = \infty$ corresponds to the usual lambda iteration. The smaller the value of γ, the more the corrections are amplified. The optimum value should be of the order of unity, while too small a value of γ will result in overcorrections and can cause divergence.

Fig. 7 shows two level populations at different stages of the iteration. The population numbers of the He II n=1 (ground state) and the He II n=2 level, relative to total helium, are plotted versus the helium number density, which serves as depth coordinate. The labels denote the number of completed iterations. The parameter γ has been decreased stepwise during the iteration, which is always a recommended procedure as corrections that are too large at the beginning may diverge. Starting from LTE (label 0), three iterations are performed with $\gamma = \infty$, then three iterations with $\gamma = 64$, and then three iterations with $\gamma = 16$. After these 9 iterations the final value of $\gamma = 4$ is retained.

Fig. 7. Run of level population after a certain number of iterations (label). The population numbers (relative to total helium) of He II n=1 (ground state) and n=2 are plotted versus helium number density, which serves as depth coordinate. A typical line-forming depth is marked "A" ($r = 2\ R_*$, $v(r) = 1/2\ v_\infty$), while "B" denotes a continuum-forming layer (optical depth in the visual about 2/3).

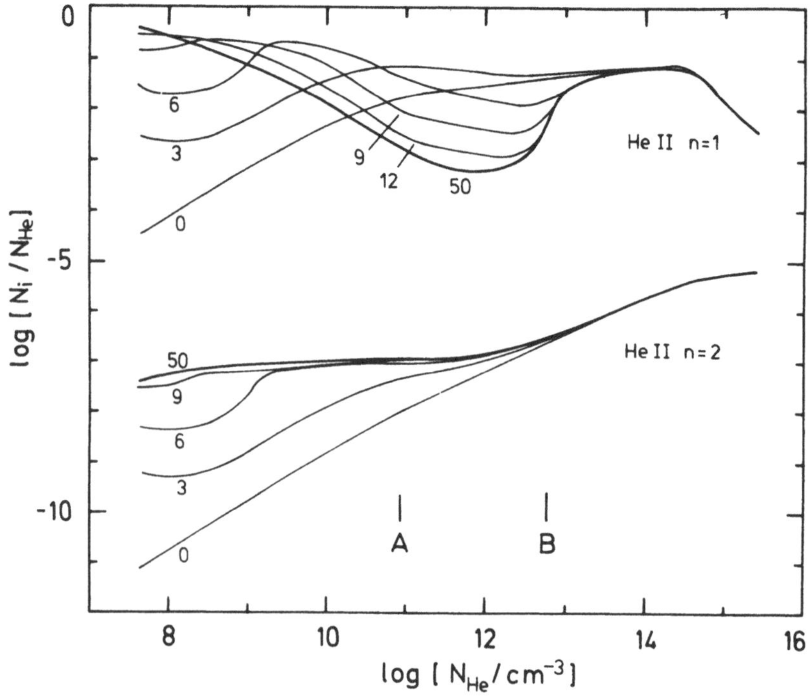

The He II ground state causes the most severe problems in atmospheres of this type, as it is only marginally coupled to the other levels, and the optical depths of the resonance transitions are extremely large. The LTE solution, which serves as the first approximation at the start of the iteration, deviates from the converged solution by some orders of magnitude. After the three lambda iterations at the beginning, the ground state still exhibits

Fig. 8. Relative corrections between subsequent iterations versus iteration number, for population numbers of the level He II n=2 at depth "A" and of the He II ground state at depth "B" (cf. Fig. 7). Different series of calculations are labelled by their parameter γ.

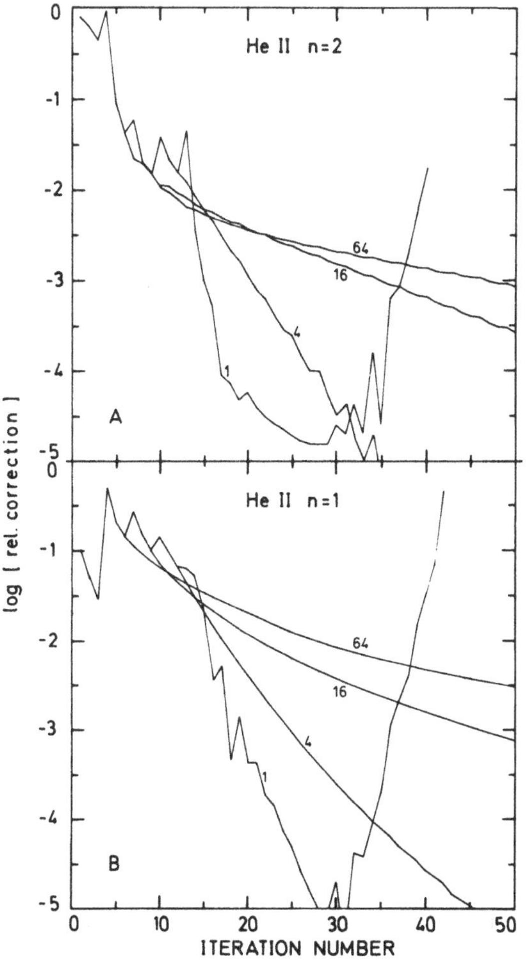

a qualitatively wrong run with density. In certain parts the corrections even point in the wrong direction. (Test calculations confirm that after many further lambda iterations the correct solution would still not be approached.) Further iteration with stepwise-decreased γ, however, quickly changes the run of the He II ground state population, and leads finally to smooth convergence.

The He II n=2 level is much easier to handle. The transition 2-3 (λ1640 Å) is not optically thick in the outer regions and would already converge in a simple lambda iteration. Hence its coupling to the ground state remains the only drawback of convergence.

The higher levels of He II, which are not presented in the figures, behave like the n=2 state.

Now we trace the rate of convergence at two representative layers (marked by A and B in Fig. 7). Fig. 8A shows the relative corrections between successive iterations for the population of He II n=2 at depth A, corresponding to a typical line-forming radius $r = 2\ R_*$; Fig. 8B describes the He II ground state at depth B, where the (visual and UV) continuum is formed. Different series of iterations are shown: The series discussed above (Fig. 7), with a final γ of 4 (=label), converges best among the presented tests. When the stepwise decrease of γ is stopped with γ= 16 or 64, respectively, slower convergence results. On the other hand, the solutions start oscillating with growing amplitude and diverge when γ is further decreased to 1.

For proper values of γ (e.g. 4 in our example), the corrections tend to align on a straight line in the logarithmic plot, i.e. they form a geometrical series. Hence, convergence is assured. We conclude from Fig. 8 that in the given example one-percent-accuracy is reached within 20 or 30 iterations.

The computing time per iteration amounts to 35 s CPU time on a CRAY-1M machine (accounting for 64 depth points and 68 impact parameter points). The main consumption arises from the formal solution of the radiation transfer in the 45 lines with 31 frequency points each. Note that this demand grows only linearly with the number of line transitions. Another 30 s are needed for every update of the continuum Eddington factors at the 89 frequency points, even when using a machine-optimized matrix inversion.

4. CONCLUSIONS

The perturbation technique, combined with the concept of core saturation, can solve the multi-level non-LTE problem by means of approximate lambda operators which are entirely local. This yields a very general iteration scheme, which can be easily combined with any available technique that yields a "formal solution" of the radiation transfer equation for the geometry under consideration. This method is applied to spherically expanding atmospheres, using the comoving-frame formalism for the line radiation transfer. Adequate approximate lambda operators are defined and tested successfully in the two-level atom case. Satisfactory convergence is also obtained in the multi-level case for atmospheres which are typical for Wolf-Rayet stars. Although up to now only the He II lines have been included, the method seems to be capable of handling up to 100 atomic levels with arbitrarily complex non-LTE coupling and is hence superior to previous methods.

Acknowledgements. It is a pleasure to thank K. Hunger, D. Husfeld, W. Schmutz, K. Werner and U. Wessolowski for inspiring ideas, fruitful discussions and encouraging collaboration.

REFERENCES

Cannon, C.J.: 1973, Astrophys. J. **185**, 621

Castor, J.I.: 1970, Mon. Not. R. astr. Soc. **149**, 111

Hamann, W.-R.: 1981, Astron. Astrophys. **93**, 353

———: 1985a, Astron. Astrophys. **145**, 443

———: 1985b, Astron. Astrophys. **148**, 364

———: 1986, Astron. Astrophys. **160**, 347

Mihalas, D.: 1978, Stellar Atmospheres, 2nd edition, San Francisco, Freeman

Mihalas, D. & Kunasz, P.B.: 1978, Astrophys. J. **219**, 635

Mihalas, D., Kunasz, P.B. & Hummer, D.G.: 1975, Astrophys. J. **202**, 465

———: 1976, Astrophys. J. **210**, 419

Rybicki, G.B.: 1972, in "Line Formation in the Presence of Magnetic Fields", eds. R.G. Athay, L.L. House, G. Newkirk Jr., Boulder: High Altitude Observatory, p. 145

Scharmer, G.B.: 1981, Astrophys. J. **249**, 720

———: 1984, in "Methods in radiative transfer", ed. W. Kalkofen, Cambridge University Press, p. 172

Scharmer, G.B. & Carlsson, M.: 1985, Journal of Computational Physics **59**, 56

Werner, K.: 1986, Astron. Astrophys. **161**, 177

Werner, K. & Husfeld, D.: 1985, Astron. Astrophys. **148**, 417

APPENDIX

In Sect. 3.1 the approximate radiation transfer has been incorporated into the statistical equations, yielding a set of equations which is local (i.e. all quantities which enter the equations refer to the considered spatial point), but non-linear in the vector of population numbers, $\mathbf{n}_{new}$, as the coefficient matrix $\mathbf{P}$ implicitly depends on these unknown variables:

$$\mathbf{n}_{new}\ \mathbf{P}(\mathbf{n}_{new}) = \mathbf{b} \quad . \tag{A1}$$

(The subscript "new" is suppressed in the following.) While the off-diagonal elements P_{ij} account for the transitions from level i to level j, the diagonal elements P_{jj} describe the losses from level j to all other levels:

$$P_{jj} = - \sum_{m \neq j} P_{jm} \quad . \tag{A2}$$

One column of $\mathbf{P}$ represents the equation of number conservation, while another equation is added for balancing the electrons. We shall not further comment on these details as they are straightforward (cf. Mihalas, 1978).

For the solution we now apply a Newton-Raphson iteration: From a current solution $\mathbf{n}^{(k)}$ the next iteration step yields $\mathbf{n}^{(k+1)}$ by

$$\mathbf{n}^{(k+1)} = \mathbf{n}^{(k)} - (\mathbf{n}^{(k)}\ \mathbf{P} - \mathbf{b})\ \mathbf{M}^{-1} \tag{A3}$$

with $\mathbf{P}$ and $\mathbf{M}$ being evaluated from the current solution $\mathbf{n}^{(k)}$. According to the Newton-Raphson algorithm, matrix $\mathbf{M}$ contains the derivatives

$$M_{ij} = \frac{\partial}{\partial n_i} \left(\sum_m n_m\ P_{mj} \right) \quad . \tag{A4}$$

Accounting for Eq. (A2) one may write

$$M_{ij} = P_{ij} + \sum_{m \neq j} D_{imj} \tag{A5}$$

with the definition

$$D_{imj} = n_m \frac{\partial P_{mj}}{\partial n_i} - n_j \frac{\partial P_{jm}}{\partial n_i} \quad . \tag{A6}$$

(Note that $D_{imj} = - D_{ijm}$.) The coefficient matrix **P** is composed of the radiative and collisional terms,

$$\mathbf{P} = - \mathbf{R} - \mathbf{C} \quad . \tag{A7}$$

We recall that the "new" radiation field (Eq. 36) must be inserted into the radiative rates **R**. The evaluation of the derivatives (Eq. A6) is straightforward.

However, with the usual formulation of the radiative rates we encountered numerical dificulties (even with 64-bit words) when calculating some Wolf-Rayet type models with extremely thick resonance lines. These problems are caused by the disproportion between the large rates in the resonance transitions and the small gains and losses which may decisively influence the solution. (Problems of this kind are also discussed by Scharmer & Carlsson, 1985.)

A suitable formulation in which these large upward and downward rates are cancelled analytically is achieved by introducing "net radiative brackets" (cf. Mihalas, 1978). Instead of Eq. (A7) we write

$$\mathbf{P} = - \mathbf{Z} - \mathbf{C} \tag{A8}$$

with the definitions

$$Z_{ul} = R_{ul} - \frac{n_l}{n_u} R_{lu} \tag{A9}$$

and

$$Z_{lu} = 0 \quad , \tag{A10}$$

denoting the lower level by subscript l and the upper level by u. For the line transitions one finds that (c f. Mihalas, 1978)

$$Z_{ul} = A_{ul} \left(1 - \frac{J_{new}}{S_{new}} \right) \quad , \tag{A11}$$

where S_{new} is the non-LTE line source function. (Analogous expressions hold for bound-free transitions.) Thus the derivative (Eq. A6) becomes

$$D_{iul} = A_{ul} \; n_u \frac{\partial}{\partial n_i} \left(J_{new} \; / \; S^L_{new} \right) \quad . \tag{A12}$$

Differentiation of the ratio yields two terms. The first one reflects that the "net radiative brackets" make the rate equations non-linear in **n**:

$$J_{new} \frac{\partial}{\partial n_i} \left(\frac{1}{S_{new}} \right) \quad . \tag{A13}$$

This term only contributes if i=l or i=u. The second term comes from the incorporation of the approximate radiation transfer:

$$\frac{1}{S_{new}} \frac{\partial J_{new}}{\partial n_i} = \frac{1}{S_{new}} \int_{core} \frac{\partial S_{new}}{\partial n_i} \varphi_x \; dx \quad . \tag{A14}$$

Note that, in a strict sense, the total (line plus underlying continuum), frequency-dependent source function appears under the integral, preventing the use of core fractions (Sect. 2.2) and contributing derivatives even when level i is not involved in the line transition under consideration.

STELLAR ATMOSPHERES IN NON-LTE: MODEL CONSTRUCTION AND LINE FORMATION CALCULATIONS USING APPROXIMATE LAMBDA OPERATORS

K. Werner, Institut für Theoretische Physik
und Sternwarte der Universität Kiel,
Leibnizstr. 15, D-2300 Kiel, Federal Republic of Germany

Abstract. We describe a method of calculating plane-parallel model atmospheres subject to the constraints of radiative, hydrostatic, and statistical equilibrium. Simple modifications make this method equally suitable for multi-level line formation calculations. It is based on the use of local, approximate lambda operators and a perturbation technique yielding exact solutions for radiative transfer problems. Up to 100 non-LTE levels can be treated, markedly exceeding the capacity of the classical complete linearization approach of Auer & Mihalas (1969) and allowing investigations of line blanketing effects in non-LTE stellar atmospheres. As an example, line formation calculations are presented for a five-level hydrogen atom with continuum in six different model atmospheres. In addition, several pure-hydrogen model atmospheres have been computed, demonstrating the power of our method.

1. INTRODUCTION

Since the presentation of the complete linearization method by Auer & Mihalas (1969), non-LTE, H-He composed model atmospheres of hot stars have been available including line opacities of hydrogen as well as those of a "mean light ion", crudely representing continuum opacities of the CNO elements of solar abundance (Mihalas 1972). Using these models and carrying out line formation calculations with different schemes (Feautrier 1964, Rybicki 1971, Auer & Heasley 1976) based on complete linearization, analyses of stars led to many satisfactory results, but also in some cases to inconsistencies, which call attention to short-comings in model and line formation calculations.

The basic difficulty is the coupling of all variables at all depth points in the model atmosphere, requiring the simultaneous solution of the corresponding equations, i.e., the radiative transfer equations, the statistical equations for all non-LTE levels and, depending on the problem in question, particular types of additional constraint equations. Since the resulting system is, in general, non-linear, self-consistency to first order is obtained by linearization and the final solution is found by iteration. The numerical difficulty is due to the large number of transfer equations needed for an adequate description of the radiation field. Even in the variable Eddington factor formalism (Auer & Mihalas 1970), by which the angle dependence of the radiation field is ignored during the linearization procedure and the number of variables greatly reduced (to one variable per frequency point, namely, the mean intensity J_ν instead of the specific intensities $I_{\nu\mu}$), a more detailed representation of the radiation field than by about 70 frequency points is not feasible. Therefore, the effects of metal line-blanketing on the temperature and pressure stratification of non-LTE atmospheres could not be investigated until very recently.

Great progress has been made by Anderson (1985) who, in analogy to the variable Eddington factor formalism, reduced the set of unknown variables by dividing the frequency spectrum into several blocks which are characterized by their total energy density of radiation. The detailed distributions within the blocks are updated by lambda iteration .

Another approach to the radiative transfer problem is the combined use of its simplified solution by approximate lambda operators and of perturbation techniques (Cannon 1973) to obtain accurate results. Scharmer (1981) used this method successfully to treat the line formation problem for a two-level atom. Up to now, this very important work has initiated a number of papers on applications of approximate lambda operators in many fields. Two-level (Scharmer & Nordlund 1982) and multi-level (Scharmer &

Carlsson 1984) line formation calculations have been carried out but, since the linearization of transfer and statistical equations in the formulation of Scharmer and coworkers is still necessary due to non-local approximate lambda operators, their method is not capable of treating more than about 20 levels in non-LTE, this being comparable to the complete linearization technique. Werner & Husfeld (1985) used Scharmer's simple (entirely local) operator as well as a perturbation technique to develop an alternative method for the line formation problem in static, plane-parallel atmospheres, which can take into account up to 100 non-LTE levels. In a similar manner, Hamann (1985, 1986) developed a successful iteration scheme for the two-level and multi-level cases in rapidly expanding atmospheres. The algorithm of Werner and Husfeld was then extended to calculate non-LTE plane-parallel model atmospheres (Werner 1986). As an alternative to Anderson's approach, this method can be used to construct atmospheric models, including about 100 non-LTE levels. Programming of this method is straightforward and it is more convenient to use than any complete linearization technique.

In the following sections we describe the method of Werner and Husfeld and present, in detail, the manner in which line formation and model atmosphere calculations are performed. We show computations for a five-level hydrogen atom with continuum in the line formation case as well as the construction of a sample of pure hydrogen atmosphere models, demonstrating the power and stability of this approach.

2. THE METHOD

At each point of the atmosphere, the constraint equations are solved simultaneously with the approximate solution of the radiative transfer equations, which - at step i in the iteration scheme - may be written as

$$J_\nu^{(i)}(\tau) =$$

$$\Lambda_\nu^*(\tau, \tau')\, S_\nu^{(i)}(\tau') + (\, \Lambda_\nu(\tau,\tau') - \Lambda_\nu^*(\tau,\tau'))\, S_\nu^{(i-1)}(\tau')) \,, \quad (1a)$$

where J_ν, S_ν, Λ_ν^*, Λ_ν denote the mean intensity, the source function, and the approximate and exact lambda operators at frequency ν, respectively. This suggests that the mean intensity be calculated by applying an approximate lambda operator to the actual source function and adding a correction (or perturbation) term,

$$\Delta J_\nu(\tau) = (\Lambda_\nu(\tau,\tau') - \Lambda_\nu^*(\tau,\tau'))\, S_\nu^{(i-1)}(\tau') , \qquad (1b)$$

which is obtained by a formal solution for the source function of the last iteration. Clearly, if this procedure converges at all (i.e. $S_\nu^{(i-1)} \longrightarrow S_\nu^{(i)}$), it converges to the exact solution of the transfer problem, independent of the choice of the approximate operator. In particular, if we choose an entirely local operator (or at least an operator allowing a one-directional flow of information, see next section), we can reduce the set of independent variables analytically by incorporating Eq.(1) into the constraint equations. The remaining system is then local, though in general non-linear , and its solution vector may be written as

$$\underline{\psi}_d = (n_1, \ldots, n_{NL}, n_H, n_e, T) , \qquad d=1, \ldots, ND , \qquad (2)$$

where d is the index of the set of ND depth points, n_i the atomic occupation number of level i of the NL bound non-LTE levels, and T the electron temperature. The system is solved by linearizing it in terms of the current estimate $\underline{\psi}^\circ_d$ and a correction $\delta\underline{\psi}_d$, as described in detail below. After this has been done at all depths, a new iteration step according to Eq.(1) is started; new correction terms are calculated for the radiation field by performing a formal solution.

To summarize, we carry out two nested iteration cycles. In the overall cycle (from now on called "Scharmer cycle"), the radiation field given by Eq.(1) is calculated. Since our equations are non-linear, they are solved depth by depth by means of a linearization procedure in an inner cycle (from now on called "Newton-Raphson cycle"). Using the finally obtained new vectors $\underline{\psi}_d$ we proceed to the next step of the outer cycle.

In analogy to the variable Eddington formalism and to Anderson's scheme, where the explicit angle or frequency dependence, respectively, of the radiation field is eliminated from the simultaneous solutions and is treated during a formal solution, we avoid an explicit consideration of the depth coupling in solving for the vectors $\underline{\psi}_d$. However, depth coupling is provided by the correction term (1b) which is obtained by formal solution on known quantities.

3. RADIATIVE TRANSFER

The shape of the approximate lambda operator must meet some special requirements, if we want to aim at reasonable convergence properties (for a detailed discussion, see Kalkofen 1985). Setting $\Lambda_\nu^* S_\nu = 0$, one obtains the simple lambda iteration with all its drawbacks if the source function is dominated by the radiation field. The reason for this behaviour is well-known (see e.g. Mihalas 1978, p. 147) : photons emitted in the wings of a line may travel large distances through the atmosphere before they are reabsorbed or they escape at the upper boundary and thus transport information about distinct parts of the medium over a long range. In contrast, core photons are trapped essentially at their place of origin, since a subsequent reabsorption is more probable. Although these core photons constitute the overwhelming portion of all line photons, the line source function quickly responds to photon leakage in the wings. The reason for this is the very large number of emission and reabsorption processes, which redistribute core photons into the wings in a short time. Performing a lambda iteration means following the scatterings of a core photon until it stands a good chance of being converted into a wing photon (or until it is thermalized). In practice, therefore, this iteration scheme is much too slow and results in stagnating corrections for the source function. However, it can be decidedly improved by ignoring the huge number of scatterings in the line core and starting from the outset with the radiation transfer in the wings. This implies that we cannot regard the radiation field in the core as a known quantity but must calculate it simultaneously, according to the radiation field in the wings. By using Scharmer's (1981) approximate lambda operator,

$$\Lambda^*_\nu(\tau, \tau') \, S_\nu(\tau') = \begin{cases} S_\nu(\tau_\nu) & \text{if } \tau_\nu > \gamma \,, \\ 0 & \text{if } \tau_\nu \leq \gamma \,, \end{cases} \tag{3}$$

in the perturbation scheme (1), the "inactive" core photons are eliminated from the transfer problem since the source function is iterated with the operator $(\Lambda_\nu - \Lambda^*_\nu)$, which vanishes rapidly with increasing optical depth, i.e., $(\Lambda_\nu - \Lambda^*_\nu)\, S_\nu << \Lambda^*_\nu\, S_\nu$ if $\tau_\nu >> 1$. Much better convergence can now be expected and, indeed, as can be seen from Eq.(15) of Hamann (this book), a formal lambda iteration can be performed, but with amplification factors to the corrections, which are obtained by physical considerations and which may become very large at great optical depths.

The parameter γ, defining the point at which the line is divided into core and wings, is of the order of unity. Hence, Eq.(3) describes Rybicki's (1972) core saturation assumption. As long as γ is not too large, the simple wing approximation is justified because lambda iteration works well in optically thin media.

While the core saturation assumption is essential for an approximate lambda operator, some modifications can still be made in the optically thin case in order to speed up convergence. Scharmer (1981) extended the Eddington-Barbier relation for the upper boundary into the atmosphere as deep as $\tau_\nu = \gamma$; the approximate operator then appears as

$$\Lambda^*_\nu(\tau, \tau')\, S_\nu(\tau') = \begin{cases} S_\nu(\tau_\nu) & \text{if} \quad \tau_\nu > \gamma \\ 1/2\, S_\nu(\bar\tau = \bar\gamma) & \text{if} \quad \tau_\nu \leq \gamma\ . \end{cases} \quad (4)$$

In principle, this approximation should be adequate for our purposes since the Eddington-Barbier relation is known to be a good estimate for static atmospheres. The optimum value for $\bar\gamma$ cannot be known in advance, because the depth dependence of the source function is unknown. But this is not a serious problem since Λ^*_ν is an approximate operator anyway, and so we simply set $\gamma = \bar\gamma$.

However, in contrast to the operator given by Eq.(3), which contains no explicit depth coupling, the operator (4) accounts for a one-directional flow of information: the radiation field in the wing frequencies is calculated from the source function of the current iteration step at a depth below the current depth. In spite of this depth coupling, our system of equations is still local if we solve it starting from the inner boundary of the atmosphere and proceed towards the upper boundary. Two conditions must be met: first, the inner boundary of the atmosphere must be deep enough to be optically thick at all frequencies ($\tau_\nu = \gamma$) and, second, $\gamma \leq \bar\gamma$, otherwise the important advantage of the one-directional coupling and locality of our system is lost.

A further refinement to the wing approximation can be introduced if we take into account that the incoming radiation field ($I_{\nu,-\mu}(\tau_\nu=0)$) at the upper boundary may be non-zero. Then the usual boundary conditions for the transport and hydrostatic equations have to be changed and we calculate the radiation field from

$$\Lambda^*_\nu(\tau, \tau')\, S_\nu(\tau') = 1/2\, S_\nu(\bar\tau_\nu = \bar\gamma) + \int_{-1}^{0} I_{\nu,\mu}(\bar\tau_\nu = 0)\, d\mu\ ,$$

where μ is the absolute value of the direction cosine, in analogy to the use of the Eddington-Barbier relation above. If the incoming radiation is not prescribed but depends on the outgoing radiation (reflected on an envelope) as

$$I_{\nu,-\mu}(\tau_\nu=0) = A_\nu I_{\nu,\mu}(\tau_\nu=0) ,$$

where A_ν denotes the albedo function, we get

$$\Lambda^*_\nu(\tau,\tau') \, S_\nu(\tau') = (1+A_\nu)/2 \quad S_\nu(\bar{\tau}_\nu = \bar{\gamma})$$

for $\tau_\nu \leq \gamma$.

In order to suppress instabilities of the iteration scheme due to continuum frequencies in the far UV, we modified the core saturation approximation (for our model calculations only):

$$\Lambda^*_\nu(\tau,\tau') \, S_\nu(\tau') = S_\nu(\tau_\nu) \, (1 - e^{-\tau_\nu/\beta}) \tag{5}$$

where β is another free parameter, of the order of 20. By this expedient we prevent overcorrections, since the factor in parantheses serves as an additional possibility for photons to escape, which means that fewer photons are trapped in the line core as compared with the pure core saturation assumption Eq.(3).

4. THE CONSTRAINT EQUATIONS

In this section we set up the constraint equations and linearize them. We restrict ourselves to the pure-hydrogen case in order to keep the presentation in a concise form (the addition of further atoms in non-LTE is straightforward and raises no problems). The emissivities η_ν and opacities χ_ν are calculated as follows (see e.g. Auer & Mihalas 1969):

$$\eta_\nu = 2h\nu^3/c^2 \sum_{i=1}^{NL} \sum_{j>i}^{NL} \sigma_{ij}(\nu) n_j g_i/g_j$$

$$+ e^{-h\nu/kT} \left(\sum_{i=1}^{NL} n_i^* \sigma_{ik}(\nu) + n_e^2 \sigma_{kk}(\nu,T) \right)$$

$$\chi_\nu = \sum_{i=1}^{NL} \sum_{j>i}^{NL} \sigma_{ij}(\nu)(n_i - n_j g_i/g_j)$$

$$\sum_{i=1}^{NL} \sigma_{ik}(\nu)(n_i - n_i^* e^{-h\nu/kT}) + n_e^2 \sigma_{kk}(\nu,T)\ (1-e^{-h\nu/kT}) \quad .$$

The LTE occupation numbers are denoted by n_i^* and $\sigma_{ij}(\nu)$ is the photon cross section for absorption (index k denotes the continuum state). The three terms of the sums in these expressions represent the contributions of bound-bound, bound-free, and free-free transitions, respectively. The source function, necessary for the calculation of the approximate radiation field, is then given by $S_\nu = \eta_\nu / \chi_\nu$. Derivatives with respect to occupation numbers and temperature are obtained in analytical form according to Mihalas et al. (1975).

4.1. Statistical equilibrium

The constraint equations of statistical equilibrium determine the level populations and are written as follows:

$$n_i \left(\sum_{\substack{j=1 \\ j \neq i}}^{NL} P_{ij} + P_{ik} \right) - \sum_{\substack{j=1 \\ j \neq i}}^{NL} n_j P_{ji} = n_e P_{ki} , \quad i=1,\ldots,NL . \qquad (6a)$$

The transition probabilities P_{ij} include radiative and collisional contributions, i.e. $P_{ij} = R_{ij} + C_{ij}$. We choose the notation

$$\underline{n} = (n_1, \ldots, n_{NL}, n_H) ,$$

where n_H is the density of the massive particles. Then we have another equation relating n_H to all bound levels :

$$\sum_{i=1}^{NL} n_i - n_H = - \sum_{j=NL+1}^{16} n_j^* - n_e \quad , \qquad (6b)$$

The second sum includes all LTE level densities n_j^* .
In short, we have to solve the equation

$$\underline{\underline{A}}\ \underline{n} = \underline{b} \quad .$$

Matrix $\underline{\underline{A}}$ and vector $\underline{b}$ are defined by equations (6):

$$A_{ij} = \begin{cases} -(R_{ji}+C_{ji}) & \text{if } j < i \\ -\left(\dfrac{n_i}{n_j}\right)^* (R_{ji}+C_{ij}) & \text{if } j > i \\ \displaystyle\sum_{m=1}^{i-1} \left(\dfrac{n_m}{n_i}\right)^* (R_{im}+C_{mi}) + \sum_{m=i+1}^{NL} (R_{im}+C_{im}) + R_{ik} + C_{ik} & \text{if } j = i \end{cases}$$

$$b_i = n_i^* (R_{ki}+C_{ik})$$

for $i,j \leq NL$, and

$$A_{NL+1,j} = 1 \qquad \text{if } j \leq NL$$

$$A_{NL+1,NL+1} = -1$$

$$b_{NL+1} = -\sum_{j=NL+1}^{16} n_j^* - n_e \ .$$

The radiative rates are computed with the radiation field Eq.(1), e.g.

$$R_{ij} = 4\pi \int \frac{\sigma_{ij}}{h\nu} \left(\Lambda_\nu^* S_\nu + \Delta J_\nu \right) d\nu$$

$$R_{ji} = 4\pi \left(\frac{n_i}{n_j}\right)^* \int \frac{\sigma_{ij}}{h\nu} e^{-h\nu/kT} \left(\Lambda_\nu^* S_\nu + \Delta J_\nu + \frac{2h\nu^3}{c^2} \right) d\nu$$

for upward and downward rates, respectively. With Λ_ν^* as defined in Eq.(4) we have

$$R_{ij} = 4\pi \int_{\text{core}} \frac{\sigma_{ij}}{h\nu} S_\nu d\nu + 4\pi \int_{\text{wing}} \frac{\sigma_{ij}}{h\nu} 1/2 \ S_\nu(\tau_\nu = \bar{\gamma}) d\nu + 4\pi \int_{\text{core+wing}} \frac{\sigma_{ij}}{h\nu} \Delta J_\nu d\nu \qquad (7a)$$

and

$$R_{ji} = 4\pi \left(\frac{n_i}{n_j}\right)^* \left[\int_{\text{core}} \frac{\sigma_{ij}}{h\nu} e^{-h\nu/kT} \left(S_\nu + \frac{2h\nu^3}{c^2} \right) d\nu \right.$$

$$+ \int_{\text{wing}} \frac{\sigma_{ij}}{h\nu} e^{-h\nu/kT} \left(1/2 \ S_\nu(\tau_\nu = \bar{\gamma}) + \frac{2h\nu^3}{c^2} \right) d\nu$$

$$\left. + \int_{\text{core+wing}} \frac{\sigma_{ij}}{h\nu} e^{-h\nu/kT} \Delta J_\nu d\nu \right] , \qquad (7b)$$

where $\int_{\text{core}}$ means the integral over optical thick frequencies. This shows the non-linearity of the statistical equations since the source function itself depends on the occupation numbers (and temperature). Therefore, we linearize by writing

$$\delta \underline{n} = \frac{\partial \underline{n}}{\partial n_e}\delta n_e + \frac{\partial \underline{n}}{\partial T}\delta T + \sum_{k=1}^{NF} \frac{\partial \underline{n}}{\partial S_k}\delta S_k .$$

The sum extends over all NF points of the chosen frequency set. Expressing the perturbations δS_k by those of temperature, electron density, and population numbers

$$\delta S_k = \frac{\partial S_k}{\partial T}\delta T + \frac{\partial S_k}{\partial n_e}\delta n_e + \sum_{i=1}^{NL+1} \frac{\partial S_k}{\partial n_i}\delta n_i$$

(with $n_{NL+1} := n_H$), we obtain, after some intermediate steps, the linearized statistical equations:

$$\begin{aligned} 0 = &- \delta \underline{n} \\ &+ \delta n_e \left[\frac{\partial \underline{n}}{\partial n_e} + \sum_{k=1}^{NF} \frac{1}{\chi_k} \frac{\partial \underline{n}}{\partial S_k} \left(\frac{\partial \eta_k}{\partial n_e} - S_k \frac{\partial \chi_k}{\partial n_e} \right) \right] \\ &+ \delta T \left[\frac{\partial \underline{n}}{\partial T} + \sum_{k=1}^{NF} \frac{1}{\chi_k} \frac{\partial \underline{n}}{\partial S_k} \left(\frac{\partial \eta_k}{\partial T} - S_k \frac{\partial \chi_k}{\partial T} \right) \right] \\ &+ \sum_{i=1}^{NL+1} \delta n_i \left[\sum_{k=1}^{NF} \frac{1}{\chi_k} \frac{\partial \underline{n}}{\partial S_k} \left(\frac{\partial \eta_k}{\partial n_i} - S_k \frac{\partial \chi_k}{\partial n_i} \right) \right] . \end{aligned} \qquad (8)$$

The derivatives of the occupation numbers are computed by

$$\frac{\partial \underline{n}}{\partial x} = \underline{\underline{A}}^{-1} \left(\frac{\partial \underline{b}}{\partial x} - \frac{\partial \underline{\underline{A}}}{\partial x} \underline{n} \right) \qquad (9)$$

for any variable x. Derivatives of $\underline{\underline{A}}$ and $\underline{b}$ may be obtained analytically.

4.2. RADIATIVE EQUILIBRIUM

At all points of the atmosphere the amount of energy emitted must equal the amount of energy absorbed. This constraint of radiative equilibrium may be written as

$$\int_0^\infty \chi_\nu (S_\nu - J_\nu) \, d\nu = 0 \quad ,$$

where χ_ν and S_ν do not include any scattering terms. Again, inserting Eq.(1) for the radiation field J_ν and using Eq.(3), for example, as the approximate lambda operator, we obtain after discretization and linearization:

$$\sum_{k_{core}} w_k(\chi_k + \delta\chi_k) \Delta J_k + \sum_{k_{wing}} w_k(\chi_k + \delta\chi_k)(S_k + \Delta J_k) = 0 \quad ,$$

where $\sum_{k_{core}}$ and $\sum_{k_{wing}}$ mean summation over core- and wing-frequency points, respectively. The w_k are quadrature weights and we define:

$$d_k := \begin{cases} 0 & \text{if} \quad \tau_k > \gamma \\ 1 & \text{if} \quad \tau_k \leq \gamma \end{cases} \quad .$$

Again, expressing the perturbations δS_k by those of temperature, electron density, and population numbers we have:

$$\sum_{k=1}^{NF} w_k \chi_k (S_k d_k - \Delta J_k) = \sum_{k=1}^{NF} w_k \Bigg[\delta n_e \left(\frac{\partial \chi_k}{\partial n_e} \Delta J_k - \frac{\partial \eta_k}{\partial n_e} d_k \right)$$

$$+ \delta T \left(\frac{\partial \chi_k}{\partial T} \Delta J_k - \frac{\partial \eta_k}{\partial T} d_k \right) \quad (10)$$

$$+ \sum_{i=1}^{NL+1} \delta n_i \left(\frac{\partial \chi_k}{\partial n_i} \Delta J_k - \frac{\partial \eta_k}{\partial n_i} d_k \right) \Bigg] .$$

If other approximate lambda operators are used, expressions analogous to Eq.(10) are easily obtained.

4.2. Hydrostatic equilibrium

The equation of hydrostatic equilibrium,

$$\frac{d}{dm}(NkT) + \frac{4\pi}{c} \int_0^\infty \frac{d(f_\nu J_\nu)}{dm} d\nu = g \quad ,$$

may be written as a difference equation connecting the depth points d and d-1, which are specified by column masses m_d and m_{d-1} (Mihalas 1978):

$$N_d kT_d - N_{d-1} kT_{d-1} + \frac{4\pi}{c} \sum_{i=1}^{NF} w_i (f_{di} J_{di} - f_{d-1,i} J_{d-1,i}) =$$

$$g \; (m_d - m_{d-1}) \quad , \qquad d = 2, \ldots, ND \; .$$

A convenient starting value is

$$N_1 k T_1 = m_1 \left(g - \frac{4\pi}{c} \sum_{i=1}^{NF} w_i \frac{\chi_{1,i}}{\rho_1} h_i J_{1,i} \right) \quad .$$

The mass density is ρ , g is the gravity and f_{di} and h_i are the variable Eddington factors which are kept fixed during the linearization procedure and which are only updated during the formal solution when the correction term Eq.(1b) is calculated. Elimination of J_{di} by means of Eq.(1) and linearization yields:

$$-N_d k T_d + N_{d-1} k T_{d-1} + g(m_d - m_{d-1})$$

$$- \frac{4\pi}{c} \sum_{i=1}^{NF} w_i \left[f_{di} \left(S_{di} a_i + \Delta J_{di}\right) - f_{d-1,i} \left(S_{d-1,i} a_i + \Delta J_{d-1,i}\right) \right]$$

$$= \delta N_d k T_d + \delta T_d k N_d + \frac{4\pi}{c} \sum_{i=1}^{NF} a_i w_i f_{di} \frac{1}{\chi_i} \times \tag{11a}$$

$$\left[\delta n_e \left(\frac{\partial \eta_i}{\partial n_e} - S_i \frac{\partial \chi_i}{\partial n_e} \right) + \delta T \left(\frac{\partial \eta_i}{\partial T} - S_i \frac{\partial \chi_i}{\partial T} \right) + \sum_{l=1}^{NL} \delta n_l \left(\frac{\partial \eta_i}{\partial n_l} - S_i \frac{\partial \chi_i}{\partial n_l} \right) \right]$$

and for the upper boundary (dropping the depth index which is d=1) with the definition

$$a_i := \begin{cases} 0 & \text{if} \quad \tau_i \leq \gamma \\ 1 & \text{if} \quad \tau_i > \gamma \end{cases} \quad :$$

$$m\left(g - \frac{4\pi}{c}\sum_{i=1}^{NF} w_i h_i \frac{\chi_i}{\rho}(S_i a_i + \Delta J_i)\right) - NkT$$

$$= \delta n_H \frac{1}{n_H}(NkT - mg) + \delta NkT + \delta TkN$$

$$+ \frac{4\pi}{c}\frac{m}{\rho}\sum_{i=1}^{NF} w_i h_i \Bigg[\delta T \left(a_i \chi_i \frac{\partial S_i}{\partial T} + (S_i a_i + \Delta J_i)\frac{\partial \chi_i}{\partial T}\right)$$

$$\partial n_e \left(a_i \chi_i \frac{\partial S_i}{\partial n_e} + (S_i a_i + \Delta J_i)\frac{\partial \chi_i}{\partial n_e}\right)$$

$$\sum_{l=1}^{NL} \delta n_l \left(a_i \chi_i \frac{\partial S_i}{\partial n_l} + (S_i a_i + \Delta J_i)\frac{\partial \chi_i}{\partial n_l}\right)\Bigg]. \qquad (11b)$$

Since we have to solve the hydrostatic equation inwards from the outer boundary to keep the integration stable, it is not possible to use the approximate lambda operator in the form of Eq.(4), which requires the solution to start from the inner boundary. No restriction of this kind, however, is encountered if the wing approximation ($\tau_\nu \leq \gamma$) $\Lambda^*_\nu S_\nu = 0$ is employed. In order not to confine ourselves to this simple approximation we did not include the hydrostatic equation in the linearized set of equations, but solve it iteratively after every iteration of the Scharmer cycle, using radiation field and Eddington factors from the formal solution. As expected, this procedure proved to be stable as long as the atmosphere was not too close to the Eddington limit. However, we emphasize that there is no basic need to keep the hydrostatic equation out of the linearization scheme.

4.4. PARTICLE CONSERVATION

The total number density N must be conserved:

$$N = n_H + n_e \quad .$$

Linearization yields

$$N - n_H - n_e = \delta n_H + \delta n_e - \delta N \quad ,$$

or, if N is assumed fixed,

$$N - n_H - n_e = \delta n_H + \delta n_e \quad . \qquad (12)$$

5.1. SOLUTION OF THE LINEARIZED EQUATIONS

As already indicated at the beginning of this section, the system of linearized equations, consisting of Eqs. (8), (10), and (12), is set up at each depth point d and may be written as

$$\underline{\underline{M}}_d \, \delta \underline{\psi}_d = \underline{c}_d \quad . \qquad (13)$$

The Matrix $\underline{\underline{M}}_d$ and the inhomogeneous vector $\underline{c}_d$ are defined as follows. The first NL+1 lines represent the statistical equations (8):

$$M_{ij} = \begin{cases} \left[\sum_{k=1}^{NF} \frac{1}{\chi_k} \frac{\partial n_i}{\partial S_k} \left(\frac{\partial \eta_k}{\partial n_j} - S_k \frac{\partial \chi_k}{\partial n_j} \right) \right] & \text{if } i \neq j \\ \left[\qquad " \qquad \right] - 1 & \text{if } i = j \end{cases}$$

$$M_{i,NL+2} = \sum_{k=1}^{NF} \frac{1}{\chi_k} \frac{\partial n_i}{\partial S_k} \left(\frac{\partial \eta_k}{\partial n_e} - S_k \frac{\partial \chi_k}{\partial n_e} \right) + \frac{\partial n_i}{\partial n_e}$$

$$M_{i,NL+3} = \sum_{k=1}^{NF} \frac{1}{\chi_k} \frac{\partial n_i}{\partial S_k} \left(\frac{\partial \eta_k}{\partial T} - S_k \frac{\partial \chi_k}{\partial T} \right) + \frac{\partial n_i}{\partial T}$$

$$c_i = 0$$

for $i \leq NL+1$. The particle conservation equation (12) is described by the next line (NL+2):

$$M_{NL+2,j} = \begin{cases} 0 & \text{if } j \leq NL \quad \text{or} \quad j=NL+3 \\ 1 & \text{if } j=NL+1 \quad \text{or} \quad j=NL+2 \end{cases}$$

$$c_{NL+2} = N - n_H - n_e \,.$$

The last line represents the radiative equilibrium equation (10):

$$M_{NL+3,j} = \begin{cases} \sum_{k=1}^{NF} w_k \left(\frac{\partial \chi_k}{\partial n_j} \Delta J_k - \frac{\partial \eta_k}{\partial n_j} d_k \right) & \text{if } j \leq NL+1 \\ \sum_{k=1}^{NF} w_k \left(\frac{\partial \chi_k}{\partial n_e} \Delta J_k - \frac{\partial \eta_k}{\partial n_e} d_k \right) & \text{if } j=NL+2 \\ \sum_{k=1}^{NF} w_k \left(\frac{\partial \chi_k}{\partial T} \Delta J_k - \frac{\partial \eta_k}{\partial T} d_k \right) & \text{if } j=NL+3 \end{cases}$$

$$c_{NL+3} = \sum_{k=1}^{NF} w_k \, \chi_k (S_k d_k - \Delta J_k)$$

These elements of $\underline{\underline{M}}$ and $\underline{c}$ are calculated with the current estimate for the solution,

$$\underline{\psi}^{o}_{d} = (n_1, \ldots, n_{NL}, n_H, n_e, T) .$$

Then system (13) is solved for the corrections

$$\delta\underline{\psi}_{d} = (\delta n_1, \ldots, \delta n_{NL}, \delta n_H, \delta n_e, \delta T)$$

which are added to $\underline{\psi}^{o}_{d}$, yielding an improved estimate for the solution $\underline{\psi}_{d}$. Then Eq.(13) is solved again and the procedure continued until convergence is reached. Since this is a Newton-Raphson iteration scheme, convergence is locally quadratic and, in general, few iterations are needed to reach the required accuracy.

At the end of this section we summarize the steps of the solution scheme:

i.) Estimate the atmospheric structure and compute source functions.

ii.) Calculate correction terms $\Delta J_\nu = (\Lambda_\nu - \Lambda^*_\nu) S_\nu$ for all frequency and depth points by formal solutions with the exact and approximate lambda operators.

iii.) Perform a depth by depth Newton-Raphson iteration by repeatedly solving the linearized set of Eqs.(13) using radiation field described by Eq.(1) (if the pressure is kept fixed during linearization: sequence arbitrary for operator (3) or going outwards starting from the inner boundary if operator (4) is used).

iv.) Solve the hydrostatic equation if not already incorporated in the linearization, and return to ii.).

These steps are iterated until a defined accuracy limit is reached.

5.2. SIMPLIFICATIONS FOR LINE FORMATION CALCULATIONS

We now assume the pressure and temperature stratification to be fixed and solve the iteration scheme only for the occupation numbers. The iteration scheme remains exactly the same as summarized in the preceding section, but the constraints of radiative and hydrostatic equilibrium are dropped and only the statistical equations plus the particle conservation equation are retained in the linearized system Eq.(13). The corresponding derivatives, with respect to temperature and total particle density, now disappear. The reduction of computing time is nevertheless only marginal. A further simplification, however, can be made which would make things easier in many applications. If we assume that the source function in the core of a transition i ⟶ j is controlled solely by n_i and n_j, we achieve detailed balancing in the core, i.e.

$$n_i R_{ij}^{core} = n_j R_{ji}^{core}.$$

Then the core integrals in Eq.(7) cancel analytically, resulting in linear statistical equations since all remaining rates are known. The system can be solved directly for the occupation numbers and hence, linearization is no longer needed. The assumption of detailed balancing in the core is justified only if there is no strong overlap of line transitions and/or if the (narrow) lines cover only a small frequency range of a given bound-free transition. At any rate, it provides at least a good first estimate for a Newton-Raphson-iteration. However, this assumption is of no use for model atmosphere calculations since non-linearity still enters into system Eq.(13) via hydrostatic and radiative equilibrium.

6. NUMERICAL TESTS

In order to investigate the algorithm outlined above we performed test computations for the multi-level line formation case and also constructed pure-hydrogen model atmospheres. The choice of input physics was exactly the same as in the complete linearization code written by Kudritzki (1976) because we intended to compare our results as closely as possible with those obtained by the "classical" method. The implemented model for the hydrogen atom consists of 16 bound levels, of which the lowest five and the proton density are allowed to depart from LTE while the remaining levels are assumed to be in LTE relative to the continuum state. Radiative

Table 1. Line formation calculations: numbers of iterations for obtaining convergence in different models and values of parameter γ. Results in case **a** are obtained by using operator (4), and in case **b** by the even more simplified operator (3). A bar means that γ is too small and hence leads to divergence.

Model			Number of iterations	
T_{eff}	log g	γ	a	b
32,000	4.1	6	–	–
		7	32	42
		8	34	44
		9	40	45
		10	39	45
40,000	4.0	3	–	–
		4	18	68
		5	20	18
		6	22	18
		7	23	19
40,000	5.0	2	–	–
		3	16	16
		4	18	18
		5	20	20
		6	22	22
50,000	4.0	5	–	–
		6	19	–
		7	19	19
		8	20	20
		9	20	20
70,000	5.5	1	–	–
		2	11	–
		3	12	14
		4	13	14
		5	13	15
100,000	7.0	0.75	20	–
		1	13	18
		2	8	9
		3	8	10
		4	8	10

transitions between the first four levels (L_α, L_β, L_γ, H_α, H_β, P_α) are explicitly included. The absorption coefficients are represented by depth-independent Doppler profiles. The exact formal solution is obtained by the standard Feautrier (1964) scheme.

6.1. MULTI-LEVEL LINE FORMATION CALCULATIONS

Hydrogen line formation has been carried out for six non-LTE model atmospheres with 32,000 K $\leq T_{eff} \leq$ 100,000 K and 4.0 $\leq \log g \leq$ 7.0 (cgs-units). The helium abundance adopted in the models is n(He/H) = 0.1 in all cases, but this quantity hardly affects our tests. The iterations were started with all hydrogen occupation numbers set in LTE .

6.1.1. Convergence properties

For each atmospheric model, we calculated the line formation for a series of different values of γ . In most cases convergence was quickly reached, with the number of iterations

Fig.1. Run of the departure coefficient of the hydrogen ground state with depth. The four curves display the departures after 1, 2, and 4 iterations as well as the converged result. The atmospheric parameters are T_{eff}= 40000 K and log g = 4, and the value of γ is 4.

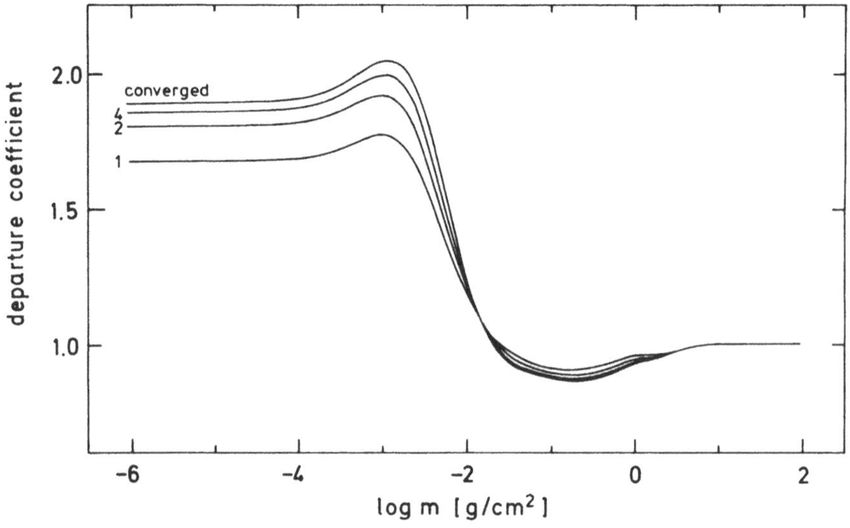

needed depending on γ, which is shown in Table 1. Convergence was assumed when the relative corrections to the occupation numbers had decreased below 10^{-3} at all depths. In some cases we followed the iteration process down to much smaller relative corrections ($<10^{-8}$) to check for the possibility of stabilizing corrections as is found in the classical lambda iteration. No problems of this kind were encountered.

In order to demonstrate the convergence properties of the iteration process in more detail, we chose the model with T_{eff}= 40000 K, log g = 4.0, γ = 4 and the operator (4). Figure 1 shows the behaviour of the ground level's departure coefficient and its dependence on the number of iterations. The depth scale is m, which is the mass in a column of unit cross section, measured inwards from the outer boundary of the atmosphere. As can be seen, the population numbers reach their final values globally. A total of 18 iterations had to be performed and comparison with the departure coefficients obtained by the complete linearization model atmosphere code (Kudritzki 1976) revealed deviations no larger than 1%.

Next, we should pay attention to those regions of the atmosphere where the line cores have very large optical depths and would therefore raise some problems in the usual lambda iteration. We examine the situation at log m=-1.025: the line core optical depths range from 30 (L_γ) to 500 (L_α). Figure 2 shows the decrease of the relative corrections of the (n=2) state during the iteration process. For comparison, it also shows the behaviour of the lambda iteration, whose corrections decrease significantly for about 10 iterations and finally cease, demonstrating dramatically the failure of this procedure. As expected, due to cancellation of the huge number of "inactive" photons by the core saturation assumption incorporated in the approximate operators, our iteration scheme is clearly superior, and the nearly linear decline of corrections (note the ordinate's logarithmic scale!) justifies a termination of the iterations when the desired accuracy is reached.

We also made calculations with the even more simplified operator given by Eq.(3). Although one would expect a somewhat slower convergence because of the omission of the approximated radiation field in the wings, roughly the same number of iterations is needed as compared with Scharmer's operator, as can also be seen in Table 1 (case b). Only at low values of γ is the behaviour different, which can be related to the instability problem (see below). Curve b) of Figure 3 shows the relative corrections to the n=2 departure coefficient in this case.

6.1.2. The value of γ and the stability of the iteration process

As already observed by Scharmer (1981), the iteration of the perturbation technique tends to become unstable if too low a value of the parameter γ is chosen. He found that a value of 1 is necessary to obtain convergence in a two-level atom with an

Fig.2. Logarithm of absolute value of relative corrections to the occupation number of the hydrogen n=2 level during the iteration process at depth log m=-1.025. The corrections in our iteration scheme (curve a) decrease almost linearly, while in the case of lambda iteration (curve b) they become fixed after about 10 iterations. Atmospheric parameters and γ have the same values as in Fig. 1.

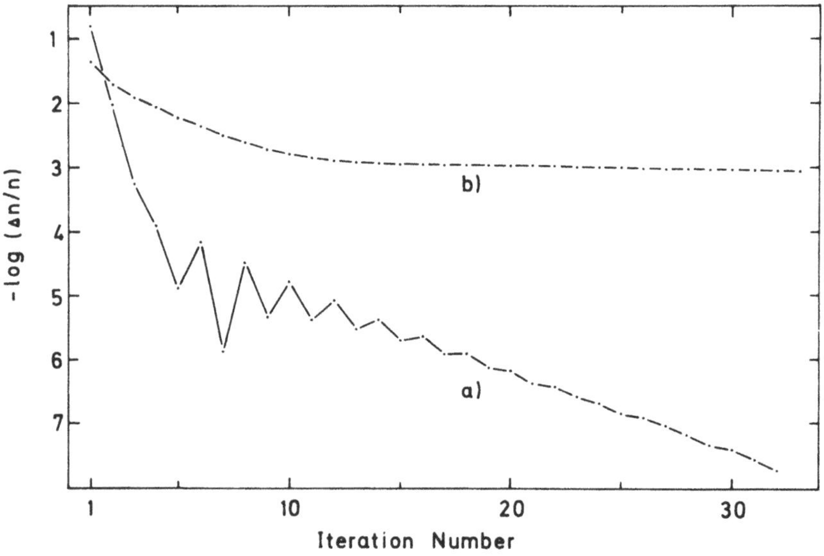

assumed Voigt profile. For a Doppler profile he had to increase the parameter to 2. In our calculations, however, we needed γ in excess of 2 in nearly all models. We were able to trace this difference down to the inclusion of bound-free transitions into the approximate transfer solution. Continua with their shallow profile function require a larger γ-value to avoid instabilities than do line transitions with their steep Doppler profiles. This finding is in agreement with the stability analysis of the Scharmer method carried out by Bond (1983), but in contrast with the results of Scharmer (1981). Consequently, we tried to improve the convergence and stability properties of the iteration scheme by excluding the continua from the simultaneous solution of approximate radiative transfer and statistical equilibrium equations and calculating them instead by means of the simple lambda iteration. This worked well for the higher temperature models ($T_{eff} \geq 40000$ K) but, for the lower temperatures, the former difficulties reappeared, due to the Lyman continuum becoming opaque. In this case it was more advantageous

Fig.3. As in Fig. 2, but a) with γ set up to 6.0, resulting in a smooth run for the relative corrections at the beginning of the iteration process. Curve b) demonstrates that the even more simplified operator in Eq.(3) with γ = 6.0 also has satisfactory convergence properties.

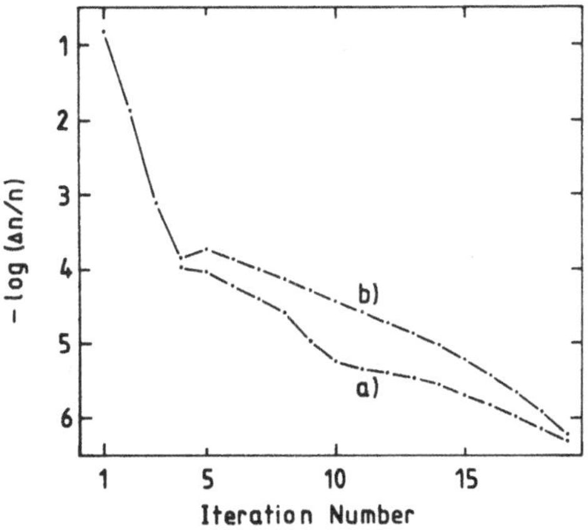

to treat all transitions (including also the continua) with the approximate lambda operators and to choose a somewhat higher value of γ. All results presented in this paper were obtained in the latter way.

Unlike Scharmer (1981), who assigns the cause of instability to the one-directional depth coupling of his wing approximation, we feel that it is the core saturation approximation which causes the destabilization at depths where even at core frequencies the existence of the outer boundary makes itself conspicuous. In an exact solution of the transfer equation, the radiation field J_ν responds (only weakly) to the photon escape, but in the simultaneous solution with the statistical equilibrium Λ^*_ν is unable to respond properly.

6.2. MODEL ATMOSPHERE CALCULATIONS

Starting from LTE models as first approximations, we computed five pure-hydrogen model atmospheres with the techniques outlined above. Their parameters cover the range from 30,000 K to 100,000 K in effective temperature T_{eff}, and 10^4 to $10^{5.5}$ (cgs-units) in gravity. Again, in order to investigate the convergence properties we tested both approximate operators (3) and (4) with different values of γ. In all cases convergence was claimed to occur whenever the relative corrections to all occupation densities as well as to the temperature, at all depths, decreased below 10^{-4}. All models were checked against those computed by the complete linearization code (Kudritzki 1976), with the result that the agreement was perfect: relative deviations in the temperature structures were within the limit of 10^{-4}. This is satisfying inasmuch as both programs are independent. In all cases, convergence was quickly reached. Table 2 shows the rate of convergence, which depends on the model parameters, the value of γ and the type of approximate lambda operator used. The overall behaviour is quite similar to that of the line formation problem. Increasing γ beyond the optimum value (where convergence is obtained within the lowest

Table 2. Model atmosphere calculations: numbers of iterations for obtaining convergence for 5 pure hydrogen model atmospheres and different values of γ. Results in case **a** are obtained by using Scharmer's operator (4) and in case **b** by the even more simplified operator (3).

Model			Number of iterations	
T_{eff}	log g	γ	a	b
30,000	4.0	4.5	–	–
		4.8	45	–
		5.	49	–
		6.	58	85
		7.	68	93
40,000	4.0	2.3	–	–
		2.5	30	–
		3.	32	45
		4.	36	52
		5.	41	56
60,000	5.0	1.7	–	–
		1.8	17	23
		2.	17	23
		2.5	20	26
		3.	21	27
		4.	24	30
80,000	5.0	1.5	23	–
		2.	15	21
		3.	19	24
		4.	21	25
		5.	23	26
		6.	24	26
100,000	5.5	1.	–	–
		1.5	19	–
		2.	14	20
		3	17	21
		4	18	23

number of iterations) worsens convergence. In the extreme case, $\gamma \longrightarrow \infty$, a lambda iteration would be performed. On the other hand, decreasing the parameter below its optimum value led to overcorrections, resulting in slower convergence (see e.g. model with T_{eff}= 80,000 K and γ = 1.5) or even in divergence. Slightly better results may be obtained if different γ's are introduced for different lines as well as for continua. Furthermore it was also found that γ may be decreased during the iteration process, thus again slightly speeding up the whole procedure. But in order to keep the presentation of our tests uniform, and since final models are calculated anyway within a few iterations, we shall leave out here detailed descriptions of these experiments. However, these refinements may become important if more complex atomic models are treated.

Fig.4. Run of temperature on a mass scale for the model with T_{eff}= 40,000 K, log g = 4.0, and γ = 2.5, shown at different stages of the iterations. Labels indicate the number of iterations performed, starting from an LTE model. Before the lines are included, a pre-converged model with pure continua in non-LTE (NLTE-C) is computed, in the first six iterations, in order to speed up the overall convergence.

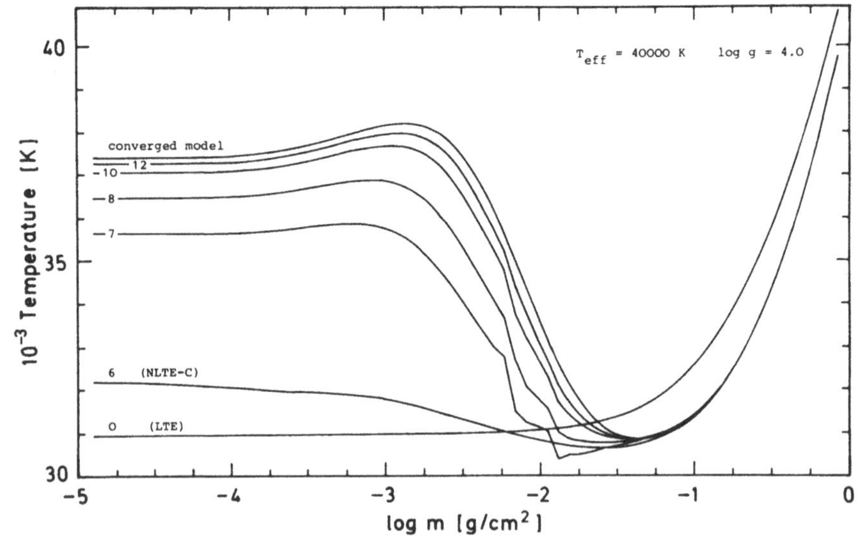

As an example, let us examine the convergence properties in more detail using the model with T_{eff}= 40,000 K and log g = 4.0, treated with γ = 2.5 and the approximate operator (4). In this case, we found the model to converge in fewer overall iterations if we first calculated a model with pure continua in non-LTE, which means that the lines are not explicitly included but are assumed to be in strict radiative detailed balance. Starting from this converged NLTE-C model as a second approximation we continued the iterations by adding the line transitions ("NLTE-L" model). This scheme saved a number of iterations. It worked, however, only in the present case for, in all other models, convergence was reached more rapidly if no NLTE-C models were computed as a second approximation, this pointing to the stability of our iteration scheme.

Fig.5. Logarithm of absolute values of relative corrections to temperature during the iteration process at three different depths: log m = −2.33 (asterisks), −0.72 (squares) and +0.22 (crosses). The corrections decrease almost linearly. The jumps at iteration number 7 are explained by the transition from the NLTE-C to the NLTE-L case. Atmospheric parameters and γ as in Fig.4.

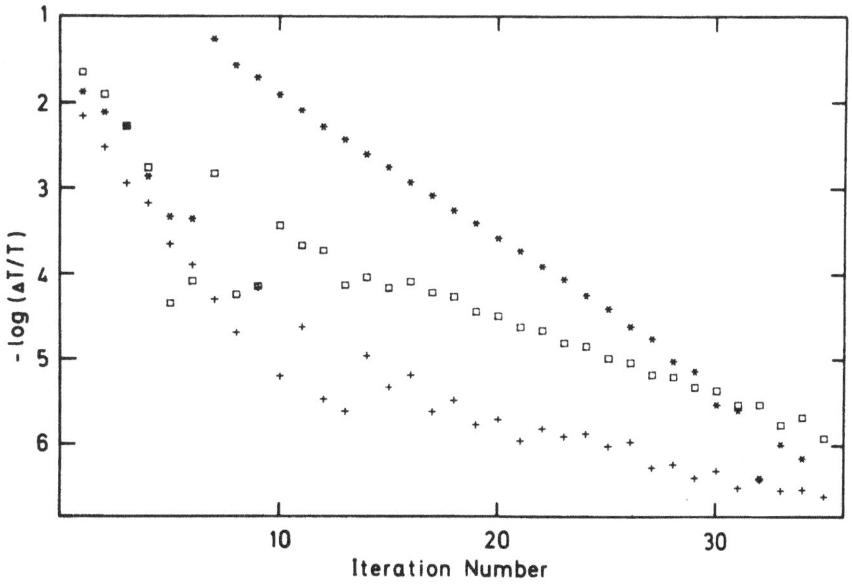

On a mass scale, Figure 4 shows the run of temperature of the starting model (LTE), of the converged model and of some semi-converged models, the numbers of iterations being marked by labels. A NLTE-C model is obtained after six iterations, with the largest relative corrections 10^{-2}, which is accurate enough to continue iterations with the lines. Clearly, convergence is global and after 30 iterations our convergence criterion was satisfied.

In Figure 5 we have plotted the logarithm of the absolute values of the relative corrections to the temperature versus the number of iterations for three different depth points. The depths in log m are $d_1 = -2.23$, $d_2 = -0.72$ and $d_3 = +0.22$. At d_1 (asterisks) all lines are optically thin except for L_α, its core having an optical depth of $\tau = 4$. At d_2 (squares) all line cores have become optically thick, the depths ranging from $\tau = 120$ (L_γ) to $\tau = 2400$ (P_α), and at d_3 (crosses) the Lyman continuum is opaque with $\tau = 6$ at the edge, while P_α has the largest optical line depth ($\tau = 5\ 10^4$). Clearly, lambda iteration would not lead to convergence at these large optical depths. However, from Figure 5 it can be seen, that the corrections in our iteration scheme decline rapidly and almost linearly in all cases; the jump at the seventh iteration arises from the inclusion of the lines at this stage. Similarly, Figure 6 displays the absolute values of relative corrections to the emergent Eddington flux at three distinct frequencies, namely at the line centers of L_α (crosses) and of L_β (asterisks), and the edge of the Lyman continuum (squares). Again, the (logarithmic) linear decline of the corrections is obvious and hence convergence is assured, which means that the accuracy of the numerical solution may be reliably estimated .

Nearly two CPU-minutes on a Cray-1M were needed to converge the above discussed model. A grid of 65 frequency-, 90 depth- and 3 angle-points was employed and the number of non-LTE levels was 6. On average four CPU-seconds are needed for one iteration cycle, somewhat more at the beginning and somewhat less near the end of the computations; later in the iteration sequence fewer Newton-Raphson iterations are needed. One exact formal solution requires about 0.2 seconds. CPU-time scales linearly with the number of depths- and frequency-points and cubically with the number of angle points (matrix inversions for formal solution) and the number of non-LTE levels (matrix inversions for linearization, see Eq.(8)). Therefore, the limit of our method is set by the number of non-LTE levels (about 100), while in principle the number of lines included raises no problem. Using Eq.(3) as an approximate operator and leaving the hydrostatic equation out of the simultaneous solution, we have no explicit depth coupling in our equations so that during one iteration step system Eq.(13) can be solved in parallel for every depth point. This implies that the turn-around time on parallel processors may be shortened considerably by roughly a factor given by the number of depth points or at least by the number of parallel working CPU's.

Fig.6. Logarithm of absolute values of relative corrections to the Eddington flux during the iteration process at three different frequencies: line centers of L_α (crosses) and L_β (asterisks) and head of the Lyman continuum (squares). Again the decline of the corrections is almost linear. Atmospheric parameters and γ are the same as in Figures 4 and 5.

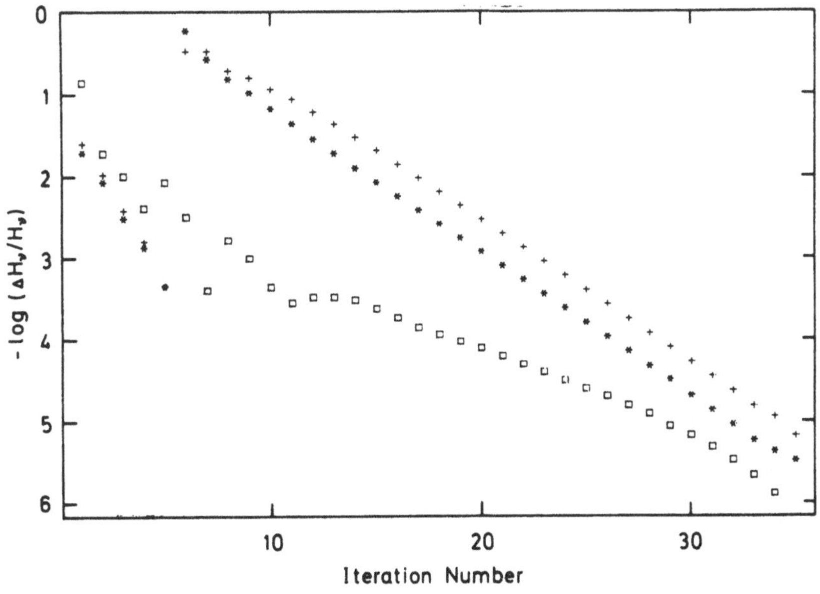

7. SUMMARY AND CONCLUSIONS

We have presented a method for solving multi-level line formation problems and for constructing non-LTE model atmospheres. It is based on the use of entirely local approximate lambda operators in combination with a perturbation technique, which simplifies the radiative transfer and yields accurate solutions of the transfer equation iteratively. Test calculations have proved the applicability and stability of this scheme. The accurate description of the radiation field in the line core by the core saturation assumption (Rybicki 1972) is essential for an approximate operator, while the radiation field in the wing does not necessarily require an approximation, since lambda iteration is known to work well in optically thin cases. Therefore, in applications where no appropriate wing approximation can be found, our scheme can still be used (see Hamann, this book).

This new method may preferably be used for problems involving complex model atoms and which cannot be solved by the classical complete linearization approach. Line formation and model atmospheres may be calculated with up to 100 non-LTE levels taken into account. Since computing time scales only linearly with the number of frequency points, a large number of opacity sources may be included, enabling one to investigate metal line blanketing effects in hot stars.

Acknowledgements. It is a pleasure to thank Prof. Dr. K. Hunger, who initiated this work, for his continued support and careful reading of the manuscript. Thanks also go to D. Husfeld (Munich) for his collaboration and to my colleagues in the Kiel institute, especially to W.-R. Hamann, for helpful discussions. Computations were carried out on the Cyber 175 and CRAY-1M machines of the Konrad-Zuse-Institut für Informationstechnik in Berlin. This work was supported by the Deutsche Forschungsgemeinschaft under grant Hu 39/23-1.

REFERENCES

Anderson,L.S.: 1985 in "Progress in Stellar Line Formation Theory", J.F.Beckman,L.Crivellari (eds.), Reidel Publishing Company, p.225

Auer,L.H.,Heasley,J.N.: 1976, Astrophs.J. **205**,165

Auer,L.H.,Mihalas,D.: 1969, Astrophys.J. **158**,641

————.1970, Monthly Notices Roy.Astron.Soc. **149**,65

Bond,D.J.: 1983, JPhys.A:Math.Gen. **14**,L239

Cannon,C.J.: 1973, Astrophys.J. **185**,621

Feautrier,P.: 1964, Compt.Rend. **258**,3189

Hamann,W.-R.: 1985, Astron.Astrophys. **148**,364

————.1986, Astron.Astrophys. **160**,347

Kalkofen,W.: 1985 in "Progress in Stellar Line Formation Theory", J.F.Beckman,L.Crivellari (eds.), Reidel Publishing Company, p.153

Kudritzki,R.P.: 1976, Astron.Astrophys. **52**,11

Mihalas,D.: 1972, "Non-LTE model atmospheres for B and O stars", NCAR TN/STR-76

————.1978, Stellar Atmospheres, 2nd ed., San Francisco, Freeman

Mihalas,D.,Heasley,J.N.,Auer,L.H.: 1975, "A Non-LTE Model Stellar Atmosphere Computer Program", NCAR-TN/STR-104

Rybicki,G.B.: 1971, J.Q.S.R.T. **11**,589

————.1972, in "Line Formation in the Presence of Magnetic Fields", ed. R.G.Athay, L.L.House, G.Newkirk,Jr. (Boulder: High Altitude Observatory), p.145

Scharmer,G.B.: 1981, Astrophys.J. **249**,720

Scharmer,G.B.,Nordlund,Å.: 1982, Stockholm Observatory Report **19**

Scharmer,G.B.,Carlsson,M.: 1984, J.Comput.Phys. **59**,56

Werner,K.: 1986, Astron.Astrophys. **161**,177

Werner,K.,Husfeld,D.: 1985, Astron.Astrophys. **148**,417

ACCELERATION OF CONVERGENCE

Lawrence Auer
Earth and Space Sciences, Los Alamos National Laboratory

ABSTRACT. We present an inexpensive method which can double or even triple the efficiency of any linearly convergent method. Successive iterations of the unaccelerated scheme are used to predict the accurate solution. As an example, the technique is applied to accelerating simple iterative solution techniques for the two-level atom non-coherent scattering problem.

1. Introduction.

It is an unfortunate computational "fact of life" that the iterative techniques which are the simplest to implement and generalize usually have comparatively poor convergence properties. In this paper we will present a powerful general acceleration technique, due to Ng (1974), which can be used to more than double the speed of any linearly convergent scheme. It is, in that sense, the vector analogue of Aitken's δ^2-process. This acceleration technique was first applied to the solution of radiation transfer problems in Buchler & Auer (1983) and Olson, Auer & Buchler (1986). The critical point is that it may be applied to accelerate **any** linearly convergent method, not just the ones discussed in those papers.

We can show that any scheme, for the iterative solution of a linear problem, which uses the information only from the preceding iteration cannot have better than linear convergence. (Radiation transfer in a two-level atom is an example of such a problem.) The iterative schemes can always be written in the form

$$\mathbf{x}^{n+1} = \mathbf{F}\cdot\mathbf{x}^{n} + \mathbf{x}^{0} \tag{1}$$

where we are using the superscript to indicate the iteration number. The matrix **F** appearing in Eq (1) is the *amplification matrix* of the iteration scheme. It does not appear explicitly in the original equations, although it is based on those equations. An example of the derivation of such a matrix is given by Eq (11). Further, note in general that one never explicitly forms the **F** matrix. Eq (1) is merely the formal representation of an iterative scheme.

Iteration with Eq (1) converges in the limit to the desired solution,

$$\mathbf{x}^{\infty} = \mathbf{F} \cdot \mathbf{x}^{\infty} + \mathbf{x}^{0} \tag{2}$$

If we define the error of the n-th iterate to be the vector $\varepsilon^n = \mathbf{x}^n - \mathbf{x}^{\infty}$, then subtracting Eq (2) from (1) we see the successive errors decrease as $\varepsilon^{n+1} = \mathbf{F} \cdot \varepsilon^n$. If we express the error vector in terms of the eigenvectors of **F**, $\mathbf{u}_i$, writing $\varepsilon^n = \Sigma\, c_i\, \mathbf{u}_i$, we have

$$\varepsilon^{n+1} = \mathbf{F} \cdot \varepsilon^n = \sum \lambda_i\, c_i\, \mathbf{u}_i \tag{3}$$

where the λ_i are the eigenvalues of the **F** matrix. This means the asymptotic behavior of the error is proportional to $(\lambda_{max})^n$. Clearly the iterative scheme will be convergent only if $|\lambda_{max}| < 1$ and the schemes represented by Eq (1) are only linearly convergent.

II. The General Acceleration Technique.

In order to achieve a higher rate of convergence we must use more information. In the Ng method one uses a succession of several iterations in order to provide this additional information. The accelerated estimate of the solution is written

$$\mathbf{x} = (1 - \sum_{m=1}^{M} \alpha_m)\, \mathbf{x}^{n} + \sum_{m=1}^{M} \alpha_m\, \mathbf{x}^{n-m} \tag{4}$$

One should note that the coefficients being used for the acceleration must sum to one in order that Eq (4) will always return $\mathbf{x}^{\infty}$ when the iterations have in fact converged. This is the origin of the $(1-\Sigma\alpha_m)$ in that equation. The problem is now to estimate the coefficients, α_m. In order to do this we implicitly apply Eq (4) after the (n-1)-th iteration. This will imply an estimate

$$\mathbf{x}' = (1 - \sum_{m=1}^{M} \alpha_m)\, \mathbf{x}^{n-1} + \sum_{m=1}^{M} \alpha_m\, \mathbf{x}^{n-m-1} \tag{5}$$

The coefficients for the acceleration are determined by making sure that the distance between **x** and **x'** is minimal in the least square sense. That is, the equations for α are chosen to minimize $r^2 = \Sigma\, w_d(x_d - x'_d)^2$, where x_d are the *components* of the estimates of the solution vector, and w_d are weights, the choice of which is discussed below. The normal equations are, formally,

$$\mathbf{A}\alpha = \mathbf{b} \tag{6}$$

which is a system of symmetric linear equations determining the M unknown acceleration coefficients , α_m, and whose elements are

$$A_{i,j} = \sum_d w_d\,(\Delta x_d^n - \Delta x_d^{n-i})\,(\Delta x_d^n - \Delta x_d^{n-j})$$

$$b_i = \sum_d w_d\,\Delta x_d^n\,(\Delta x_d^n - \Delta x_d^{n-i})$$

where we have used

$$\Delta\mathbf{x}^k = \mathbf{x}^k - \mathbf{x}^{k-1}$$

The iteration scheme with acceleration consists of applying the basic iterative technique to generate M+2 successive estimates of the solution vector, which are stored temporarily. Eq (6) is then solved for the α_m, and Eq (4) is used to generate the accelerated vector.

While the general acceleration based on Eq (6) is straightforwardly coded, pseudocode for the second order, M=2, case is given in the appendix. As we will see, this provides adequate acceleration and minimizes the amount of storage required.

III. An Example: Radiative Transfer in a Two-Level Atom.

Having presented the acceleration technique in a general form, let us now concentrate on its application to the solution of radiative transfer problems, in particular, to the solution of the two-level atom using diagonal approximate operators. The questions with the greatest importance affecting efficiency are 1) the choice of weights and 2) the optimal order to use in the acceleration.

Formally, the *two-level atom problem* is the solution of the integral equation

$$\mathbf{S} = (1 - \varepsilon)\,\bar{\mathbf{J}} + \varepsilon\,\mathbf{B} = (1 - \varepsilon)\,\Lambda(\mathbf{S}) + \varepsilon\,\mathbf{B} \tag{7}$$

for $S(\tau)$. In general, the continuous problem is replaced with a discretized one by using a basis set to interpolate the τ-variation of $S(\tau)$ implicitly. The problem is, thus, reduced to solving a set of linear equations for $S_d = S(\tau_d)$. While the direct solution may be relatively easily found, it is often more efficient (definitely so for multidimensional problems!) to use an iterative technique to obtain the solution. In the iterative schemes we will consider here, we will use an approximation to the true Λ-operator

$$\bar{\mathbf{J}} = \Lambda^*(\mathbf{S}) + \Delta\bar{\mathbf{J}} \tag{8}$$

where by definition

$$\Delta\bar{\mathbf{J}} = \Lambda\,(\mathbf{S}) - \Lambda^*(\mathbf{S}) \tag{9}$$

and $\Delta\,\bar{\mathbf{J}}$ is to be determined iteratively. The critical point is that if the approximation to the Λ-operator is *diagonal*,

$$\bar{J}_d = \Lambda^*_d\, S_d + \Delta\bar{J}_d \tag{10}$$

the iterative scheme for S is especially simple. At each grid point we have

$$S_d^{n+1} = \frac{(1-\varepsilon)\,\Delta\bar{J}_d^{\,n} + \varepsilon B_d}{1 - (1-\varepsilon)\,\Lambda^*_d} \tag{11}$$

which involves only scalar operations. With the new estimates of S_d we use the accurate *explicit* Λ-operator to evaluate $\Delta\,\bar{\mathbf{J}}$ and continue to iterate until convergence. It is important to note that although the approximate Λ^*-operator in Eq (8) is local, it is *implicit* and this suffices to overcome the notoriously poor convergence of explicit Λ-iteration. Further, Eq (11) is equally valid in two or three dimensions. The only change necessary when treating multidimensional radiative transfer is to develop an appropriate method for evaluation of the explicit formal solution of the transfer equations. This is unfortunately not true of the powerful multipoint method developed by Scharmer (1983).

The two approximate operators we will consider in the remainder of this paper are the *implicit diagonal* and *core-wing operators*. For their implicit diagonal operator Olson, Auer & Buchler (1985) noted that if one used

$$\Lambda_d^* = \Lambda_{dd} \tag{12}$$

i.e., the diagonal of the true discretized Λ-operator, $\bar{J}_d = \Sigma \Lambda_{dd'} S_{d'}$, then iteration using Eq (11) will converge at a rate roughly equal to the maximum of the sum of the off-diagonal elements in the rows of $\Lambda_{dd'}$, which is always less than unity. To derive the core-wing approximate operator, Rybicki (1972) used the physical fact that for a frequency and direction at which the optical depth to the boundary is very large, the intensity saturates to the local value of the source function; that is, $I_{\nu\omega} \approx S_\nu$ if $\tau_{\nu\omega} >> 1$, and $\tau_{\nu\omega}$ is the monochromatic optical depth to the nearest boundary in the direction ω. The core-wing approximation assumes that the only contribution to $\bar{J}$ comes from frequency-angle values for which $\tau_{\nu\omega} > \gamma$, with γ a tunable parameter, and assumes saturation for these quadrature points. The diagonal approximation is, accordingly,

$$\Lambda_d^* = \frac{1}{4\pi} \int d\omega \int d\nu \, \phi_{\nu\omega} \Theta(\tau_{\nu\omega}) \tag{13}$$

where the integrand is

$$\begin{aligned} \Theta(\tau) &= 1, \quad \tau > \gamma \\ &= 0, \quad \tau \le \gamma, \end{aligned}$$

with τ being the optical depth to the nearest boundary.

Iteration using either of these approximations for Λ^* is simple and both are easily generalized to multiple dimensions. Unfortunately, a price has been paid in order to use a diagonal operator. While the difficulty with Λ-iteration at depth has been avoided, the information must still perform the equivalent of a random walk through the logical grid, that is, over the discrete index, d, instead of the continuous variable, τ. This is shown by applying these methods to the solution of the two-level atom problem in a semi-infinite atmosphere with $\varepsilon = 10^{-4}$, using a logarithmic grid and 5 points per decade.

The relatively poor performance of iteration using either Eq (12) or (13) is demonstrated for this strongly scattering test problem in Fig 1a and 2a. Both approximations after 15 iterations are still far from the direct solution, which is marked with stars (*). Fortunately, both schemes are effectively accelerated by the scheme described in this paper. Fig 1b and 2b show the convergence using Eq(4) with M=2 and the relative weighting described in the next paragraph. Now, after 15 iterations the implicit diagonal scheme has converged to graphical accuracy and the core-wing scheme, although not yet converged, has also been greatly improved.

There are strong *a priori* reasons for thinking that weighting will be important. Consider solution for $\bar{J}$ in a two-level atom. The largest values in $\bar{J}$ occur at depth where the field has saturated to the Planck function and the smallest values are at the boundary. The errors in $\bar{J}$ at depth would dominate the sum of the residuals, so in order to achieve relatively uniform convergence in the estimates of $\bar{J}$ at all optical depths, we could give the points at depth less weight. One simple form for evaluating the weights that overcomes this problem is to use relative weighting for the n+1 iteration

$$w_d = \frac{1}{S_d^n} \tag{14}$$

where S^n is the last estimate of the solution.

The overall convergence properties of the iteration, varying both the order of the extrapolation and the weighting, are given for the test case in this table:

Accelerated Convergence, $\tau=10^5$, $\varepsilon=10^{-4}$

	Implicit Diagonal		Core-Wing	
Acceleration	Unweighted	Weighted	Unweighted	Weighted
None	74		>100	
1	37	27	95	93
2	23	19	41	36
3	24	24	34	28
4	20	17	23	20

Here the total number of formal solutions needed to achieve a solution with relative accuracy everywhere better than 10^{-3} is given. The order of acceleration used, M, is given in the first column. The parameters chosen for this example represent a very difficult, highly scattering case. If a less perverse scattering parameter, such as $\varepsilon=10^{-2}$, were used, a much smaller number of iterations would be necessary but the power of the acceleration would not be as apparent. Fortunately, although the number of iterations will depend on the particular problem being considered, the conclusions we can draw from these data are of general applicability.

Our basic conclusion is that any order of acceleration yields improved convergence, but second order (M=2) acceleration is significantly better than the lowest order in all cases. There is, unfortunately, no definitive statement one can make about the use of yet higher orders. The optimal choice of M depends on the basic iterative method being used. In the sample problem, for example, M=4 appears to be optimal for the core-wing approximation and M=2 for the implicit diagonal. In general, the number of iterations is at first dramatically reduced by relatively low order extrapolation but then the reduction is a slow and, in fact, not even necessarily a monotonic function of the order. On the other hand, higher order does not significantly increase the number of iterations and, as in the core-wing case, may markedly decrease the work. Accordingly, we suggest for other iterative schemes a safe, although not necessarily optimal, approach is to use M=2 and, if possible, investigate the consequences of the use of higher M.

It appears that, while weighting does improve convergence, the effect is not dramatic. The use of Eq (14) corresponds to the elimination of a few extrapolation cycles and thus does not represent an improvement by a very significant factor. Still, since the use of Eq (14) always does improve the convergence and the computational cost is negligible, its use is recommended.

IV. Concluding Comment.

This method belongs in the "tool-kit" of everyone trying to do low order iterative solution of radiative transfer problems. Pseudocode for the second order method is given in the appendix. It is remarkably simple to code. The computing needed to perform the accelerations is absolutely negligible relative to the cost of even a single formal solution, yet can achieve a dramatic reduction in the number of iterations required.

V. Acknowledgement.

We gratefully acknowledge the support of the Department of Energy during this work.

References

Buchler, J. R., & Auer, L . H., 1973. In *Proceedings of the 2nd International Conference and Workshop on the Radiative Properties of Hot Dense Matter*, ed J. Davis. World Scientific, Singapore.

Ng, K. C. 1974. *J. Chem. Phys.*, **61**, 2680.

Olson, G. L., Auer, L. H. & Buchler, J. R. 1986. *J. Quant. Spectrosc. Rad. Transfer* , **35**, 441.

Rybicki, G. B. 1972. in *Line Formation in Magnetic Fields*, ed R. G. Athay, L. H. House, G. Newkirk, Jr. (Boulder: National Center for Atmospheric Research) 146.

Scharmer, G. B. 1983. *Astron. Astrophys.*, **117**, 83.

Appendix: Pseudocode for Second Order Acceleration.

Let y0(i), i=1,...,n be the latest iterate, and y1(i), y2(i), y3(i) the preceding ones. The acceleration is accomplished by

```
1) Initialize
   a1 ← 0;   b1 ←0;   b2 ← 0;   c1 ← 0;   c2 ← 0

2) For i = 1 to n

wt ← 1, or wt ← 1 / abs(y0(i))   (depending on weighting being used)
d0 ← y0(i) - y1(i);      d1← y0(i) - 2*y1(i) + y2(i);   d2← y0(i) - y1(i) - y2(i) + y3(i)
a1← a1 + wt*d1*d1;       b1← b1+wt*d1*d2;               c1 ← c1+wt*d0*d1
                         b2 ← b2+wt*d2*d2;              c2 ← c2+wt*d0*d2

3) Compute Coefficients
   a ← (b2*c1 - b1*c2) / (b2*a1 - b1*b1)
   b ← (a1*c2 - b1*c1) / (b2*a1 - b1*b1)

4) Accelerate,
   For i = 1 to n:        y(i) ← (1-a-b) *y0(i) + a*y1(i) + b*y2(i)
```

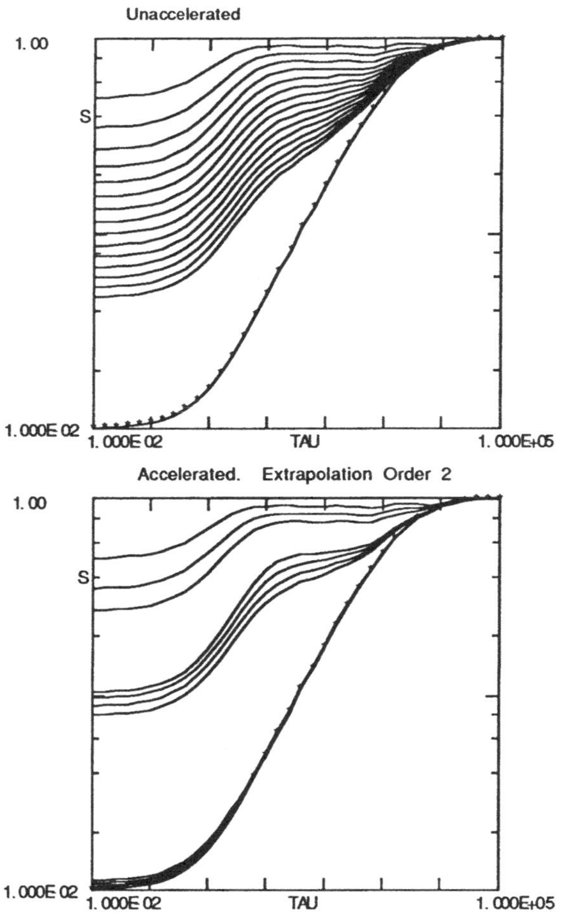

Fig 1a,b. Implicit Diagonal Operator, 15 iterations for $\varepsilon = 10^{-4}$. The exact solution has been marked with *. Note, the accelerated case has already converged to graphical accuracy.

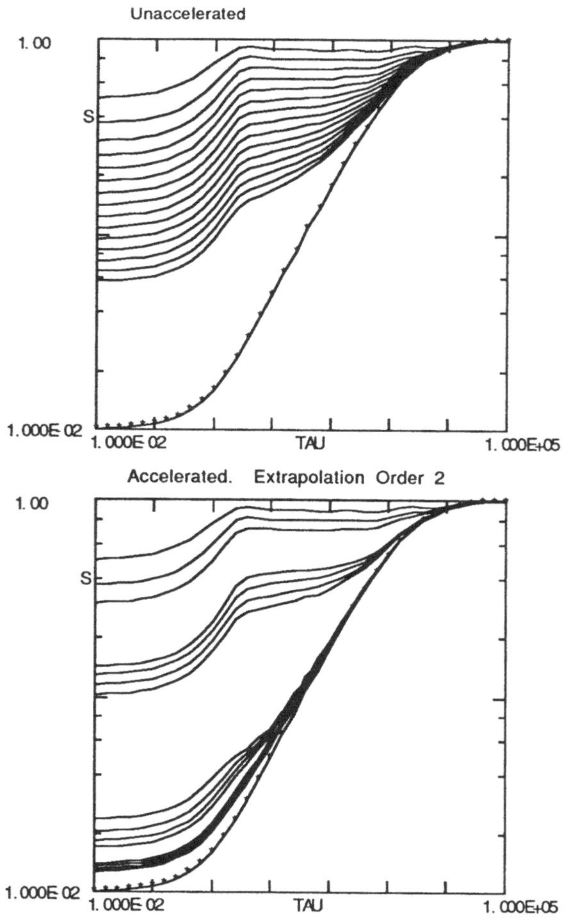

Fig 2a,b. Core Wing Approximate Operator. While convergence is not as good for this method, acceleration is clearly effective. In fact, if fourth order acceleration is used, the two are comparable.

LINE FORMATION IN A TIME-DEPENDENT ATMOSPHERE

W. Kalkofen
Harvard-Smithsonian Center for Astrophysics, Cambridge, USA

ABSTRACT: We describe an operator perturbation method for solving radiative transfer problems in moving media; the transfer equation is solved in the quasi-static approximation but the structure of the medium may be time-dependent. The characteristic feature of the method is that it solves transfer problems by separating the calculation into two parts: that of the error made by a solution in satisfying the constraint of statistical equilibrium, and that of corrections to the solution. The error calculation uses the exact operator of the transfer problem, the correction calculation, an approximate operator. Like Scharmer's method, the basic formulation of this method is based on integral equations. Both calculate the error of the solution with the aid of the Feautrier equation. But unlike Scharmer's, the present method employs the Feautrier equation also for determining corrections to the solution. It accomplishes this by exploiting a relation between the operators of the direct integral and the direct differential equation formulations.
The error is calculated with dense sets of angle and frequency points that provide an accurate formal solution of the transfer equation; but the correction calculation uses one angle and only those few frequency points that are necessary for describing the gross structure of the solution.
High speed and efficiency in time-dependent problems are achieved by two stratagems: the symmetric (*i.e.,* static) profile function is used in the approximate transfer operator, which thus needs to be calculated and inverted only at the beginning of a time-dependent calculation; and the starting solution at any time step after the first is the converged solution of the preceding time step. High accuracy is guaranteed by the Feautrier equation.
The properties of this method are illustrated in two problems of a line in a moving medium, and a variant using integral equations in the correction calculations is discussed.

1. INTRODUCTION

The transfer of radiation in a spectral line through a medium whose microscopic state is determined by the line radiation field has many characteristic length scales. They describe the properties of individual photons as well as their joint effect on the structure of the gas: The vast majority of the photons travel the relatively short distance of the monochromatic photon mean-free-path at the center of the line; but the structure of the medium, and in particular that of the source function of the line in question, is controlled by photons travelling in the line wings and is characterized by the thermalization length, which can differ by many order of magnitude from the mean-free-path of typical line photons in a highly dilute gas.

In the well-known and frequently studied idealization of a Doppler-broadened line formed in complete redistribution, where scattered line photons retain no memory of the frequency or direction in which they were absorbed, the thermalization length is related to the collision parameter ϵ by $\tau_{therm} \simeq \epsilon^{-1}$ (Avrett & Hummer 1965), where $\sqrt{\pi}\times\tau$ is distance measured in units of the photon mean-free-path at line center and ϵ is the probability of a collisional de-excitation. Typical values of ϵ are 10^{-2} or smaller; stronger lines tend to have smaller values of ϵ. Thus, there is a large disparity between the distance travelled by typical line photons and the distance over which the line photons affect the structure of the gas.

The physical characteristic of widely differing length scales in non-LTE line transfer is paralleled by the mathematical property of stiffness in the equations describing the transfer (*cf.* Dahlquist & Björck 1974, Stoer & Burlisch 1980, Press *et al.* 1986). To describe this property, consider the iterative solution of a line transfer problem consisting of the radiative transfer equation for a line and of the equations of statistical equilibrium of the two combining states of the atom under study. Suppose that the transfer equation is solved first for the line intensity given an assumed dependence of the source function on depth, and this dependence is then determined given the just computed line radiation field. To effect the coupling between radiation field and source function, information must be transmitted over distances of the order of a thermalization length. Since almost all photons are transported in the core of the line where information about the state of the gas is carried only over short distances of the order of the monochromatic photon mean-free-path at line center, the number of iterations necessary to couple the radiation field to the structure of the atmosphere, and therefore the number of iterations necessary to solve the line transfer problem, is of the order of the ratio of the thermalization length to the photon mean-free path at line center, *i.e.*, of the order of $1/\epsilon$ (this estimate applies to Doppler-broadening in a static gas; similar expressions hold for other types of line broadening, and also for coherent scattering, *cf.* Avrett & Hummer 1965, Hummer & Rybicki 1971). For dilute media, where the scattering parameter is very small compared to unity, this can be a very large number. This so-called Λ-iteration is then bound to fail.

In successful numerical schemes, the equations of radiative transfer and of the structure of the gas, *i.e.*, the equations of statistical equilibrium for the atom under study, are based on a single mathematical statement of the transfer problem, either an integral equation for the line source function (Avrett & Hummer 1965, Avrett & Loeser 1966, Athay & Skumanich 1967) or a set of coupled differential equations for the monochromatic intensity (Feautrier 1964, Cuny 1967, Auer & Mihalas 1968, 1969). These equations are solved in a single step in the linear problems considered here, where the gross structure of the atmosphere specifying the thermal source term, the collision parameter, and the absorption profile is prescribed on the optical depth scale. In this approach, no convergence difficulties arise.

These "direct" methods require matrix inversions: in the integral equation method the order of the single matrix is equal to the number of depth points, N_τ, and in the differential equation method there are N_τ matrices whose order is equal to the number of angle-frequency points, $N_{\mu\nu}$ ($\equiv N_\mu \times N_\nu$). These methods become very time-consuming when either N_τ or $N_{\mu\nu}$ is large since the solution time scales as $N_\tau{}^3$ and $N_\tau \times N_{\mu\nu}^3$, respectively. This is the case, for example, in time-dependent problems in which the gross structure of the atmosphere is obtained from hydrodynamic equations describing a flowing gas: The number of depth points is large since the scale is set by the slowest process, not usually line transitions; the number of frequency points is large since the bandwidth of a line in a moving medium is increased over that in a static medium by twice the flow shift (and both wings of the line must be treated in a moving medium); and the number of angle points must be large as well since the profile function is anisotropic.

Following an approach described by Cannon (1973, 1984, 1985), Scharmer proposed a very efficient numerical scheme for the solution of such problems using integral equations (*cf.* Scharmer 1981, 1983, 1984; Scharmer & Nordlund 1982; Scharmer & Carlsson 1985), in which the exact operator describing the radiative transfer and the constraint of statistical equilibrium is perturbed about a simpler approximate operator. The simplification results from a very simple description of the radiative transfer, which amounts to an application of the Eddington-Barbier relation to the interior of the medium. As a consequence, the construction of the *approximate* integral operator takes very much less time than that of the *exact* integral operator, and the solution of the resulting matrix equation takes very much less time as well: because of the simplification in the description of the transfer, the resulting matrix, ordinarily coupling all depth points of the atmosphere, has nearly triangular structure in Scharmer's approach (for a line formed in a semi-infinite medium). Therefore the inversion time scales only as N_τ^2.

In Scharmer's operator perturbation approach, the transfer problem is split into a calculation of the *error* with which a given solution satisfies the constraint equations of statistical equilibrium, and a calculation of *corrections* to the solution. The latter is carried out with his approximate operator, but the error, which becomes the driving term in the correction equation, is calculated with the help of the exact operator. Since this is a calculation of the intensity for *known* source function, the equations are no longer coupled. Hence individual differential equations for the specific intensities may be used in this phase of the solution. Scharmer accomplishes this by means of the Feautrier equation, which ensures that the converged solution of the transfer problem has the high accuracy of the Feautrier equation (*i.e.*, second-order, or, with the Hermite scheme of Auer 1976, fourth-order).

The approach described in the present paper is like Scharmer's, with the transfer problem formulated in terms of integral equations. But it then uses a relation between the integral and the differential operators for the direct solution of the transfer problem to restate the problem in terms of differential equations.

The essential characteristic of this particular operator perturbation approach is that two angle-frequency sets are employed. The first set describes the behavior of individual photons; this set is dense, with many mesh points; it is used in the error calculation. The second set describes the overall structure of the atmosphere; it is sparse,with very few points, and it is used in the calculation of corrections to the solution. The dense set ensures the accuracy of the converged solution, the sparse set accounts for the speed of the method.

The paper is structured as follows: In section 2 we consider the equations for the differential and integral equation approaches to the solution of line transfer problems and derive a relation between the corresponding operators describing the *direct* solution of the equations (*i.e.*, in a single step). In section 3 we discuss the operator perturbation approach based on integral equations and rephrase the equations in terms of differential equations only. Section 4 contains the numerical results of calculations using approximate operators with very few frequency points. In section 5 we consider a variation of the basic approach concerning the correction equations and discuss possible limitations of the method;

and in the final section we summarize the method and the results.

2. THE BASIC EQUATIONS

We write the transfer equation of a spectral line with background continuum for an atmosphere with plane-parallel stratification as a second-order differential equation,

$$\mu^2 \frac{d^2}{d\tau_{\nu\mu}^2} J_{\nu\mu}(\tau) = J_{\nu\mu}(\tau) - \tilde{S}_{\nu\mu}(\tau) \quad , \tag{2.1}$$

with appropriate boundary conditions (Feautrier 1964, Auer 1967, Mihalas 1978), where $J_{\nu\mu}(\tau)$ is the specific monochromatic intensity,

$$J_{\nu\mu}(\tau) = \frac{1}{2}\left[I_{\nu,|\mu|}(\tau) + I_{-\nu,|-\mu|}(\tau)\right] \quad , \tag{2.2}$$

averaged along a ray defined by the frequencies ν and $-\nu$ and the directions μ and $-\mu$ (*i.e.*, in opposite line wings and directions), $\tilde{S}_{\nu\mu}(\tau)$ is the total source function at frequencies of the line,

$$\tilde{S}_{\nu\mu}(\tau) = (1 - \rho_{\nu\mu}(\tau))S(\tau) + \rho_{\nu\mu}(\tau)B(\tau) \quad , \tag{2.3}$$

comprising the contributions of the line transition S and of the background continuum B, and $\rho_{\nu\mu}$ is the opacity ratio for the background continuum,

$$\rho_{\nu\mu}(\tau) = \frac{\kappa_c(\tau)}{\kappa_l(\tau)\varphi_{\nu\mu}(\tau) + \kappa_c(\tau)} \quad . \tag{2.4}$$

The total opacity, consisting of the line and continuum contributions $\kappa_l\varphi_{\nu\mu}$ and κ_c, is used to define the monochromatic optical depth $\tau_{\nu\mu}$ in terms of the geometrical depth x, *i.e.*, $d\tau_{\nu\mu} = (\kappa_l\varphi_{\nu\mu} + \kappa_c)dx$; τ is a reference optical depth.

The equations of statistical equilibrium for a line formed in complete redistribution (*cf.* Thomas 1957) in a static or slowly time-varying medium are written in the form of the source function equation,

$$S(\tau) = (1 - \epsilon(\tau))\bar{J}(\tau) + \epsilon(\tau)\mathcal{B}(\tau) \quad , \tag{2.5}$$

where ϵ describes the probability of a collisional transition (or radiative transition other than a direct downward transition in the line under discussion, *cf.* Athay 1972), $\mathcal{B}$ is the Planck function at the electron temperature (or a suitable generalization to describe the strength of the sources emitting photons into the line. Note that the Planck function $\mathcal{B}$ is often equal to the source term B of the background continuum), and $\bar{J}$ is the mean integrated intensity,

$$\bar{J}(\tau) = \int_0^1 d\mu \int_{-\infty}^{\infty} d\nu \varphi_{\nu\mu}(\tau) J_{\nu\mu}(\tau) \quad , \tag{2.6}$$

where $\varphi_{\nu\mu}$ is the normalized absorption profile, which is isotropic in a static atmosphere but depends on direction μ in a moving medium.

In order to facilitate the manipulation of the operators we simplify the notation by dropping all subscripts and arguments and write the equations containing operators in a more compact form. The transfer equation (2.1) is then

$$DJ = J - \tilde{S} \quad , \tag{2.7}$$

with the differential operator D, and the integral expression (2.6) is

$$\bar{J} = \Omega J \quad , \tag{2.8}$$

with the integral operator Ω.

For given line and continuum opacities κ_l and κ_c, collision probability ϵ, and the source terms B and $\mathcal{B}$, the *transfer problem* is defined by the expressions for the total source function (2.3), the line source function (2.5), and the mean integrated intensity (2.8), and by the equation of radiative transfer (2.7). Note that these equations are linear in the line source function and in the intensity. The reason for the linearity is that the problem has been stated in terms of the optical depth; thus the question of solving for the populations in the individual energy levels is not addressed. The equations could therefore be solved directly, without iterations. The perturbation equations will, however, require iterations. They are, nevertheless, much faster to solve than the equations for a direct solution. For a direct solution, the four equations are combined either into a single differential equation, or into a single integral equation. Even though we will not use a direct solution method, we need to derive the respective equations since the approach detailed below depends on a relation between the operators occurring in them.

To derive the equation for the *direct differential method* we insert equations (2.3), (2.5), and (2.8) into the differential equation (2.7) for the specific monochromatic intensity, resulting in

$$TJ = \psi \quad , \tag{2.9}$$

where T is a block tridiagonal difference operator,

$$T = (1 - D) - (1 - \rho)(1 - \epsilon)\Omega \quad , \tag{2.10}$$

and 1 is the Kronecker-δ. The inhomogeneous term, ψ, given by

$$\psi = (1 - \rho)\epsilon\mathcal{B} + \rho B \quad , \tag{2.11}$$

is the driving term of the differential equation. The matrix T has the well-known tridiagonal block structure (*cf.* Mihalas 1978, page 156), and the vectors J and ψ have analogous structure, with the angle-frequency blocks of ψ obtained by repeating the elements of the depth-dependent vectors within the frequency blocks.

Since the inhomogeneous term ψ and the matrix T are assumed known, the linear system (2.9) of coupled equations could be solved directly, in a single step, for the specific monochromatic intensity $J_{\nu\mu}$. This solution would also contain the emergent specific intensity, $I^{+}_{\nu\mu}(0) = 2J_{\nu\mu}(0)$.

In the *direct integral method*, the equations (2.3), (2.7), and (2.8) are inserted into the expression (2.5) for the line source function, resulting in

$$\mathcal{L}S = \varphi \quad , \tag{2.12}$$

where $\mathcal{L}$ is the integral operator,

$$\mathcal{L} = 1 - (1 - \epsilon)\Omega(1 - D)^{-1}(1 - \rho) \quad , \tag{2.13}$$

and φ the inhomogeneous term,

$$\varphi = (1 - \epsilon)\Omega(1 - D)^{-1}\rho B + \epsilon\mathcal{B} \quad . \tag{2.14}$$

The integral equation (2.12) is a complete statement of the transfer problem and, again, the solution of this system of coupled equations solves the transfer problem, yielding $S(\tau)$; the energy loss rate $\bar{J}(\tau) - S(\tau)$ can be directly computed by solving the local equation (2.5), and the mean intensity $J_{\nu\mu}(\tau)$ and the emergent intensity $I^{+}_{\nu\mu}(0)$ are obtained by solving the scalar difference equations (2.7) for known source function.

In addition to the transfer problem with background continuum described by the equations (2.9) or (2.12), we consider also the transfer problem without background continuum. The differential equation (2.9) then becomes

$$T_0 J = \epsilon\mathcal{B} \quad , \qquad \rho = 0 \quad ; \tag{2.15}$$

and the integral equation (2.12),

$$LS = \epsilon \mathcal{B} \quad , \qquad \rho = 0 \quad ; \tag{2.16}$$

with the operators

$$T_0 = (1 - D_0) - (1 - \epsilon)\Omega_0 \quad , \tag{2.17}$$

and

$$L = 1 - (1 - \epsilon)\Omega_0(1 - D_0)^{-1} \quad . \tag{2.18}$$

The operators T_0 and L are *exact* operators for the line transfer problem *without* background continuum; they may also be viewed as *approximate* operators for the transfer problem *with* background continuum. But in addition to the neglect of the continuous opacity, they may differ from the operators T and $\mathcal{L}$ in the number of grid points for which they are constructed; so they would be approximate even for the line without background continuum. What makes T_0 and L interesting here is that they can be related by an equation that is *exact* provided a background continuum is absent.

In order to derive the relation between the operators T_0 and L we may suppose solving equation (2.15) for the specific monochromatic intensity J, evaluating the integral (2.8) for the mean integrated intensity $\bar{J}$, and then finding the line source function S from equation (2.5). By comparing this resulting expression for the line source function S from the differential equation with the formal solution of the integral equation (2.16), *i.e.*, with $S = L^{-1}\epsilon\mathcal{B}$, and equating the operators acting on the "thermal" source term $\epsilon\mathcal{B}$, we obtain the equation

$$L^{-1} = 1 + (1 - \epsilon)\Omega_0 T_0^{-1} \quad , \tag{2.19}$$

which is the desired relation between the approximate integral and differential operators of the transfer problem.

3. SOLUTION BY OPERATOR PERTURBATION

We solve the integral equation (2.12) of the transfer problem by means of operator perturbation (Cannon 1973, 1984, 1985; Scharmer 1981, 1984; for the equivalence of the two methods, *cf.* Kalkofen 1984), using the approach and notation of Kalkofen (1985): Following Cannon, we write the exact integral operator $\mathcal{L}$ of the transfer problem (2.12) as an approximate operator L plus the correction operator $\mathcal{L} - L$ and expand the line source function S in a series,

$$S^{(n)} = \sum_{i=0}^{n} s^{(i)}, \quad S = S^{(\infty)} \quad . \tag{3.1}$$

Inserting the expressions for the operator and for the source function into the "integral equation" (2.12) we obtain a hierarchy of perturbation equations,

$$Ls^{(n+1)} = \mathcal{E}^{(n)} \quad , \tag{3.2}$$

where the driving term is given by

$$\mathcal{E}^{(n)} = \varphi - \mathcal{L}S^{(n)} \quad . \tag{3.3}$$

Note that the driving term of the integral equation (3.2) is identical to the error made by the nth-order source function in the equation (2.12) of statistical equilibrium.

A typical estimate at the beginning of a calculation is to assume that the zeroth-order source function $S^{(0)}$ is equal to zero; the error $\mathcal{E}^{(0)}$ is then equal to the driving term φ of the exact integral equation (2.12). Now, if the approximate operator L were equal to the exact operator $\mathcal{L}$, the approximate equation (3.2) would be identical to the exact equation (2.12), and the first-order source function $S^{(1)} = s^{(1)}$ would solve the problem (2.12); of course, even if $S^{(0)}$ were different from zero, $S^{(1)} = S^{(0)} + s^{(1)}$ would be the exact solution of the problem, but no advantage would be gained from this perturbation method since in addition to solving the equation with the exact integral operator, a non-zero error term would have to be evaluated as well. In fact, the operator perturbation has an advantage over a direct method only if the calculation of both the error $\mathcal{E}^{(n)}$ and of the correction $s^{(n)}$ can be executed very efficiently, a task we accomplish by exploiting the relation between the differential and integral operators of the problem. We consider first the evaluation of the error term.

The full expression for the error $\mathcal{E}^{(n)}$ is given by

$$\begin{aligned} \varphi - \mathcal{L}S^{(n)} = {} & \epsilon B + (1-\epsilon)\Omega(1-D)^{-1}\rho B \\ & - S^{(n)} + (1-\epsilon)\Omega(1-D)^{-1}(1-\rho)S^{(n)} \quad . \end{aligned} \tag{3.4}$$

It is easy to see that the calculation can be broken down into several parts. First, note that the monochromatic Λ-operator $(1-D)^{-1}$ acts only on known functions. It is therefore convenient to define intensities by means of differential equations.

Thus, the specific monochromatic intensity due to the background continuum may be defined by *scalar* differential equations (*i.e.*, separate equations at the individual angle-frequency points),

$$(1-D)J_c = \rho B \quad , \tag{3.5}$$

and the corresponding mean integrated intensity by sums over the angle-frequency components of the intensity with the integration weights Ω,

$$\bar{J}_c = \Omega J_c \quad . \tag{3.6}$$

Equations (3.5) and (3.6) are solved only in the first iteration (of each time step, if the problem is time-dependent). The (scalar) differential equations for the intensity due to the line transition are

$$(1-D)J_l^{(n+1/2)} = (1-\rho)S^{(n)} \quad ; \tag{3.7}$$

the half-integral index of the intensity $J_l^{(n+1/2)}$ is to indicate that the line intensity is obtained from the line source function $S^{(n)}$ by applying the monochromatic Λ-operator; this operation improves the quality of J, but by less than a full iteration step (by very much less if the value of ϵ is very small). The corresponding mean integrated intensity is again obtained with the aid of the integral operator Ω, *i.e.*, by summing over the angle-frequency components of the line intensity,

$$\bar{J}_l^{(n+1/2)} = \Omega J_l^{(n+1/2)} \quad . \tag{3.8}$$

These intensities can now be used to define the Λ-iterated source function,

$$S^{(n+1/2)} = (1-\epsilon)(\bar{J}_l^{(n+1/2)} + \bar{J}_c) + \epsilon B \quad . \tag{3.9}$$

The error term may therefore be written as

$$\mathcal{E}^{(n)} = S^{(n+1/2)} - S^{(n)} \quad . \tag{3.10}$$

Thus, the driving term of the integral equation (3.2) for the source function correction $s^{(n+1)}$ is given by the difference between the nth-order source function $S^{(n)}$ and the source function $S^{(n+1/2)}$ obtained from it by means of a Λ-iteration.

It is interesting to note that ordinary Λ-iteration would solve the integral equation (2.12) by generating the source function $S^{(n+1/2)}$ from the source function $S^{(n)}$ repeatedly. Since the integrated Λ-operator $\Omega(1-D)^{-1}$ acts mainly via the short-range interaction of the radiative transfer whereas the scale length of the scattering problem is the thermalization length, *i.e.*, the characteristic length of the operator $1-(1-\epsilon)\Omega(1-D)^{-1}$, Λ-iteration will fail to converge when scattering is very important, *i.e.*, when the value of ϵ is very small. The present iteration method will converge, however, provided these long-range scattering interactions are contained in the approximate integral operator L.

We calculate the correction term $s^{(n+1)}$ of the source function from equation (3.2), in which we replace the integral operator L by its equivalent (2.19) involving the differential operator T_0,

$$s^{(n+1)} = L^{-1}\mathcal{E}^{(n)} = \left[1 + (1-\epsilon)\Omega_0 T_0^{-1}\right]\mathcal{E}^{(n)} \quad . \tag{3.11}$$

Again we see that the equation can be solved by breaking it into several parts: First, we define a correction intensity by means of the differential equation

$$T_0 j^{(n+1)} = \mathcal{E}^{(n)} \quad , \tag{3.12}$$

and the corresponding mean integrated intensity by means of the integral expression

$$\bar{j}^{(n+1)} = \Omega_0 j^{(n+1)} \quad . \tag{3.13}$$

The correction equation (3.11) then becomes

$$s^{(n+1)} = \mathcal{E}^{(n)} + (1-\epsilon)\bar{j}^{(n+1)} \quad . \tag{3.14}$$

Thus the transfer problem (2.12) for the direct calculation of the source function S by means of a single integral equation is now written in the form of a perturbation expansion for the iterative calculation of the source function series (3.1); the most important parts of the perturbation approach are equations (3.7), (3.8), and (3.10) for the error $\mathcal{E}^{(n)}$ of the conservation equation (2.12); and equations (3.12), (3.13), and (3.14) for the correction terms $\bar{j}^{(n+1)}$ and $s^{(n+1)}$. Even though the perturbation equations are solved repeatedly, the calculations can be carried out very rapidly. This is obviously true for the error term, which requires the solution of the scalar (*i.e.*, uncoupled) transfer equations (3.7) for the monochromatic line intensity $J_l^{(n+1/2)}$ and the frequency integration for the corresponding mean integrated intensity $\bar{J}_l^{(n+1/2)}$; the computer time for these operations scales linearly with the numbers of depth, frequency, and angle points. The correction calculation (3.12) for the "intensity" $j^{(n+1)}$, on the other hand, scales as the third power of the number of angle-frequency points; as a differential equation it is linear in the number of depth points and, of course, linear in the number of iterations that are necessary to satisfy a convergence criterion.

The *error* calculation with the operators D and Ω requires a set of frequency and angle points that is sufficiently dense to insure good accuracy; for the *correction* calculation with the operators D_0, Ω_0, and hence T_0, a sparse set of angle and frequency points is adequate. Since the approximate differential operator T_0 must describe the long-range interactions due to scattering, this sparse set must contain frequencies for which the photon mean-free-path is of the order of the thermalization length. In a moving medium this may imply a range of frequency values even for a depth-independent scattering parameter ϵ. Typical grid points are the frequency at line center and near the frequency for which the photon mean-free-path matches the thermalization length. Thus, like the core saturation approximation (Rybicki 1972), the sparse set emphasizes radiative transfer in the line wings. The number of angle points may be quite small; a single μ-value may suffice. Furthermore, if the approximate operator refers to a static medium, only one line wing needs to be treated.

It is interesting to note that the equations (3.12) and (3.13) need not be solved as differential equations. Under some conditions it can become advantageous to transform them into the equivalent integral equation for the mean integrated intensity $\bar{J}^{(n+1)}$ and to solve this integral equation (*cf.* section 5).

4. NUMERICAL RESULTS

The perturbation method was tested in two cases patterned after transfer problems with macroscopic flow suggested by Scharmer (1984, p 196ff). In one test, the atmosphere is a model of the solar chromosphere with a line shifted by a periodic, vertical velocity field with an amplitude of three Doppler widths; in the second test, the atmosphere is isothermal and the velocity amplitude is one Doppler width. In both problems the collision parameter ϵ has the value 0.01, the line is Doppler-broadened, and background emission is neglected (a very small background opacity is included in order to prevent the line opacity from becoming very small). In the chromospheric problem, there are about $1\frac{1}{2}$ wave trains in the layers between $\tau = 10^{-1}$ and $\tau = 10^{3.4}$, where the source function saturates to the Planck function. The time steps are separated by one twentieth of the wave period. We discuss first the isothermal problem.

The line transfer problem in an isothermal, semi-infinite atmosphere with constant collision parameter is a modification by a flow field of the standard non-LTE problem. The thermalization depth of a Doppler-broadened line in the corresponding static atmosphere with $\epsilon = 0.01$ is of the order of $\tau = 100$; the characteristic scale length of the flow-shifted line may not be very much larger than $\delta\tau \simeq 100$, although in general this depends on the detailed depth dependence of the velocity and the Planck function. If the atmosphere were static, the critical bandwidth for the calculation would extend from line center to about 2.5 Doppler widths into one line wing, *i.e.*, to a frequency at which the photon mean-free-path is about as long as the characteristic length. In the moving medium with the velocity amplitude of one Doppler width, the corresponding band extends from -3.5 to $+3.5$ Doppler widths. In the error calculation we cover this band width with 30 frequency points in the half profile, *i.e.*, with 59 frequency points spread over the whole profile, and thus with 236 angle-frequency points in this calculation for 4 angle points.

The line transfer problem is solved at several levels of approximation. The fewest simplifications are introduced into the operator in a procedure that is analogous to the variable Eddington factor technique (*cf.* Mihalas 1978), where the approximate operator uses only a single angle point rather than the full set; the matrix is computed with the direction-dependent, asymmetric profile function appropriate for a moving medium. The savings in solution time compared to the direct solution by means of differential equations is then due to the smaller size of the matrix that couples the angle-frequency components of the intensity; this time scales as the cube of the order of the coupling matrix, thus, in this example, as the cube of the number of angle points, *i.e.*, the perturbation calculation is more efficient by factor of 4^3 (for a single iteration).

At the next level, the approximate operator is based on a static atmosphere. This results in further savings of three kinds. The most immediate one is that the absorption profile is symmetric, requiring frequency points in only one line wing. In addition, fewer frequency points are needed to cover all significant frequencies that provide coupling between distant parts of the atmosphere. And finally, the most important savings in a time-dependent problem, the time-consuming inversion of the matrix operators is carried out only at the beginning of the first iteration. The order of the correction matrices can then be reduced further by cutting the number of frequency points; in such a wider frequency mesh, care must be taken that frequencies controlling the overall structure of the solution be included in the set.

We solved the line transfer problem for the asymmetric profile with 59 frequency points between $-3.5\Delta\nu_D$ and $3.5\Delta\nu_D$, and for symmetric profiles with 26, 13, 11, and 5 frequency points between 0 and $2.5\Delta\nu_D$. In all cases we started the iterations with zero source function at the first time step.

The astonishing result of the tests carried out with these different approximate operators is that one cannot point to any significant systematic differences in the rates of convergence. But there are enormous differences in the computer effort.

The least efficient operator by far is the one constructed in analogy to the variable Eddington factor technique. In addition to being expensive in itself, it must be updated at every time step. In contrast, the matrices for the static medium are constructed and inverted only once.

At the first time step and after the fifth iteration (*cf.* TABLE 1), all cases give corrections below 1% (a typical requirement for the maximal allowed error). The accuracy is better than 1% after the fourth iteration, but not in all cases could one have judged the accuracy of the numerical solution from the magnitude of the corrections with enough confidence to terminate the iterations after the fourth. At the second time step (out of 20 per period), two iterations suffice to reach an accuracy of 1%.

The equations for the isothermal atmosphere converge at an acceptable rate even when only two angle-frequency points (frequency points at 1 and $2\Delta\nu_D$) are used in the construction of the approximate operator (compared to 236 angle-frequency points for the exact operator!). Only one additional iteration is needed both at the first and at subsequent time steps. But since the cost of solving the problem by means of this operator perturbation method is in the iterations (because of the error calculations), and also since one has the feeling that this sparse operator must be less robust than operators more generously supplied with, perhaps redundant, frequency points, the case of two points will not be discussed further in this comparison.

The time advantage of the most efficient matrix (5 frequency points, spaced $\frac{1}{2}\Delta\nu_D$ apart) over the least efficient (the variable Eddington factor analogue) is a factor of about 50 in the first time step, and a factor of 80 in the second and, generally, subsequent time steps. With this code, which is designed

TABLE 1
Isothermal Atmosphere, with $\Delta\nu_{vel} = \Delta\nu_D$
Maximal Relative Corrections and Computation Times

iteration	max. corr. $\delta S/S$	%time (correction)	%time (error)	time(s) (total)
		time step 1		
1	—	100	0	0.6
2	0.24402	29	71	2.2
3	0.04663	19	81	3.9
4	−0.01389	15	85	5.6
5	0.00430	14	86	7.2
6	−0.00094	12	88	9.0
		time step 2		
1	−0.07200	6	94	1.7
2	−0.01411	6	94	3.2
3	0.00407	6	94	5.0
4	−0.00125	6	94	6.6
5	0.00039	6	94	8.3

to test the method, not optimized for speed of computation, the time for solving the equations with 5 frequency points in the correction calculation is about 3s on a VAX11-780 (at the first time step, the calculation takes about twice as long, *cf.* TABLE 1, last column). This is not much longer than it takes to compute and store the (full) profile function for the error calculation; with a Voigt profile, the times for calculating the profile function and for solving the transfer problem would have been comparable (*cf.* Scharmer 1984, his TABLE 1, for similar conclusions). Most of the time is spent in the formal solution (*cf.* TABLE 1, column 4), which determines the error of a given source function in satisfying the conservation equations. At the first time step, 15% of the solution time is needed for solving the coupled equations for the corrections to the source function (TABLE 1, column 3, after the fourth iteration), and at later times only 6%, which is less than the time needed to calculate the profile function, even for the case of pure Doppler-broadening.

Because of the large fraction of time spent in determining the formal solution of the transfer problem, the overall timing of the method grows linearly with the numbers of frequency, angle, and depth points of the error calculation and with the number of iterations. It is therefore not essential to be able to solve the correction equations with the lowest possible number of frequency points; it is much more important to choose a set that minimizes the number of iterations necessary for convergence.

In the problem under discussion, no further improvements are to be expected from a more careful selection of frequency points since the two iterations that are required at most time steps for an accuracy of 1% are probably close to the minimum necessary for this case. The exception is the first time step, where no good source function estimate is available and hence the starting solution is $S(\tau) = 0$. But the number of iterations is larger when the collision parameter is decreased or when the velocity amplitude is increased. Another calculation that also deserves attention is that of the absorption profile. For Voigt functions or more complicated broadening profiles, that part of the calculation can absorb a large fraction of the total time needed to solve a problem. This will not change the overall timing of the solution, however, which remains linear in the grid sizes of all the variables.

In the line transfer calculation for the model chromosphere we study the properties of the iteration scheme in a case that requires more iterations than the preceding problem. We consider a line formed in a velocity field with an amplitude of three Doppler widths. Because of the larger velocity, the bandwidth in this lab frame formulation spans the approximate frequency range from $-6\Delta\nu_D$ to $+6\Delta\nu_D$. We perform the error calculation with 67 frequency points covering both wings of the profile; five points per Doppler width between line center and $5\Delta\nu_D$ in either wing, and four frequency points per Doppler width between $5\Delta\nu_D$ and $7\Delta\nu_D$. Only a single angle point is used in this demonstration, both for the error and for the correction.

For the correction calculations we use 5 frequency points in the symmetric profile. Again, because of the symmetry, the profile function and the associated matrices are calculated and inverted only at the beginning of the first time step and stored for later use.

The iterations yield fairly rapidly decreasing maximal corrections of the source function S that alternate in sign, implying overshoot in the corrections δS (*cf.* TABLE 2, columns 2 and 4). An accuracy of better than 1% is reached after 7 iterations; *i.e.*, the eighth iteration has a value of the maximal relative correction below 0.01. Since the ratio of consecutive, maximal corrections is nearly constant (approximately -1.8), convergence can be sped up by assuming that

TABLE 2
Maximal Relative Corrections for the Chromospheric Model

	t = 0		t = Period/4	
iteration	unmodified	accelerated	unmodified	accelerated
1	—	—	−0.26564	−0.26618
2	0.48784	0.48784	0.08129	0.08151
3	−0.19757	−0.19757	−0.05073	−0.03748
4	0.09740	0.09740	0.02705	0.00643
5	−0.05627	−0.04116	−0.01493	−0.00161
6	0.03070	0.00805	0.00797	0.00027
7	−0.01772	−0.00226	−0.00428	
8	0.00998	−0.00031	0.00228	
9	−0.00572		−0.00122	
10	0.00325		0.00065	
11	−0.00185			
12	0.00105			
13	−0.00060			

this ratio applies equally at all depth points and is constant for all subsequent iterations. Thus, dividing the corrections to the source function by 1-r, where r is the ratio of consecutive maximal corrections (which typically occur at the same or at a neighboring depth point), and beginning this damping of δS when the maximal relative corrections have dropped below 10%, the number of iterations is reduced (*cf.* TABLE 2, columns 3 and 5). The difference between unmodified and accelerated convergence is clearly exhibited by the rate at which the maxima of $|\delta S/S|$ drop to a value below 0.001: 13 iterations *vs.* 8 at the first time step, and 7 *vs.* 3 at the sixth.

With the accuracy of the solution set at 1%, 2 iterations are saved by this strategy at the first time step; and, depending on the criterion used for inferring that the error must be less than 1%, one or two iterations are saved at the sixth time step. At other times, the speed-up may occur when the corrections are already too small to translate into savings.

5. DISCUSSION

The operator perturbation method of this paper is well-suited to the solution of radiative transfer problems in moving media. Its timing is very favorable, in part, because of the small number of frequency points used in the calculation of corrections to the solution and the resulting preponderance of the error calculation. However, depending on the temperature and velocity structure of the atmosphere, one may need more frequency points and, in addition, one may want to construct the approximate operator for the anisotropic profile function. Because of the rapid growth rate of the solution time with the number of angle-frequency points, the solution of the correction equations may threaten to take so long that the perturbation method becomes as expensive as the direct differential equation method. In order to address such problems we consider an alternative formulation of the correction equations and ask under what conditions a modified perturbation method remains faster than a direct solution method.

The speed of the perturbation method in time-dependent problems is due to two features in particular: the expansion of the exact integral operator about an approximate operator appropriate for the symmetric profile function of a static atmosphere, and the use of the converged solution of the preceding time step as the starting solution of the current time step. Unless the time steps are very large, the preceding time will always provide a much better zeroth-order solution than does $S(\tau) = 0$. But the operator for the static atmosphere need not be a close approximation to the exact operator.

Flow can modify the characteristic length of the solution. Thus, a velocity disturbance might alter the interaction within the atmosphere, coupling distant regions between which there is no significant interaction in the static case and at the same time decoupling otherwise coupled regions. For example, a periodic velocity field of large amplitude effectively decouples regions one half wavelength apart. If that distance is of the order of the thermalization length, the static operator couples layers that are only weakly interacting in the moving medium. Thus, an approximate operator constructed with the static profile function may show a qualitatively incorrect behavior and therefore have poor numerical properties, resulting in slow convergence or perhaps even failure to converge. For sufficiently large velocities, the co-moving frame formulation may be the only viable, albeit expensive, alternative. But there may be intermediate cases of moderately large flow velocity for which the perturbation method of this paper, suitably modified, might provide a more efficient alternative. While it will still be advantageous to start the calculation at a new time with the converged solution of the preceding time, thus retaining one of the features that account for the speed of the method, one might be forced to a profile function that reflects the flow. This might necessitate a frequent updating of the approximate operator, even if that were not necessarily at every time step.

In a moving medium, the absorption profile is anisotropic and therefore requires frequency points in both line wings. Since the growth rate of the computation time for the differential equations in the correction calculation is proportional to the third power of the number of angle-frequency points, it does not take many frequency points in an approximate operator constructed with the anisotropic profile function to make the correction calculation more expensive than the error calculation. It is therefore interesting to ask whether the basic procedure described in this paper needs to be modified to retain a timing advantage over a direct solution method.

The modification to be considered concerns the method of solution of the set of coupled differential equations (3.12) and of the integral (3.13). Instead of solving them as differential equations for the specific, monochromatic intensity $j^{(n+1)}$ we transform them into the equivalent integral equation for the mean integrated intensity $\bar{j}^{(n+1)}$. Using the definitions (3.13) and (2.17), equations (3.12) can be cast into the form

$$(1 - D_O)j^{(n+1)} = (1 - \epsilon)\bar{j}^{(n+1)} + \mathcal{E}^{(n)} \quad , \tag{5.1}$$

from which we obtain the integral equation

$$\lambda \bar{j}^{(n+1)} = E^{(n)} \quad . \tag{5.2}$$

The integral operator λ is given by

$$\lambda = 1 - \Omega_O(1 - D_O)^{-1}(1 - \epsilon) \quad ; \tag{5.3}$$

(note the superficial resemblance of the integral operators λ and L, equation 2.18). The inhomogeneous term of the integral equation (5.2) is

$$E^{(n)} = \Omega_O(1 - D_O)^{-1}\mathcal{E}^{(n)} \quad . \tag{5.4}$$

This solution method requires the monochromatic integral operators $(1 - D_O)^{-1}$, which are obtained from the corresponding tridiagonal difference operators $(1 - D_O)$ in approximately $N_\tau{}^2$ operations. From the solution of equation (5.2) for the mean integrated intensity $\bar{j}^{(n+1)}$ we obtain the correction term $s^{(n+1)}$ by using equation (3.14). The correction equations in the form (5.2) are referred to as *modified.*

The *direct* method that serves as the comparison is the differential equation method. It can be solved as a differential equation in the Feautrier approach or as an integral equation in the Rybicki approach (Rybicki 1971; *cf.* also Auer 1984 or Mihalas 1978), where the differential equation is the starting formulation for the integral equation method, with which it shares the structure of the equations to be solved and the dependence of the timing on the grid sizes of the independent variables.

First, we note that we need not concern ourselves with the case where the differential equations are solved as differential equations, for which the computation time scales as $N_\tau \times N_{\mu\nu}{}^3$, because in that case the perturbation method as described above is expected to be very much faster than a direct method and hence need not be modified.

In a moving medium the number of frequency points tends to be large, both line wings must be treated, and several angle points are necessary to account for the anisotropy of the profile function. As a consequence the computation time for the differential equation tends to be longer than that of the corresponding integral equation. Then the *direct* method of choice for solving the differential equation is to transform it into the corresponding integral equation, *i.e.*, to use the Rybicki method of solution.

The integral equation method consists of two parts, the construction of the frequency-integrated operator and the solution of the resulting system of coupled equations. The construction of $N_{\mu\nu}$ matrices, with $N_\tau{}^2$ elements each, scales as $N_{\mu\nu} \times N_\tau{}^2$, the solution of the integral equation scales as $N_\tau{}^3$. Thus there are two cases to consider, depending on the relative times of the two parts. Of these, the case where the matrix inversion time is longer does not concern us since then the direct method is faster than the modified perturbation method since the latter scales the same way and, in addition, requires iterations. Thus we need consider only the case where the construction time is the longer. Then $N_{\mu\nu} \times N_\tau{}^2$ is larger than $N_\tau{}^3$, *i.e.*, the number of angle-frequency points exceeds the number of depth points, $N_{\mu\nu} > N_\tau$.

In the modified perturbation method the construction of the operator λ (eq. 5.3) scales as $N^*_{\mu\nu} \times N_\tau{}^2$, where $N^*_{\mu\nu}$ is the number of angle-frequency points of the approximate operator, and the solution of equation (5.2) scales as $N_\tau{}^3$. If we confine ourselves to the case where the correction calculation is carried out for a single angle point we distinguish two cases, where either $N^*_\nu \times N_\tau{}^2$ or $N_\tau{}^3$

dominates the solution time of the modified perturbation method; here N_ν^* is the number of frequency points used for the approximate operator.

If the set-up time of the integral operator $\Omega_0(1-D_0)^{-1}$ dominates, *i.e.*, $N_\nu^* > N_\tau$, the modified perturbation solution has a time advantage provided $N_\nu^* < N_\mu N_\nu$; but if the solution time of equation (5.2) dominates, *i.e.*, $N_\nu^* < N_\tau$, the modified perturbation solution is favored provided $N_\tau < N_\mu N_\nu$. In the first example discussed in this paper, where $N_\nu = 60$ (for both wings of the profile) and $N_\mu = 4$, and if N_ν^* is as large as N_τ, the modified perturbation solution is faster by a factor of about six if parts of the frequency-integrated matrix are stored (from an $L \times U$ decomposition, for example; *cf.* Scharmer & Nordlund 1982), or otherwise if the number of iterations is smaller than about 6. For the problem we solved, where two iterations sufficed for convergence, there is thus still a speed advantage of the modified perturbation method, but the gain is not very large.

Although the modification of the perturbation method yields computation times that are not as short as those of the approach discussed in the body of the paper, it does have a desirable feature: it allows a gradual transition from the modified perturbation method to the direct differential equation approach. This transition can be useful in developing numerical code, for example, or to study the behavior of the approximate operator when the number of frequency points is reduced gradually, or to track down numerical difficulties encountered in the solution of the equations. We note parenthetically that another approach to trouble-shooting in case of convergence difficulties is to determine the eigenvalues of the matrix in the expansion for the source function corrections (Kalkofen 1985) that result from combining equations (3.1), (3.2), and (3.3).

One may ask what we have achieved by taking the correction equations (3.12) to (3.14), which were obtained from their integral equation form (3.2) and (3.3), and transforming them back into integral equations (5.2, 5.4, 3.14). First note that the integral equation (3.2) for the source function correction $s^{(n+1)}$ is different from the integral equation (5.2) for the intensity correction $\bar{j}^{(n+1)}$. The main difference is that in equation (3.2) the matrix elements of L would ordinarily be calculated starting from the integral for the monochromatic intensity in terms of the source function, whereas in equation (5.2) the elements of λ are obtained from the monochromatic second-order difference operator $(1-D_0)$. The elements of λ thus come from a very accurate depth integration of the transfer equation, with all the approximations in the angle and frequency quadratures (yet accurately maintained normalization of the profile function and the corresponding integration weights Ω_0). If the stress in the approximate operator is on an accurate depth integration, the way we have chosen is much more efficient than using the original integral equation. The latter lends itself more easily to approximations of the depth integration, but without reducing the angle-frequency set for the correction calculation; that is the essence of Scharmer's approximation.

6. SUMMARY

We have described an operator perturbation method for the solution of line transfer problems with complete frequency redistribution of photons over the line. The basic equations, consisting of the equation of radiative transfer and the conservation equations of statistical equilibrium, are formulated in terms of an integral equation for the source function. The operator of this equation is expanded about a simpler, approximate integral operator as suggested by Cannon, resulting in a hierarchy of integral equations for corrections to the source function; these equations are similar to those of Scharmer's method. The equations separate a transfer problem into the calculation of the error with which a provisional solution satisfies the conservation equations, and the calculation of corrections to the solution; the error in the constraint equations becomes the driving term in the correction equation.

The distinctive feature of the method is the use of two separate sets of angle and frequency points: a dense set consisting of many grid points for the calculation of the error, and a sparse set with very few points for the calculation of corrections to the solution. The dense set guarantees that the quadrature error in the frequency integration is negligible, the sparse set must insure that the overall structure of the solution is correctly reproduced. It must therefore contain frequency points for which the monochromatic photon mean-free-path is of the order of the thermalization length of the medium; the sparse set is responsible for the high speed of the method.

The integral equation for the error is transformed into a set of scalar differential equations for the specific monochromatic intensity, which are solved for known source function. These equations are Feautrier equations with second-order accuracy, or Hermite equations with fourth-order accuracy. The integral equation for the correction to the solution is transformed into a set of coupled differential equations; this transformation is performed with the help of an operator relation obtained from the direct statements of the transfer problem in either integral or differential form. We also discuss a method of solution that transforms the coupled differential equations back into integral equations; this form may be useful when the differential velocity is very large.

High speed in time-dependent problems with velocity amplitudes of the order of the sound velocity is achieved by approximate operators based on a static medium, for which the operators and their inverses are computed only at the beginning of a time-dependent calculation, and by the use of the converged solution of the preceding time step as the starting solution at the current time step.

The numerical properties of the method are shown in the case of a single line formed in a time-dependent atmosphere in statistical equilibrium: In an isothermal problem with a periodic velocity field of one Doppler width, two iterations suffice for an accuracy of better than 1% (except for the first time step, where three iterations are necessary), and in a chromospheric problem with a periodic velocity field having an amplitude of three Doppler widths, two or three iterations suffice for convergence. For the latter case, an accelerated iteration procedure is discussed that is useful in this case of oscillatory convergence.

In the demonstration problems, almost all the computation time is used for calculating the error, only a negligible amount being needed for solving the coupled differential equations for the corrections to the solution. Therefore the solution time of the method scales linearly with the number of grid points of all the independent variables: depth, angle, and frequency. Hence the optimal choice of grid points for the sparse set used in the correction calculation minimizes the number of iterations necessary for convergence.

ACKNOWLEDGMENTS

I thank Gene Avrett for comments on the manuscript.

REFERENCES

Athay, R. G. 1972, *Radiation Transport in Spectral Lines*, D. Reidel Publishing Co., Dordrecht, Holland.

Athay, R. G. & Skumanich, A. 1967, *Ann. d'Astroph.*, **30**, 669.

Auer, L. H., 1967, *Astrophys. J. (Letters)*, **150**, L53.

———— 1976, *J. Quant. Spectr. Rad. Transf.*, **16**, 931.

———— 1984. *Methods in Radiative Transfer*, W. Kalkofen ed., Cambridge University Press, Cambridge, 79.

Auer, L. H. & Mihalas, D. 1968, *Astrophys. J.*, **151**, 311.

———— 1969, *Astrophys. J.*, **158**, 641.

Avrett, E. H. & Hummer, D. G. 1965, *Monthly Notices Roy. Astron. Soc.*, **130**, 295.

Avrett, E. H. & Loeser, R. 1966, *Smithsonian Astrophys. Obs. Spec. Rept.*, **201**, 98.

Cannon, C. J., 1973, *J. Quant. Spectrosc. Rad. Transfer*, **13**, 627.

———— 1984, *Methods in Radiative Transfer*, W. Kalkofen ed., Cambridge University Press, Cambridge, 157.

———— 1986, *The Transfer of Spectral Line Radiation*, Cambridge University Press, Cambridge.

Cuny, Y. 1967, *Ann. d'Astroph.*, **30**, 143.

Dahlquist, G. & Björck, Å. 1974, *Numerical Methods*, Prentice-Hall, Englewood Cliffs, N.J.

Feautrier, P. 1964, *Compt. Rend. Acad. Sci. Paris*, **258**, 3189.

Hummer, D. G. & Rybicki, G. B. 1971, *Ann. Rev. Astron. Astroph*, **9**, 237.

Kalkofen, W. 1984, *Methods in Radiative Transfer*, W. Kalkofen ed., Cambridge University Press, Cambridge, 427.

———— 1985, *Progress in Stellar Spectral Line Formation Theory*, J. E. Beckman & L. Crivellari eds., D. Reidel Publishing Co., Dordrecht, Holland.

Mihalas, D. 1978, *Stellar Atmospheres*, Second Edition, W. H. Freeman & Co., San Francisco.

Press, W. H., Flannery, B. P., Teukolsky, S. A. & Vetterling, W. T. 1986, *Numerical Recipes*, Cambridge University Press, Cambridge.

Rybicki, G. B. 1971, *J. Quant. Spectrosc. Rad. Transfer*, **11**, 589.

———— 1972, *Line Formation in the Presence of Magnetic Fields*, R. G. Athay, L. L. House & G. Newkirk, Jr. eds., Boulder: High Altitude Observatory, p145.

Scharmer, R. 1981, *Astrophys. J.*, **249**, 720.

———— 1983, *Astron. Astroph.*, **117**, 83.

———— 1984, *Methods in Radiative Transfer*, W. Kalkofen ed., Cambridge University Press, Cambridge, 173.

Scharmer, G & Carlsson, M. 1985, *J. Comp. Phys.*, **59**, 56.

Scharmer, G. & Nordlund, Å. 1982, *Stockholm Obs. Rep.*, **19**.

Stoer, J. & Burlisch, R. 1980, *Introduction to Numerical Analysis*, Springer-Verlag, New York.

ITERATIVE SOLUTION OF MULTILEVEL TRANSFER PROBLEMS

Eugene H. Avrett and Rudolf Loeser
Harvard-Smithsonian Center for Astrophysics, Cambridge, USA

ABSTRACT: We show how the "equivalent two-level atom" equations can be used to solve multilevel transfer problems. Iterative convergence can be very slow if the coupled equations are not formulated in a way that directly requires the solution to be self consistent. We give an example in which convergence is obtained after a few iterations or requires more than 100 iterations, depending on how the equations are formulated. We also examine the accuracy of using the escape-probability approximation in multilevel problems instead of solving the transfer equations, and find that large errors can occur when this approximation is used for all line transitions. In multilevel cases, the approximation should not be used for the weaker line transitions.

1. INTRODUCTION

Calculations of the transfer of line radiation generally involve atoms or ions with large numbers of energy levels and coupled radiative transitions. Various methods are available for obtaining the solution of the basic two-level case, *i.e.*, solving the combined statistical equilibrium and radiative transfer equations for a given line transition. See Feautrier (1964), Avrett and Hummer (1965), Avrett (1965), Athay and Skumanich (1967), Hummer (1968), Jefferies (1968), Avrett and Loeser (1969), Hummer and Rybicki (1971), Avrett (1971), Athay (1972), Kalkofen (1974), Mihalas (1978), Scharmer (1981, 1984), Avrett and Loeser (1984), and Cannon (1985).

Some methods are available that can be applied directly to multilevel problems, solving all line transitions simultaneously. See Auer and Mihalas (1969), Auer (1984, and references therein), Scharmer and Carlsson (1985), Schönberg and Hampe (1986), and Carlsson (1986).

Alternatively, any of the methods developed for the basic two-level case can be applied separately to each line transition of a multilevel atom. This is the so-called equivalent two-level approach. See Cuny (1967), Avrett (1968), Avrett and Kalkofen (1968), Vernazza, Avrett and Loeser (1973), Mihalas and Kunasz (1978), Gouttebroze and Leibacher (1980), Vernazza, Avrett and Loeser (1981), Lites and Skumanich (1982), Lemaire and Gouttebroze (1983), Skumanich and Lites (1985), and Catala and Kunasz (1987).

An advantage of the equivalent two-level approach is that it can be applied to highly complex cases with very many levels and transitions. In some applications, however, we find that only when the equations are formulated with sufficient care will rapid iterative convergence be obtained, instead of slow convergence, or no convergence at all.

The formulation of the equivalent two-level equations was discussed in some detail by Avrett (1968). Since the conference proceedings containing that paper are no longer generally available, we give here an updated account of the essential method and results.

2. STATISTICAL EQUILIBRIUM EQUATIONS

The statistical equlibrium equation for level m of an $\mathcal{N}$-level atom is

$$n_m \left(\sum_{\substack{\ell=1 \\ \neq m}}^{\mathcal{N}} P_{m\ell} + P_{m\kappa} \right) = \sum_{\substack{\ell=1 \\ \neq m}}^{\mathcal{N}} n_\ell P_{\ell m} + n_\kappa P_{\kappa m} \,, \tag{1}$$

where n_a is the number of atoms per unit volume in level a and P_{ab} is the number of transitions from a to b, per atom in level a, per unit time. The subscript κ refers to the next higher stage of ionization.

We can eliminate n_κ using the continuum equation,

$$n_\kappa \sum_{\ell=1}^{\mathcal{N}} P_{\kappa\ell} = \sum_{\ell=1}^{\mathcal{N}} n_\ell P_{\ell\kappa} \,. \tag{2}$$

Then it follows that

$$n_m \sum_{\substack{\ell=1 \\ \neq m}}^{\mathcal{N}} (P_{m\ell} + p_{m\ell}) = \sum_{\substack{\ell=1 \\ \neq m}}^{\mathcal{N}} n_\ell (P_{\ell m} + p_{\ell m}), \tag{3}$$

where

$$p_{ab} = P_{a\kappa} P_{\kappa b} / \sum_{\ell=1}^{\mathcal{N}} P_{\kappa\ell} \,. \tag{4}$$

The bound-bound rate coefficients are given by

$$\left. \begin{aligned} P_{m\ell} &= A_{m\ell} + B_{m\ell} \bar{J}_{m\ell} + C_{m\ell} \\ P_{\ell m} &= B_{\ell m} \bar{J}_{m\ell} + C_{\ell m} \end{aligned} \right\} m > \ell \,, \tag{5}$$

where the Einstein A and B coefficients are related by

$$A_{m\ell} = (2h\nu_{m\ell}^3 / c^2) B_{m\ell} \,, \tag{6}$$

and

$$g_m B_{m\ell} = g_\ell B_{\ell m} \,. \tag{7}$$

The collisional excitation and deexcitation rates satisfy the equation

$$n_m^* C_{m\ell} = n_\ell^* C_{\ell m} \,, \tag{8}$$

where

$$\frac{n_m^*}{n_\ell^*} = \frac{g_m}{g_\ell} e^{-h\nu_{m\ell}/kT} , \quad m > \ell. \tag{9}$$

If we let $Z_{ab} = C_{ab} + p_{ab}$, the statistical equilibrium equation (3) may be written as

$$\begin{aligned} n_m &\left[\sum_{\ell=1}^{m-1} (A_{m\ell} + B_{m\ell}\bar{J}_{m\ell} + Z_{m\ell}) + \sum_{\ell=m+1}^{\mathcal{N}} (B_{m\ell}\bar{J}_{\ell m} + Z_{m\ell}) \right] \\ &= \sum_{\ell=1}^{m-1} n_\ell (B_{\ell m}\bar{J}_{m\ell} + Z_{\ell m}) + \sum_{\ell=m+1}^{\mathcal{N}} n_\ell (A_{\ell m} + B_{\ell m}\bar{J}_{\ell m} + Z_{\ell m}) . \end{aligned} \tag{10}$$

This is the single-rate statistical equilibrium equation, based on the values of $\bar{J}$. ($\bar{J}$ is defined explicitly in the next section.)

The bound-bound radiative rates can be expressed as net rates as follows:

$$n_m (A_{m\ell} + B_{m\ell}\bar{J}_{m\ell}) - n_\ell B_{\ell m}\bar{J}_{m\ell} = n_m A_{m\ell}\rho_{m\ell} \,, \tag{11}$$

where the net rate coefficient $\rho_{m\ell}$ is given by

$$\rho_{m\ell} = 1 - \frac{\bar{J}_{m\ell}}{S_{m\ell}} \,. \tag{12}$$

Here $S_{m\ell}$ is the line source function

$$S_{m\ell} = \frac{2h\nu_{m\ell}^3/c^2}{\left(\dfrac{g_m}{g_\ell}\dfrac{n_\ell}{n_m} - 1 \right)} \,. \tag{13}$$

Equation (10) then can be written as

$$\begin{aligned} n_m &\left[\sum_{\ell=1}^{m-1} (A_{m\ell}\rho_{m\ell} + Z_{m\ell}) + \sum_{\ell=m+1}^{\mathcal{N}} Z_{m\ell} \right] \\ &= \sum_{\ell=1}^{m-1} n_\ell Z_{\ell m} + \sum_{\ell=m+1}^{\mathcal{N}} n_\ell (A_{\ell m}\rho_{\ell m} + Z_{\ell m}) . \end{aligned} \tag{14}$$

This is the net-rate statistical equilibrium equation, based on the values of ρ.

For simplicity we now consider the equations for a three-level atom. The corresponding results for the $\mathcal{N}$-level case are given in Section 12.

In order to solve for the line source function S_{21} we write the level 2 and 3 net-rate equations as

$$\begin{aligned} &\frac{n_2}{n_1}(A_{21}\rho_{21} + Z_{21} + Z_{23}) - \frac{n_3}{n_1}(A_{32}\rho_{32} + Z_{32}) = Z_{12}\,, \\ &-\frac{n_2}{n_1}Z_{23} + \frac{n_3}{n_1}(A_{31}\rho_{31} + Z_{31} + A_{32}\rho_{32} + Z_{32}) = Z_{13}\,. \end{aligned} \tag{15}$$

We solve this pair of equations for n_2/n_1, substitute the result on the right side of equation (13), and obtain

$$S_{21} = \frac{\bar{J} + \epsilon_{21}\mathcal{B}_{21}}{1 + \epsilon_{21}}\,, \tag{16}$$

where

$$\epsilon_{21} = \epsilon_{21}^{a} - \beta_{21}\epsilon_{21}^{b}\,, \tag{17}$$

$$\mathcal{B}_{21} = B_{21}(\epsilon_{21}^{b}/\epsilon_{21})(1 - \beta_{21})\,, \tag{18}$$

$$\beta_{21} = \exp(-h\nu_{21}/kT)\,, \tag{19}$$

$$B_{21} = \frac{2h\nu_{21}^3/c^2}{\exp(h\nu_{21}/kT) - 1}\,, \tag{20}$$

and where

$$\epsilon_{21}^{a} = \frac{1}{A_{21}}\left[Z_{21} + \frac{Z_{23}(A_{31}\rho_{31} + Z_{31})}{A_{31}\rho_{31} + Z_{31} + A_{32}\rho_{32} + Z_{32}}\right], \tag{21}$$

and

$$\epsilon_{21}^{b} = \frac{g_1}{g_2\beta_{21}A_{21}}\left[Z_{12} + \frac{Z_{13}(A_{32}\rho_{32} + Z_{32})}{A_{31}\rho_{31} + Z_{31} + A_{32}\rho_{32} + Z_{32}}\right]. \tag{22}$$

In order to solve for the line source function S_{31} we follow the same procedure, also using the level 2 and 3 net-rate equations, and obtain results identical to equations (16) – (20) except that the 21 subscripts are replaced by 31, and where

$$\epsilon_{31}^{a} = \frac{1}{A_{31}}\left[Z_{31} + \frac{(A_{32}\rho_{32} + Z_{32})(A_{21}\rho_{21} + Z_{21})}{A_{21}\rho_{21} + Z_{21} + Z_{23}}\right], \tag{23}$$

and

$$\epsilon_{31}^{b} = \frac{g_1}{g_3\beta_{31}A_{31}}\left[Z_{13} + \frac{Z_{12}Z_{23}}{A_{21}\rho_{21} + Z_{21} + Z_{23}}\right]. \tag{24}$$

For the line source function S_{32} we use the level 1 and 3 equations and obtain the analogous parameters

$$\epsilon_{32}^{a} = \frac{1}{A_{32}}\left[Z_{32} + \frac{(A_{31}\rho_{31} + Z_{31})Z_{12}}{Z_{12} + Z_{13}}\right], \tag{25}$$

and

$$\epsilon_{32}^{b} = \frac{g_2}{g_3\beta_{32}A_{32}}\left[Z_{23} + \frac{(A_{21}\rho_{21} + Z_{21})Z_{13}}{Z_{12} + Z_{13}}\right]. \tag{26}$$

In all three cases, note that the ϵ^a and ϵ^b parameters for a given transition depend on the ρ values corresponding to the other transitions.

3. THE RADIATIVE TRANSFER EQUATIONS

The transfer equation for the absorption and emission of line radiation has the simple form

$$\frac{dI_\nu}{ds} = -\frac{h\nu_{m\ell}}{4\pi}\varphi_\nu\Big[(n_\ell B_{\ell m} - n_m B_{m\ell})I_\nu - n_m A_{m\ell}\Big], m > \ell\,, \tag{27}$$

provided that the absorption and emission have the same <u>uncorrelated</u> dependence on frequency. The frequency profile φ_ν is normalized according to

$$\int \varphi_\nu d\nu = 1\,, \tag{28}$$

where the integral extends over the line. The specific intensity I_ν is the radiant energy along ds per unit area, time, frequency, and solid angle. The mean intensity J_ν is the angle-average of I_ν, and the quantity $\bar{J}$ in equations (5), (10), and (12) is defined by

$$\bar{J} = \int \varphi_\nu J_\nu d\nu\,. \tag{29}$$

The transfer equation (27) does not include the effects of absorption and emission by other processes in the given frequency band. (For a discussion of these effects, see Avrett (1965), Hummer (1968), or Athay (1972).) We assume that the line is broadened only by Doppler motions so that

$$\varphi_\nu = \frac{1}{\sqrt{\pi}\,\Delta\nu_D}\exp\left[-\left(\frac{\nu - \nu_{m\ell}}{\Delta\nu_D}\right)^2\right], \tag{30}$$

where

$$\Delta\nu_D = \frac{\nu_{m\ell}}{c}\sqrt{\frac{2kT}{M}}\,. \tag{31}$$

At the line center ($\nu = \nu_{m\ell}$) the absorption coefficient (in cm^{-1}) is given by

$$\kappa_{m\ell} = \frac{h\nu_{m\ell}}{4\pi} \frac{1}{\sqrt{\pi}\,\Delta\nu_D}(n_\ell B_{\ell m} - n_m B_{m\ell}). \tag{32}$$

The line-center optical depth τ measured inward from the boundary of the atmosphere in the normal direction is given by

$$d\tau = -\kappa_{m\ell}\,\mu^{-1}\,ds\,, \tag{33}$$

where μ is the cosine of the angle between the direction of I_ν (along ds) and the outward normal. Then the transfer equation (27) becomes

$$\mu\frac{dI_\nu}{d\tau} = \exp\left[-\left(\frac{\nu - \nu_{m\ell}}{\Delta\nu_D}\right)^2\right](I_\nu - S)\,, \tag{34}$$

where $S = S_{m\ell}$ is given by equation (13). This equation reduces to the standard form

$$\mu\frac{dI_\nu}{d\tau_\nu} = I_\nu - S. \tag{35}$$

when we introduce the monochromatic optical depth

$$d\tau_\nu = \exp\left[-\left(\frac{\nu - \nu_{m\ell}}{\Delta\nu_D}\right)^2\right]d\tau. \tag{36}$$

Given a set of monochromatic depths τ_{ik}, $i = 1, 2, \ldots, N$, corresponding to the frequency ν_k, the mean monochromatic intensity J_{ik} (at depth i and frequency ν_k) can be expressed as

$$J_{ik} = \sum_{j=1}^{N} W^{\Lambda}_{ijk} S_j\,, \tag{37}$$

where the "lambda-operator" weighting coefficients depend only on the τ_{ik} values, and on the manner in which S is assumed to vary from one optical depth value to the next. Equation (29) for $\bar{J}$ then can be written as

$$\bar{J}_i = \sum_{j=1}^{N} \overline{W}^{\Lambda}_{ij} S_j\,, \tag{38}$$

where

$$\overline{W}^{\Lambda}_{ij} = \int \varphi_\nu W^{\Lambda}_{ijk}\,d\nu\,. \tag{39}$$

The numerical procedures we use to evaluate W_{ijk}^{Λ} and $\overline{W}_{ij}^{\Lambda}$ are described later in Section 13. In this paper we consider only the simple case of a plane-parallel semi-infinite atmosphere with no inward radiation at the boundary.

4. THE COMBINED EQUATIONS

In Section 2 the statistical equilibrium equations were used to obtain

$$S = \frac{\bar{J} + \epsilon \mathcal{B}}{1 + \epsilon} \tag{40}$$

for transitions 21, 31, and 32. These equations apply to each depth i in the atmosphere. Equation (38), for each of these transitions, gives the dependence of $\bar{J}$ at depth i on the values of S at all other depths j.

Equations (38) and (40) may be combined, giving

$$S_i = \frac{\sum_{j=1}^{N} \overline{W}_{ij}^{\Lambda} S_j + \epsilon_i \mathcal{B}_i}{1 + \epsilon_i} . \tag{41}$$

Given the values of ϵ_i and $\mathcal{B}_i$ we can solve the set of these equations for $i = 1, 2, \ldots, N$ to obtain the values of S_i. Once the S_i are known, it follows from equations (12) and (40) that

$$\rho_i = \epsilon_i \left(\frac{\mathcal{B}_i}{S_i} - 1 \right) . \tag{42}$$

Note that ϵ and $\mathcal{B}$ for the 21 transition depend on ρ for the 31 and 32 transitions, that ϵ and $\mathcal{B}$ for the 32 transition depend on ρ for the 21 and 31 transitions, etc. Thus from equations (41) and (42) we can determine ρ_i for any one transition if we are given the values of ρ_i for the other two transitions. For simplicity we use a given set of ρ_i values for all three transitions to determine improved ρ_i values for all three transitions, (rather that using each improved ρ_i to determine the ρ_i values for the next transition). Starting with all $\rho_i = 0$ we sometimes find that after five or ten such iterations the successive values of S are identical to an accuracy of six or more significant figures.

5. ESCAPE-PROBABILITY APPROXIMATION

Instead of starting with all $\rho_i = 0$, we can use an escape-probability formula, such as

$$\rho_i = \frac{1}{2 + 4\tau_i\sqrt{\pi}\ln\tau_i}\,, \tag{43}$$

as given by Canfield, McClymont, and Puetter (1984). Here τ is the line-center optical depth for the given transition. Note that if we adopt equation (43) for each ρ with no further correction, it is unnecessary to solve the radiative transfer equations: the rate equations (15) give n_2/n_1 and n_3/n_1 which in turn determine S_{21}, S_{31}, and S_{32}. In Section 10 we compare a solution obtained in this way with the corresponding exact solution.

6. DIFFICULTIES WITH THE NET-RATE FORMULATION

In optically thick regions of the atmosphere it is necessary to deal with the radiative rates as net-rates, since cancellation tends to occur between pairs of single radiative rates. Formulations that disregard such cancellation often do not converge properly.

In optically thin regions the use of net radiative rates can lead to difficulties. For example, if all of the Z terms in the net-rate equation (15) were neglected in comparison to the $A\rho$ terms, these equations would reduce to the indeterminate form

$$\begin{aligned} &\frac{n_2}{n_1}A_{21}\rho_{21} + \frac{n_3}{n_1}A_{32}\rho_{32} = 0\,,\\ &\frac{n_3}{n_1}(A_{31}\rho_{31} + A_{32}\rho_{32}) = 0\,, \end{aligned} \tag{44}$$

suggesting that $n_2 = n_3 = 0$. However, the single-rate equations (10) in the same case reduce to

$$\begin{aligned} &\frac{n_2}{n_1}(A_{21} + B_{21}\bar{J}_{21} + B_{23}\bar{J}_{32}) - \frac{n_3}{n_1}(A_{32} + B_{32}\bar{J}_{32}) = B_{12}\bar{J}_{21}\,,\\ &-\frac{n_2}{n_1}B_{23}\bar{J}_{32} + \frac{n_3}{n_1}(A_{31} + B_{31}\bar{J}_{31} + A_{32} + B_{32}\bar{J}_{32}) = B_{13}\bar{J}_{31}\,, \end{aligned} \tag{45}$$

which are well defined.

Also, the ρ values obtained from equation (42) can be negative, which in some cases can give negative values of n_2/n_1 and n_3/n_1 from the net-rate equations (15). Negative values of these intrinsically positive ratios can result from an inconsistent set of ρ values at an early stage in the iterative calculation. No such difficulties can occur in the solution of the single-rate equations (10) since every $\bar{J}$ is positive.

7. AN ALTERNATIVE ρ CALCULATION

The basic relationship between $\bar{J}$ and S is given by equation (40) from which it follows that

$$\bar{J} = (1 + \epsilon)S - \epsilon\mathcal{B}, \tag{46}$$

After solving the simultaneous equations (41) for S, the corresponding $\bar{J}$ can be obtained from equation (46). We can also determine $\bar{J}$ by an integration over S (equation 38), but the same result normally is obtained by equation (46).

As explained in Section 4, one basic iterative procedure is to start with values of ρ_{21}, ρ_{31}, and ρ_{32}, solve the three sets of equations (41) for S_{21}, S_{31}, and S_{32}, and then obtain new ρ values from equation (42). An alternative to this last step is to 1) obtain $\bar{J}$ from each S using equation (46), 2) solve the single-rate equations (10) for n_2/n_1 and n_3/n_1, 3) use these number density ratios in equation (13) to determine internally-consistent values of S'_{21}, S'_{31}, and S'_{32}, and 4) determine the corresponding ρ' values from

$$\rho' = 1 - \frac{\bar{J}}{S'} . \tag{47}$$

The substitution of $\bar{J}$ from equation (46) gives

$$\rho' = 1 - \frac{S}{S'} + \epsilon\left(\frac{\mathcal{B}}{S'} - \frac{S}{S'}\right) . \tag{48}$$

This equation reduces to equation (42) when S and S' are identical.

In optically thick regions where $\rho \ll 1$, equation (48) gives almost meaningless results, except when the iterative solution has converged so that S and S' are essentially the same. In optically thick regions, ρ should be calculated by equation (42).

In optically thin regions, ρ is not necessarily small compared with unity, and it is often more appropriate to use equation (48) instead of equation (42). However, the use of equation (48), even when restricted to optically thin regions, tends to produce an oscillatory iterative behavior. For this reason we obtain ρ at small optical depths by means of a linear combination of the two equations, *i.e.*,

$$\rho_i = (1 - w_i)\rho_i^\circ + w_i\rho'_i , \tag{49}$$

where ρ_i° is given by equation (42), and where w_i typically has the value 0.5 in the boundary region of the atmosphere where τ_i for the strongest line is no greater than about 10. Specifically, we let

$$w_i = \begin{cases} w, & i \leq i^\star - 3\,, \\ (2 - i + i^\star)w/5, & i^\star - 3 < i < i^\star + 2\,, \\ 0, & i \geq i^\star + 2\,, \end{cases} \tag{50}$$

where $i^\star$ is such that $\tau_{i^\star -1} \leq \tau^\star < \tau_{i^\star}$, and where we normally choose the values $w = 0.5$ and $\tau^\star = 10$. Here we have one common w_i set for all transitions, based on the τ_i values of the strongest line (the 21 transition in this case).

Solutions based on this alternative ρ calculation have better iterative convergence properties than those based on $\rho = \rho^\circ$ at all depths, as shown in the following section.

8. A THREE-LEVEL HYDROGEN-TYPE CALCULATION

Here we give numerical solutions for a simplified three-level case in which the frequencies, statistical weights, and Einstein A coefficients are those corresponding to the three lowest levels of atomic hydrogen. These values are $\nu_{21} = 2.47 \times 10^{15}$ Hz, $\nu_{31} = 2.93 \times 10^{15}$ Hz, $g_1 = 2$, $g_2 = 8$, $g_3 = 18$, $A_{21} = 4.68 \times 10^8$ s^{-1}, $A_{31} = 5.54 \times 10^7$ s^{-1}, and $A_{32} = 4.39 \times 10^7$ s^{-1}. For simplicity all bound-free transitions are ignored. Then each Z_{ab} is equal to C_{ab} rather than $C_{ab} + p_{ab}$. The absorption profiles are determined by equations (30) and (31). We choose a plane-parallel, semi-infinite, isothermal atmosphere with $T = 5000$ K. The line source functions are assumed to be frequency independent as in equation (13), and we do not include absorption or emission due to any processes other than the 21, 31, and 32 transitions.

We consider two cases. In Case I the collisional de-excitation rates are constant with depth and are given by $C_{21} = C_{31} = C_{32} = 10^5$ s^{-1}. In Case II these rates are given by

$$C_{21} = C_{31} = C_{32} = 10^5\,[1 - 0.99\exp(-0.1\tau_{21})]\ \mathrm{s}^{-1}\,. \tag{51}$$

We show that the second of these two cases cannot easily be solved by the basic procedure described in Section 4.

The data given above are sufficient to allow us to determine S_{21}, S_{31}, and S_{32} as functions of line-center optical depth τ_{21}, τ_{31}, and τ_{32}. The three optical depth scales are related to each other as follows. From equation (32), the ratio of the line-center absorption coefficients $\kappa_{m\ell}$ and $\kappa_{m'\ell'}$ depends on ratios of number densities according to

$$\frac{\kappa_{m\ell}}{\kappa_{m'\ell'}} = \frac{n_\ell}{n_{\ell'}} \left\{ \frac{(g_m/g_\ell)(A_{m\ell}/\nu^3_{m\ell})[1-(g_\ell/g_m)(n_m/n_\ell)]}{(g_{m'}/g_{\ell'})(A_{m'\ell'}/\nu^3_{m'\ell'})[1-(g_{\ell'}/g_{m'})(n_{m'}/n_{\ell'})]} \right\}. \tag{52}$$

We choose κ_{21} and a set of geometrical depths to obtain a convenient set of τ_{21} values. The corresponding values of τ_{31} and τ_{32} then are determined from the calculated number density ratios. In all the calculations reported here we have used the set of 53 depths $\tau_{21} = 0,\ 1\times10^{-3},\ 2\times10^{-3},\ 5\times10^{-3},\ 1\times10^{-2},\ 2\times10^{-2},\ \ldots,\ 5\times10^{13}, 1\times10^{14}$.

In Section 7 we described how each ρ_i is determined from ρ_i° and ρ'_i. Note that we obtain the basic $\rho_i = \rho_i^\circ$ iterative procedure described in Section 4 when $w = 0$ in equation (50). We now give solutions for Case I (constant collision rates) and Case II (depth-dependent collision rates) with $w = 0$ and $w = 0.5$.

For Case I, we find similar iterative convergence for the procedures with $w = 0$ and $w = 0.5$. Rapid convergence occurs with either procedure. After 10 iterations there are no changes greater than a few parts in 10^4. The three independently-computed source functions are consistent with each other when

$$\xi = \left(\frac{2h\nu^3_{21}/c^2}{S_{21}} + 1\right)\left(\frac{2h\nu^3_{32}/c^2}{S_{32}} + 1\right)\left(\frac{2h\nu^3_{31}/c^2}{S_{31}} + 1\right)^{-1} \tag{53}$$

is equal to unity; in the tenth iteration we find that $|\xi - 1|$ does not exceed 6×10^{-4} at any depth. For all three transitions, ρ_i° and ρ'_i are the same to within one part in 10^4, except at very large optical depths ($> 10^6$). These tests indicate that the solution is converged and self-consistent.

The iterative values of ξ at $\tau_{21} = 0$, 0.1, 1, 10, 100 are listed in Table 1. The solution we obtain is given in Table 2. The $w = 0$ and $w = 0.5$ procedures give the same solution in ten iterations to within one part in 10^4.

Now consider Case II in which the collision rates decrease by two orders of magnitude in the surface region (see equation 51). Case II is a better characterization than Case I of an atmosphere having an outward decrease of the electron number density.

When we use the $w = 0$ procedure, the iterative solution of Case II converges very slowly; Table 3 shows the iterative behavior of ξ obtained with $w = 0$ at the depths $\tau_{21} = 0$, 0.1, 1, 10, 100. After 15 iterations the surface value

Table 1

Iterative Behavior of ξ (equation 53)
for Case I (constant collision rates) with $w = 0$,
for five selected values of τ_{21}.

Iteration	$\tau_{21} = 0$	$\tau_{21} = 0.1$	$\tau_{21} = 1$	$\tau_{21} = 10$	$\tau_{21} = 100$
1	0.017	0.019	0.028	0.045	0.050
2	0.348	0.355	0.382	0.442	0.584
3	0.840	0.853	0.901	0.989	0.994
4	0.924	0.924	0.926	0.950	0.984
5	0.988	0.986	0.985	1.000	0.999
6	0.991	0.990	0.989	0.994	0.999
7	1.001	1.000	0.998	1.000	1.000
8	0.999	0.999	0.998	0.999	1.000
9	1.001	1.001	1.000	1.000	1.000
10	1.000	1.000	1.000	1.000	1.000

of ξ is 0.412. Extrapolation suggests that at least 100 iterations would be required to obtain $\xi = 1$ at the surface with this method.

We obtain rapid convergence for Case II using the $w = 0.5$ procedure, as shown in Table 4. Here we have used $\tau^* = 10$ (see equation 50). Figure 1 shows a comparison of the iterative behavior of ξ at $\tau = 0$ for Case II using $w = 0$ and $w = 0.5$.

The solution for Case II, obtained with $w = 0.5$, is given in Table 5.

9. INTERPRETATION

Figure 1 clearly shows that the method described in Section 7 with $w = 0.5$ has much better characteristics than the method with $w = 0$ when we solve multilevel cases with collision rates that decrease towards the surface. The two methods have similar characteristics when the collision rates are constant with depth.

As explained in Section 7, the method with $w = 0.5$ is based in part on the set of internally-consistent source functions S'_{21}, S'_{31}, and S'_{32}, and is thus forced to approach an internally-consistent solution, corresponding to $\xi = 1$

Table 2

The Source Functions for Case I.

τ_{21}	S_{21}/B_{21}	τ_{31}	S_{31}/B_{31}	τ_{32}	S_{32}/B_{32}
0	0.0150	0	0.00319	0	0.211
1.0(-2)	0.0154	1.6(-3)	0.00321	2.50(-13)	0.207
1.0(-1)	0.0175	1.6(-2)	0.00333	2.69(-12)	0.188
1	0.0309	1.6(-1)	0.00411	3.93(-11)	0.131
1.0(1)	0.100	1.6	0.00844	1.06(-9)	0.0835
1.0(2)	0.337	1.6(1)	0.0240	3.64(-8)	0.0703
1.0(3)	0.789	1.6(2)	0.0533	9.75(-7)	0.0668
1.0(4)	1.011	1.6(3)	0.0677	1.53(-5)	0.0662
1.0(5)	1.022	1.6(4)	0.0685	1.67(-4)	0.0662
1.0(6)	1.025	1.6(5)	0.0690	1.69(-3)	0.0666
1.0(7)	1.025	1.6(6)	0.0712	1.69(-2)	0.0687
1.0(8)	1.025	1.6(7)	0.0860	1.69(-1)	0.0830
1.0(9)	1.022	1.6(8)	0.172	1.69	0.167
1.0(10)	1.013	1.6(9)	0.505	1.67(1)	0.495
1.0(11)	1.003	1.6(10)	0.909	1.65(2)	0.905
1.0(12)	1.000	1.6(11)	0.993	1.63(3)	0.993
1.0(13)	1.000	1.6(12)	1.000	1.63(4)	1.000
$B_{21} = 1.122(-11)$		$B_{31} = 2.264(-13)$		$B_{32} = 1.756(-5)$	

in Figure 1. This accounts for the behavior shown in Figure 1 for Case II, but does not explain why $w = 0$ gives equally rapid convergence when the collision rates are constant with depth (Case I).

We surmise that Case I hides a basic difficulty that becomes apparent only when the collision rates vary with depth. When these rates decrease toward the boundary there is a corresponding increase in the thermalization length (Jefferies 1960, Avrett and Hummer 1965, Athay 1972). The thermalization length is essentially the optical distance τ_{th} from the boundary where S thermalizes, *i.e.*, approaches $\mathcal{B}$ (see equation 40). In the present case τ_{th} is of order ϵ^{-1}. The decreasing collision rates toward the boundary tend to reduce the coupling between the three source functions, causing greater difficulty than when the thermalization length is the same in the boundary region as it is deep in the atmosphere.

Table 3

Iterative Behavior of ξ
for Case II (depth-dependent collision rates) with $w = 0$,
for five selected values of τ_{21}.

Iteration	$\tau_{21} = 0$	$\tau_{21} = 0.1$	$\tau_{21} = 1$	$\tau_{21} = 10$	$\tau_{21} = 100$
1	0.016	0.018	0.027	0.045	0.050
2	0.145	0.166	0.258	0.435	0.584
3	0.325	0.375	0.598	0.963	0.995
4	0.309	0.360	0.588	0.917	0.984
5	0.342	0.401	0.663	0.988	0.999
6	0.342	0.406	0.681	0.979	0.999
7	0.358	0.426	0.720	0.996	1.000
8	0.362	0.435	0.741	0.993	1.000
9	0.372	0.450	0.767	0.999	1.000
10	0.377	0.459	0.787	0.998	1.000
11	0.386	0.472	0.808	1.000	1.000
12	0.392	0.482	0.825	0.999	1.000
13	0.399	0.494	0.842	1.000	1.000
14	0.405	0.504	0.857	1.000	1.000
15	0.412	0.515	0.870	1.000	1.000

10. ESCAPE-PROBABILITY SOLUTIONS

Here we compare the solution for Case II given in Table 5 with the solution obtained by the use of equation (43) for ρ as a function of τ for the 21, 31, and 32 transitions. In this case, the iterative solution converges completely in three iterations since we only need to determine successively the number densities from ρ_{21}, ρ_{31}, and ρ_{32} in order to redetermine τ_{21}, τ_{31}, and τ_{32}; only a few such iterations are required.

In Tables 6, 7, and 8 we compare ρ^e and S^e obtained in this way with ρ and S obtained in Section 8. The escape-probability formula for ρ is known to be inaccurate for small optical depths. Table 6 shows that S^e_{21} and S_{21} agree to within a factor of 2 only for τ_{21} greater than ≈ 100. Table 7 shows that in the case of the 32 line, such agreement occurs for τ_{32} greater than ≈ 10.

Table 4

Iterative Behavior of ξ for Case II with $w = 0.5$, for five selected values of τ_{21}.

Iteration	$\tau_{21} = 0$	$\tau_{21} = 0.1$	$\tau_{21} = 1$	$\tau_{21} = 10$	$\tau_{21} = 100$
1	0.016	0.018	0.027	0.045	0.050
2	1.924	2.108	2.828	3.706	0.636
3	1.283	1.320	1.395	1.381	1.013
4	0.600	0.610	0.636	0.644	0.938
5	1.077	1.087	1.123	1.134	0.980
6	1.043	1.046	1.053	1.053	1.000
7	0.937	0.940	0.945	0.947	0.995
8	1.009	1.010	1.014	1.014	0.998
9	1.006	1.006	1.006	1.006	1.000
10	0.992	0.993	0.993	0.993	0.999
11	1.001	1.001	1.001	1.002	1.000
12	1.001	1.001	1.001	1.001	1.000
13	0.999	0.999	0.999	0.999	1.000
14	1.000	1.000	1.000	1.000	1.000
15	1.000	1.000	1.000	1.000	1.000

Note, however, that τ_{31} is of order 10^9 at the depth where $\tau_{32} = 10$. Thus the error in S^e_{32} for $\tau_{32} < 10$ will also occur in S^e_{31} for $\tau_{31} < 10^9$, as shown in Table 8. These results indicate that if the escape-probability formula for ρ is used for all transitions, large errors can occur at $\tau \gg 1$ in some of the strong lines due to the use of the ρ formula for weak lines. See the comments to this effect by Rybicki (1984, p. 48).

Much better results are obtained when we use the escape-probability formula only for the 21 and 31 transitions, and solve the combined line transfer and statistical equilibrium equations to obtain ρ for the 32 transition. The results in this case are given in Table 9. Here we use the notation ρ^*_{32}, τ^*_{32}, and S^*_{32} to distinguish these results from the values ρ_{32}, τ_{32}, and S_{32} that were obtained by solving the combined line transfer and statistical equilibrium equations for all three transitions.

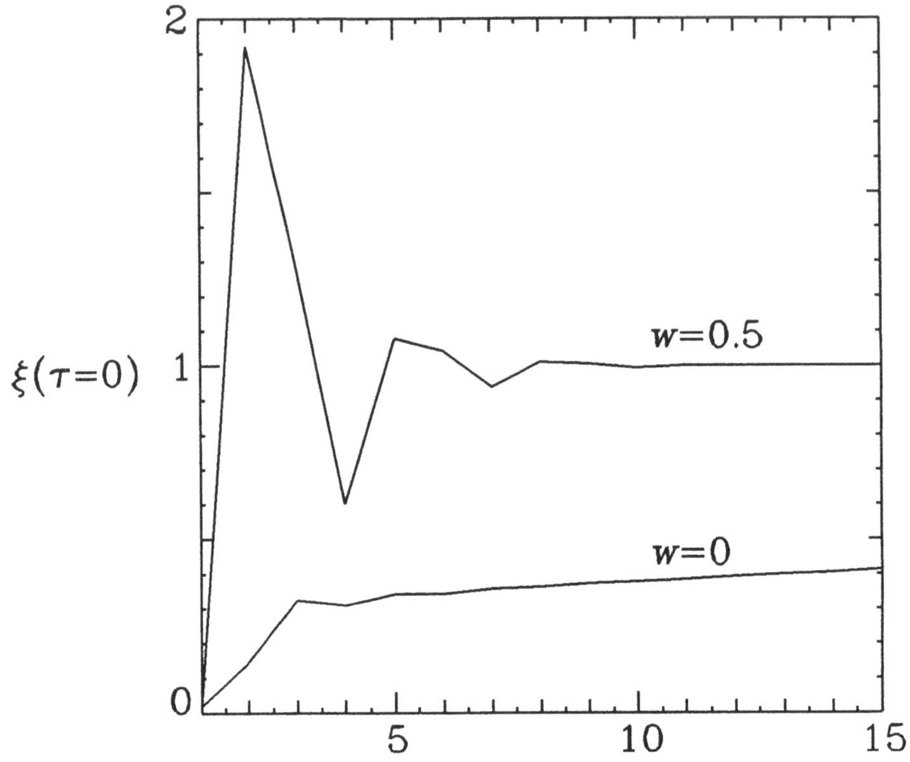

Figure 1. Calculated values of $\xi(\tau = 0)$ for Case II as a function of iteration number.

Now S^e_{31} and S_{31} agree to within a factor of 2 for τ_{31} greater than ≈ 100, compared with the substantial errors in Table 8. The solution given in Table 9 is easier to calculate than the one in Table 5 and may be an acceptable first approximation for some purposes.

11. MULTIPLET PROBLEMS

It is useful to consider the equations given in Section 2 for ϵ^a_{21}, ϵ^b_{21}, ϵ^a_{31}, and ϵ^b_{31} when $A_{32} = 0$, and when the collision rates between levels 2 and 3 are very large. When $\nu_{21} \approx \nu_{31}$ the equations then describe two transitions of a multiplet with a common lower level and with strong collisional coupling between the upper levels.

Table 5

Solution for Case II.

τ_{21}	S_{21}/B_{21}	τ_{31}	S_{31}/B_{31}	τ_{32}	S_{32}/B_{32}
0	0.0138	0	0.00181	0	0.130
1.0(-2)	0.0141	1.6(-3)	0.00182	2.30(-13)	0.128
1.0(-1)	0.0161	1.6(-2)	0.00193	2.48(-12)	0.118
1	0.0286	1.6(-1)	0.00267	3.63(-11)	0.0921
1.0(1)	0.0957	1.6	0.00743	1.00(-9)	0.0768
1.0(2)	0.336	1.6(1)	0.0239	3.59(-8)	0.0704
1.0(3)	0.788	1.6(2)	0.0533	9.74(-7)	0.0668
1.0(4)	1.011	1.6(3)	0.0677	1.53(-5)	0.0662
1.0(5)	1.022	1.6(4)	0.0685	1.67(-4)	0.0662
1.0(6)	1.025	1.6(5)	0.0690	1.69(-3)	0.0666
1.0(7)	1.025	1.6(6)	0.0712	1.69(-2)	0.0687
1.0(8)	1.025	1.6(7)	0.0860	1.69(-1)	0.0830
1.0(9)	1.025	1.6(8)	0.172	1.69	0.167
1.0(10)	1.013	1.6(9)	0.505	1.67(1)	0.495
1.0(11)	1.003	1.6(10)	0.909	1.65(2)	0.905
1.0(12)	1.000	1.6(11)	0.993	1.63(3)	0.993
1.0(13)	1.000	1.6(12)	1.000	1.63(4)	1.000
$B_{21} = 1.122(-11)$		$B_{31} = 2.264(-13)$		$B_{32} = 1.756(-5)$	

In the limit as C_{23} and C_{32} approach infinity, but where $C_{23}/C_{32} = n_3^*/n_2^*$, we obtain

$$\epsilon_{21}^a = \frac{1}{A_{21}}\left[Z_{21} + \frac{n_3^*}{n_2^*}(A_{31}\rho_{31} + Z_{31})\right], \tag{54}$$

$$\epsilon_{21}^b = \frac{g_1}{g_2\beta_{21}A_{21}}(Z_{12} + Z_{13}), \tag{55}$$

$$\epsilon_{31}^a = \frac{1}{A_{31}}\left[Z_{31} + \frac{n_2^*}{n_3^*}(A_{21}\rho_{21} + Z_{21})\right], \tag{56}$$

and

$$\epsilon_{31}^b = \frac{g_1}{g_3\beta_{31}A_{31}}(Z_{13} + Z_{12}). \tag{57}$$

These equations are well defined in the limit of infinite collisional coupling between levels 2 and 3, and the numerical solution for S_{21} and S_{31}

Table 6

Escape-Probability Solution for the 21 Transition.

τ_{21}	ρ_{21}	ρ^e_{21}	S_{21}	S^e_{21}	S^e_{21}/S_{21}
0	3.27(-4)	5.00(-1)	1.55(-13)	4.85(-17)	3.13(-4)
1.0(-2)	3.33(-4)	5.00(-1)	1.58(-13)	5.34(-17)	3.38(-4)
1.0(-1)	4.05(-4)	5.00(-1)	1.81(-13)	9.66(-17)	5.34(-4)
1	8.32(-4)	5.00(-1)	3.21(-13)	5.06(-16)	1.58(-3)
1.0(1)	1.32(-3)	9.13(-3)	1.07(-12)	1.68(-13)	1.57(-1)
1.0(2)	4.39(-4)	6.56(-4)	3.76(-12)	2.83(-12)	7.53(-1)
1.0(3)	6.42(-5)	5.37(-5)	8.85(-12)	9.21(-12)	1.04
1.0(4)	3.13(-6)	4.65(-6)	1.13(-11)	1.13(-12)	1.00
1.0(5)	5.46(-7)	4.16(-7)	1.15(-11)	1.15(-11)	1.00
1.0(6)	4.08(-8)	3.79(-8)	1.15(-11)	1.15(-11)	1.00
1.0(7)	2.44(-9)	3.51(-9)	1.15(-11)	1.15(-11)	1.00
1.0(8)	5.94(-10)	3.29(-10)	1.15(-11)	1.15(-11)	1.00
1.0(9)	6.01(-11)	3.10(-11)	1.15(-11)	1.15(-11)	1.00
1.0(10)	-2.15(-12)	2.94(-12)	1.14(-11)	1.14(-11)	1.00
1.0(11)	2.16(-12)	2.80(-13)	1.12(-11)	1.12(-11)	1.00
1.0(12)	2.17(-12)	2.68(-14)	1.12(-11)	1.12(-11)	1.00
1.0(13)	2.17(-12)	2.58(-15)	1.12(-11)	1.12(-11)	1.00

presents no difficulty. Results for cases of this type have been discussed by Avrett (1966) and by Avrett and Kalkofen (1968).

However, there are ways of formulating the equations for ϵ^a and ϵ^b that are not well defined in the limit as C_{23} and C_{32} approach infinity. For example, we could use only the first of the pair of equations (15) (*i.e.*, the level-2 statistical equilibrium equation) to determine n_2/n_1 in equation (13). In the same limiting case, equation (13) then becomes

$$S_{21} = \frac{2h\nu_{21}^3/c^2}{\left[\dfrac{g_3}{g_1}\dfrac{n_1}{n_3}\exp(-h\nu_{32}/kT) - 1\right]}, \tag{58}$$

which reduces to $S_{21} = S_{31}$ as $\nu_{32} \to 0$. Thus the equations reduce to a statement that S_{21} and S_{31} are the same, as the result of the strong collisional coupling, but do not constitute well-defined equations for computing S_{21} and S_{31}.

Table 7

Escape-Probability Solution for the 32 Transition.

τ_{32}	τ_{32}^e	ρ_{32}	ρ_{32}^e	S_{32}	S_{32}^e	S_{32}^e/S_{32}
0	0	5.22(-1)	5.00(-1)	2.28(-6)	8.56(-5)	3.75(1)
2.30(-13)	7.47(-17)	5.14(-1)	5.00(-1)	2.24(-6)	8.56(-5)	3.82(1)
2.48(-12)	1.03(-15)	4.76(-1)	5.00(-1)	2.08(-6)	8.56(-5)	4.12(1)
3.63(-11)	3.81(-14)	3.26(-1)	5.00(-1)	1.62(-6)	8.56(-5)	5.28(1)
1.00(-9)	8.16(-11)	1.92(-1)	5.00(-1)	1.35(-6)	2.70(-6)	2.00
3.59(-8)	2.04(-8)	1.18(-1)	5.00(-1)	1.24(-6)	3.84(-7)	3.10(-1)
9.74(-7)	9.66(-7)	7.13(-2)	5.00(-1)	1.17(-6)	1.74(-7)	1.49(-1)
1.53(-5)	1.53(-5)	6.31(-2)	5.00(-1)	1.16(-6)	1.56(-7)	1.34(-1)
1.67(-4)	1.67(-4)	6.27(-2)	5.00(-1)	1.16(-6)	1.55(-7)	1.34(-1)
1.69(-3)	1.69(-3)	6.23(-2)	5.00(-1)	1.17(-6)	1.55(-7)	1.32(-1)
1.69(-2)	1.69(-2)	6.02(-2)	5.00(-1)	1.21(-6)	1.55(-7)	1.28(-1)
1.69(-1)	1.69(-1)	4.91(-2)	5.00(-1)	1.46(-6)	1.55(-7)	1.06(-1)
1.69	1.69	2.22(-2)	9.33(-2)	2.93(-6)	7.98(-7)	2.72(-1)
1.67(1)	1.67(1)	4.53(-3)	4.96(-3)	8.70(-6)	8.30(-6)	9.54(-1)
1.65(2)	1.65(2)	4.65(-4)	3.79(-4)	1.59(-5)	1.62(-5)	1.02
1.63(3)	1.63(3)	3.35(-5)	3.18(-5)	1.74(-5)	1.74(-5)	1.00
1.63(4)	1.63(4)	1.97(-6)	2.78(-6)	1.76(-5)	1.76(-5)	1.00

This example demonstrates that one should avoid the use of the statistical equilibrium equation for the upper level alone (or for the lower level alone) particularly when strong collisional coupling occurs. As in Section 2 for the three-level case, we must use a combination of the equations for more than one level.

In the following section we describe ways of handling the statistical equilibrium equations in the general $\mathcal{N}$-level case.

Table 8

Escape-Probability Solution for the 31 Transition.

τ_{31}	ρ_{31}	ρ_{31}^e	S_{31}	S_{31}^e	S_{31}^e/S_{31}
0	-4.03(-1)	5.00(-1)	4.09(-16)	4.56(-18)	1.11(-2)
1.6(-3)	-3.97(-1)	5.00(-1)	4.13(-16)	5.02(-18)	1.22(-2)
1.6(-2)	-3.58(-1)	5.00(-1)	4.37(-16)	9.07(-18)	2.08(-2)
1.6(-1)	-1.86(-1)	5.00(-1)	6.04(-16)	4.75(-17)	7.86(-2)
1.6	1.45(-2)	1.03(-1)	1.68(-15)	5.26(-16)	3.13(-1)
1.6(1)	3.82(-3)	5.26(-3)	5.41(-15)	1.26(-15)	2.33(-1)
1.6(2)	4.71(-4)	3.92(-4)	1.21(-14)	1.86(-15)	1.54(-1)
1.6(3)	3.69(-5)	3.25(-5)	1.53(-14)	2.05(-15)	1.34(-1)
1.6(4)	1.99(-6)	2.84(-6)	1.55(-14)	2.07(-15)	1.34(-1)
1.6(5)	2.70(-7)	2.55(-7)	1.56(-14)	2.07(-15)	1.33(-1)
1.6(6)	2.73(-8)	2.34(-8)	1.61(-14)	2.07(-15)	1.29(-1)
1.6(7)	-4.14(-10)	2.17(-9)	1.95(-14)	2.07(-15)	1.06(-1)
1.6(8)	2.07(-10)	2.03(-10)	3.90(-14)	1.07(-14)	2.74(-1)
1.6(9)	7.06(-11)	1.92(-11)	1.14(-13)	1.09(-13)	9.56(-1)
1.6(10)	3.92(-11)	1.82(-12)	2.06(-13)	2.09(-13)	1.01
1.6(11)	3.59(-11)	1.74(-13)	2.25(-13)	2.25(-13)	1.00
1.6(12)	3.56(-11)	1.67(-14)	2.26(-13)	2.26(-13)	1.00

12. $\mathcal{N}$-LEVEL CALCULATIONS OF ϵ AND $\mathcal{B}$

In Section 2 we derived the equations

$$S = \frac{\bar{J} + \epsilon\mathcal{B}}{1 + \epsilon}, \tag{59}$$

$$\epsilon = \epsilon^a - \beta\epsilon^b, \tag{60}$$

and

$$\mathcal{B} = B(\epsilon^b/\epsilon)(1 - \beta), \tag{61}$$

and we gave explicit expressions for ϵ^a and ϵ^b in the three-level case. To obtain results for the 21 transition we used the level 2 and 3 net-rate equations (15). The same equations were used to obtain results for the 31 transition, but we used the level 1 and 3 equations for the 32 transition. In all of these cases the equation corresponding to the lower level of the particular transition is excluded.

Table 9

Solution for Case II Based on ρ_{21}^e, ρ_{31}^e, and ρ_{32}^* (see Section 10).

τ_{21}	τ_{32}	τ_{32}^*	S_{21}^e/S_{21}	S_{31}^e/S_{31}	ρ_{32}^*/ρ_{32}	S_{32}^*/S_{32}
0	0	0	3.15(-4)	7.81(-3)	1.88	0.258
1.0(-2)	2.30(-13)	7.63(-17)	3.38(-4)	8.52(-3)	1.91	0.262
1.0(-1)	2.48(-12)	1.06(-15)	5.35(-4)	1.46(-2)	2.06	0.283
1.	3.63(-11)	3.90(-14)	1.58(-3)	5.52(-2)	3.01	0.363
1.0(1)	1.00(-9)	8.18(-11)	1.57(-1)	2.86(-1)	2.91	1.83
1.0(2)	3.59(-8)	2.04(-8)	7.50(-1)	7.72(-1)	1.21	1.03
1.0(3)	9.74(-7)	9.64(-7)	1.04	1.04	0.98	0.999
1.0(4)	1.53(-5)	1.53(-5)	9.93(-1)	9.94(-1)	1.00	1.00
1.0(5)	1.67(-4)	1.67(-4)	1.00	1.00	1.00	1.00
1.0(6)	1.69(-3)	1.69(-3)	1.00	1.00	1.00	1.00
1.0(7)	1.69(-2)	1.69(-2)	1.00	1.00	1.00	1.00
1.0(8)	1.69(-1)	1.69(-1)	1.00	1.00	1.00	1.00
1.0(9)	1.69	1.69	1.00	1.00	1.00	1.00
1.0(10)	1.67(1)	1.67(1)	1.00	1.00	1.00	1.00
1.0(11)	1.65(2)	1.65(2)	1.00	1.00	1.00	1.00
1.0(12)	1.63(3)	1.63(3)	1.00	1.00	1.00	1.00
1.0(13)	1.63(4)	1.63(4)	1.00	1.00	1.00	1.00

In the general $\mathcal{N}$-level case all $\mathcal{N}-1$ equations can be used exclusive of the one corresponding to the lower level.

In order to solve for the UL transition, we arrange the terms in the level-U net-rate equation so that

$$\frac{n_U}{n_L}(A_{UL}\rho_{UL} + Z_{UL} + \cdots) + \cdots = Z_{LU}\,. \tag{62}$$

There are $\mathcal{N}-2$ other net-rate equations if we exclude the one for level L. These other equations can be used to eliminate the $\mathcal{N}-2$ unknowns n_m/n_L, $m = 1, 2, \ldots, \mathcal{N}$, $m \neq U, L$. As a result we obtain an equation of the form

$$\frac{n_U}{n_L}(\rho_{UL} + \epsilon_{UL}^a) = \frac{n_U^*}{n_L^*}\epsilon_{UL}^b\,, \tag{63}$$

and it follows that

$$S_{UL} = \frac{\bar{J}_{UL} + \epsilon_{UL}\mathcal{B}_{UL}}{1 + \epsilon_{UL}}\,, \tag{64}$$

where ϵ_{UL} and $\mathcal{B}_{UL}$ are related to ϵ^a_{UL} and ϵ^b_{UL} as in equations (60) and (61). The results in Section 2 correspond to this procedure when $\mathcal{N} = 3$. Note that ϵ^a_{UL} and ϵ^b_{UL} depend on the values of $\rho_{m\ell}$ for all transitions $m\ell \neq UL$ but not on the number densities, since the unknowns n_m/n_L, $m \neq U, L$ are eliminated as described above.

Equation (63) also can be derived by another procedure that is simpler and often just as reliable as the procedure just described. The net-rate equations for the pair of levels U and L, $U > L$, can be written as

$$\frac{n_U}{n_L}(A_{UL}\rho_{UL} + Z_{UL} + X_{UL}) = Z_{LU} + \frac{n_1}{n_L}V_U\,, \tag{65}$$

and

$$Z_{LU} + X_L = \frac{n_U}{n_L}(A_{UL}\rho_{UL} + Z_{UL}) + \frac{n_1}{n_L}V_L\,. \tag{66}$$

The explicit expressions for X_U, V_U, X_L, and V_L depend on the unknown ratios n_m/n_L, $m \neq U, L$, and are given in Appendix B of Vernazza, Avrett, and Loeser (1981). It then follows that

$$\epsilon^a_{UL} = \frac{1}{A_{UL}}\left(Z_{UL} + \frac{X_U V_L}{V_U + V_L}\right), \tag{67}$$

and

$$\epsilon^b_{UL} = \frac{n^*_L}{n^*_U A_{UL}}\left(Z_{LU} + \frac{X_L X_U}{V_U + V_L}\right). \tag{68}$$

In the $\mathcal{N} = 3$ case this procedure gives results that are identical with those based on equations (62) and (63). For $\mathcal{N} > 3$ the two procedures differ in that equations (67) and (68) contain a dependence on number densities that must be evaluated as part of the iterative solution.

Our experience with $\mathcal{N} > 3$ cases is that the procedure based on equations (62) and (63) usually gives optimum results, but that special problems sometimes arise for which the procedure based on the level U and L equations is preferable.

13. DETERMINATION OF THE WEIGHTING COEFFICIENTS

In Section 3 we indicated that the transfer equation

$$\mu \frac{dI}{d\tau} = I - S \tag{69}$$

can be solved to express the mean intensity at any one depth in terms of S at other depths. (Here we have omitted frequency subscripts for simplicity.) Given the optical depths τ_i, $i = 1, 2, \ldots, N$, we can write

$$J_i = \sum_{j=1}^{N} W_{ij}^{\Lambda} S_j \,. \tag{70}$$

The weighting coefficients W_{ij}^{Λ} can be determined from the τ_i values by any of several different methods. These include 1) solving the transfer equation (69) implicitly for I or J in terms of S by a difference-equation method such as that of Feautrier (1964); see Mihalas (1978) and Auer (1984), and 2) deriving equation (70) from the integral expression of I or J in terms of S; see Avrett and Loeser (1969, 1984). It is not our purpose here to intercompare the results of difference-equation or integral methods, but only to document the procedure we used to obtain the results given in Section 8.

The integral expression for J in terms of S is

$$J(\tau) = \frac{1}{2} \int_0^{\infty} \mathrm{E}_1(|t - \tau|) S(t)\, dt \,. \tag{71}$$

Given a particular τ_i, $i \neq 1, N$, we assume that $S(t)$ in the central interval $\tau_{i-1} \leq t \leq \tau_{i+1}$ is the quadratic function having the values S_{i-1}, S_i, and S_{i+1} at τ_{i-1}, τ_i, and τ_{i+1}, respectively, and that $S(t)$ outside this central interval varies linearly in every successive adjoining interval, having the value S_j at each τ_j. For $i = 1$ or N, $S(t)$ is assumed to vary linearly between τ_i and τ_2 or τ_{N-1}, respectively. For $t > \tau_N$, $S(t)$ is the linear function having the values S_{N-1} at τ_{N-1} and S_N at τ_N.

At large τ, equation (71) has the property

$$\lim_{\tau \to \infty} J(\tau) = S(\tau) + \frac{1}{3} S''(\tau) \,, \tag{72}$$

which is satisfied by our choice of a quadratic representation of $S(t)$ at each value of τ. Our choice of linear segments outside each central interval $\tau_{i-1} \leq t \leq \tau_{i+1}$ gives almost the same numerical accuracy as the use of quadratic segments here as well. Linear segments also provide greater numerical stability when the temperature changes substantially from one depth to the next.

Given this representation of S, equation (70) can be derived from equation (71), where the "lambda operator" weighting coefficients W_{ij}^{Λ} depend analytically on the values of τ_i, $i = 1, 2, \ldots, N$. Note that the angle integration of I to get J is treated exactly here, and not by numerical quadrature.

Restoring the frequency subscripts by the use of the indices k (so that J_{ik} corresponds to the frequency ν_k), we obtain equation (37), and then integrate over frequency to obtain equation (38).

In general, the Doppler width $\Delta\nu_D$ in equation (31) is not constant but depends on depth. Let $\langle\Delta\nu_D\rangle$ be the value at some reference depth and define $x = (\nu - \nu_{m\ell})/\langle\Delta\nu_D\rangle$. Also let $\eta = \langle\Delta\nu_D\rangle/\Delta\nu_D$. Equation (29) then becomes

$$\bar{J}_i = \frac{2}{\sqrt{\pi}}\eta_i \int_0^\infty \exp(-\eta_i^2 x^2) J_i(x)\, dx\,. \tag{73}$$

This integral is replaced by a sum of discrete terms such that

$$\int_0^\infty f(x)\, dx = \sum_{k=1}^{K} A_k f(x_k)\,. \tag{74}$$

Then

$$\bar{J}_i = \frac{2}{\sqrt{\pi}}\eta_i \sum_{k=1}^{K} A_k \exp(-\eta_i^2 x_k^2)\, J_{ik}\,. \tag{75}$$

Substitution of equation (37) then gives equation (38) where

$$\overline{W}_{ij}^{\Lambda} = \frac{2}{\sqrt{\pi}}\eta_i \sum_{k=1}^{K} A_k \exp(-\eta_i^2 x_k^2)\, W_{ijk}^{\Lambda}\,. \tag{76}$$

All of the numerical results given in this paper are based on a piece-wise linear representation of $f(x)$ in equation (74), and on the set of 24 values x_k = 0, 0.1, 0.2, 0.3, 0.4, 0.5, 0.6, 0.7, 0.8, 0.9, 1.1, 1.3, 1.5, 1.8, 2.1, 2.4, 2.8, 3.4, 3.9, 4.5, 6, 8, 15, 50.

14. NORMALIZATION

The profile function φ_ν has the normalization $\int \varphi_\nu d\nu = 1$. Thus in equation (73) we have

$$\frac{2}{\sqrt{\pi}}\eta_i \int_0^\infty \exp(-\eta_i^2 x^2)\, dx = 1\,. \tag{77}$$

It is important to preserve this normalization when the integral is replaced by a sum of discrete terms. Thus, let

$$U_i = \frac{2}{\sqrt{\pi}}\eta_i \sum_{k=1}^{K} A_k \exp(-\eta_i^2 x_k^2)\,. \tag{78}$$

In general, U_i will differ from unity. The properly normalized form of equation (76) is then

$$\overline{W}_{ij}^{\Lambda} = \frac{2}{\sqrt{\pi}}\frac{\eta_i}{U_i} \sum_{k=1}^{K} A_k \exp(-\eta_i^2 x_k^2) W_{ijk}^{\Lambda}\,. \tag{79}$$

Now when

$$W_{ijk}^{\Lambda} \to \begin{cases} 1, & j = i, \\ 0, & j \neq i, \end{cases} \tag{80}$$

we also obtain

$$\overline{W}_{ij}^{\Lambda} \to \begin{cases} 1, & j = i, \\ 0, & j \neq i, \end{cases} \tag{81}$$

which leads to the corresponding result $S \to B$ in equation (41).

From equation (41) it can be seen that without this normalization, the frequency quadrature must be accurate to one part in ϵ^{-1}, but that with the normalization, the frequency quadrature does not require higher accuracy than other parts of the calculation.

The numerical solution for Case I, specified in Section 8, also was given by Avrett (1968), but the procedure used at that time did not include this normalization. As a consequence, the surface values of S_{21}, S_{31}, and S_{32} reported there differ from the values in Table 2 by the factors 0.77, 0.83, and 1.08, respectively.

15. SUMMARY

We anticipate that the "equivalent two-level atom" method for solving multilevel transfer problems will continue to be one of the principal methods for solving complex cases that involve many coupled radiative transitions. Here we have shown that the convergence properties of this type of method can depend critically on how the equations are formulated. One illustrative example we give in Section 8 is a problem that might need at least 100 iterations to reach convergence when formulated in a straightforward manner, but that converges rapidly when the formulation directly requires the solution to be internally consistent.

We give explicit numerical results for examples with three atomic levels, and show how the formulation is generalized to atomic configurations with an arbitrary number of levels and radiative transitions.

Also, we include solutions based on the escape-probability approximation, and show that in multilevel problems the use of this approximation can lead to larger errors than in the two-level case.

REFERENCES

Athay, R. G. 1972, *Radiation Transport in Spectral Lines* (Dordrecht: Reidel), 263pp.

Athay, R. G., and Skumanich, A. 1967, An integral equation for the line source function and its numerical solution, *Ann. d'Astrophys.*, **30**, 669 - 676.

Auer, L. H. 1984, Difference equations and linearization methods for radiative transfer, in *Methods in Radiative Transfer*, ed. W. Kalkofen (Cambridge: Cambridge University Press), 237 - 279.

Auer, L. H., and Mihalas, D. 1969, Non-LTE model atmospheres. III. A complete linearization method, *Astrophys. J.*, **158**, 641 - 655.

Avrett, E. H. 1965, Solutions of the two-level line transfer problem with complete redistribution, *Smithsonian Astrophys. Obs. Spec. Rept.* **174**, 101 - 140.

__________ 1966, Source function equality in multiplets, *Astrophys. J.*, **144**, 59 - 75.

__________ 1968, Questions of consistency and convergence in the solution of multilevel transfer problems, in *Resonance Lines in Astrophysics* (Boulder, Colorado: National Center for Atmospheric Research), 27 - 63.

__________ 1971, Solution of non-LTE transfer problems, *J. Quant. Spectrosc. Radiat. Transfer*, **11**, 511-529.

Avrett, E. H., and Hummer, D. G. 1965, Non-coherent scattering II: Line formation with a frequency independent source function, *Mon. Not. Roy. Astron. Soc.*, **130**, 295-331.

Avrett, E. H., and Kalkofen, W. 1968, Transfer of line radiation by multilevel atoms, *J. Quant. Spectrosc. Radiat. Transfer*, **8**, 219-250.

Avrett, E. H., and Loeser, R. 1969, Formation of line and continuum spectra. I. Source function calculations, *Smithsonian Astrophys. Obs. Spec. Rept.* **303**, 99pp.

__________ 1984, Line transfer in static and expanding spherical atmospheres, in *Methods in Radiative Transfer*, ed. W. Kalkofen (Cambridge: Cambridge University Press), 341 - 379.

Canfield, R. C., McClymont, A. N., and Puetter, R. C. 1984, Probabilistic radiative transfer, in

Methods in Radiative Transfer, ed. W. Kalkofen (Cambridge: Cambridge University Press), 101 - 129.

Cannon, C. J. 1985, *The Transfer of Spectral Line Radiation* (Cambridge: Cambridge University Press), 541 pp.

Carlsson, M. 1986, A computer program for solving multi-level non-LTE radiative transfer problems in moving or static atmospheres, *Uppsala Astron. Obs. Rept.* **33**, 156 pp.

Catala, C., and Kunasz, P. B. 1987, Line formation in the winds of Herbig Ae/Be stars, *Astron. Astrophys.*, (in press).

Cuny, Y. 1967, Détermination exacte de la structure stationaire d'une atmosphère d'hydrogène. Problème hors de l'éqilibre thermodynamique locale, *Ann. d' Astrophys.*, **30**, 143 - 183.

Feautrier, P. 1964, Sur la résolution de l'équation de transfert, *C. R. Acad. Sc. Paris*, **258**, 3189 - 3191.

Gouttebroze, P., and Leibacher, J. W. 1980, Solar atmospheric dynamics. I. Formation of optically thick chromospheric lines, *Astrophys. J.*, **238**, 1134 - 1151.

Hummer, D. G. 1968, Non-coherent scattering - III. The effect of continuous absorption on the formation of spectral lines, *Mon. Not. Roy. Astron. Soc.*, **138**, 73 - 108.

Hummer, D. G., and Rybicki, G. B. 1971, The formation of spectral lines, *Ann. Rev. Astron. Astrophys.*, **9**, 237 - 270.

Jefferies, J. T. 1960, Source function in a non-equilibrium atmosphere. VII. The interlocking problem, *Astrophys. J.*, **132**, 775 - 789.

——— 1968, *Spectral Line Formation* (Waltham, MA: Blaisdell), 298 pp.

Kalkofen, W. 1974, A comparison of differential and integral equations of radiative transfer, *J. Quant. Spectrosc. Radiat. Transfer*, **14**, 309 - 316.

Lemaire, P., and Gouttebroze, P. 1983, Magnesium II line formation: the contribution of high atomic levels to the resonance lines, *Astron. Astrophys.*, **125**, 241 - 245.

Lites, B. W., and Skumanich, A. 1982, A model of a sunspot chromosphere based on OSO 8 observations, *Astrophys. J. Suppl.*, **49**, 293 - 316.

Mihalas, D. 1978, *Stellar Atmospheres* (San Francisco: Freeman), 632 pp.

Mihalas, D., and Kunasz, P. B. 1978, Solution of the co-moving frame equation of transfer in spherically symmetric flows. V. Multilevel atoms, *Astrophys. J.*, **219**, 635 - 653.

Rybicki, G. B. 1984, Escape probability methods, in *Methods in Radiative Transfer*, ed. W. Kalkofen (Cambridge: Cambridge University Press), 21 - 64.

Scharmer, G. B. 1981, Solution to radiative transfer problems using approximate lambda operators, *Astrophys. J.*, **249**, 720 - 730.

——— 1984, Accurate solutions to non-LTE problems using approximate lambda operators, in *Methods in Radiative Transfer*, ed. W. Kalkofen (Cambridge: Cambridge University Press), 173 - 210.

Scharmer, G. B., and Carlsson, M. 1985, A new method for solving multi-level non-LTE problems, in *Progress in Stellar Spectral Line Formation Theory*, ed. J. E. Beckman and L. Crivellari (Dordrecht: Reidel), 189 - 198.

Schönberg, K., and Hampe, K. 1986, Multilevel line formation in the comoving frame: accurate solution using an approximate Newton-Raphson operator, *Astron. Astrophys.*, **163**, 151 - 158.

Skumanich, A., and Lites, B. W. 1985, Radiative transfer diagnostics: understanding multi-level transfer calculations, in *Progress in Stellar Spectral Line Formation Theory*, ed. J. E. Beckman and L. Crivellari (Dordrecht: Reidel), 175 - 187.

Vernazza, J. E., Avrett, E. H., and Loeser, R. 1973, Structure of the solar chromosphere. I. Basic computations and summary of the results, *Astrophys. J.*, **184**, 605 - 631.

——— 1981, Structure of the solar chromosphere. III. Models of the EUV brightness components of the quiet sun, *Astrophys. J. Suppl.*, **45**, 635 - 725.

AN ALGORITHM FOR THE SIMULTANEOUS SOLUTION OF THOUSANDS OF TRANSFER EQUATIONS UNDER GLOBAL CONSTRAINTS

Lawrence S. Anderson
Ritter Observatory, The University of Toledo, Toledo, Ohio 43606, U.S.A.

ABSTRACT: I derive an algorithm for the solution of many equations of transfer coupled by integral constraints such as statistical, radiative, and hydrostatic equilibrium. The formalism uses complete linearization, variable Eddington factors, and the Feautrier transfer algorithm to reduce non-linear systems to a set of matrix equations for corrections to dependent variables which is tridiagonal in depth. A new feature makes use of the redundancy of spectral information to reduce the number of dependent radiation densities from the thousands required to resolve the spectrum to at most about one hundred. The angular moment equations of transfer are integrated over (possibly non-contiguous) regions of frequency, within which all photons experience similar probabilities for destruction, creation, and scattering to other regions. The formalism linearizes with respect to the total radiation densities in each region. The detailed *distribution* of radiation within each region is Λ-iterated via formal solutions of the transfer equations for each resolved frequency. In this respect the algorithm resembles the Eddington factor formalism used to reduce angular dependency. In linear systems (for example radiative transfer in the line of a two-level atom) the algorithm essentially reduces to Cannon's perturbation techniques with different definitions for his course grid weights. The method is particularly well suited for the study of nonLTE line blanketing in planar and spherical geometries.

1. BACKGROUND AND MOTIVATION

The classical model atmosphere problem in astrophysics consists of solving the coupled set of transfer equations for all photon frequencies and rate equations for all atomic transitions under the global constraints of radiative and hydrostatic equilibrium. The solution remains elusive. All the elementary physics has been understood for two decades now, with the exception of scattering with partial redistribution. However, the problem has not been solved for want of the computing ability to perform the immense number of operations required. The emergent spectrum from any plasma containing a healthy mix of all the naturally occurring elements shows features from upwards of 10^6 to 10^7 bound-bound transitions. To accurately solve the rate equations which determine the populations of atoms in individual electronic states and hence the photon absorption and emission rates one needs to resolve these features over the range of frequencies contributing to radiative equilibrium. A typical one-dimensional calculation must solve these resolved transfer equations at about 100 depth points and in 6 propagation directions. Even this effort is small compared to the calculation of the radiation, temperature, and pressure-dependent opacities and emissivities. Finally, the propensity for radiation to scatter long distances with redistribution in frequency and angle greatly complicates the problem by coupling all depths, frequencies, and angles in a highly non-local manner.

Until recently, this computational immensity left us at an impasse. On the one hand, if we wished to include the thermodynamic effect of *all* atomic transitions present in the real medium we had to resort to statistical methods to reduce the complexity of the opacity spectrum, such as the use of opacity distribution functions (hereafter ODFs; first used by Labs 1951, but brought to ultimate fruition by Kurucz 1979), or opacity sampling (Sneden, Johnson, and Krupp 1976). While these statistical methods are very powerful, their use required local thermodynamic equilibrium (LTE) for the atomic statistics. On the other hand, if we wished to remove the LTE constraint, the need to follow the life history of photons in detail limited us to only a few transitions (e.g. the pioneering work of Auer and Mihalas culminating in the models of Mihalas 1972 and the code of Mihalas, Heasley, and Auer 1975).

Unfortunately, at least for hot stars, the differences between LTE models with few transitions and nonLTE models with few transitions are at least as great as the differences between LTE models with few transitions and LTE models with many transitions. Self-consistent, nonLTE models of cool stars in radiative equilibrium have not been attempted, so for these conditions the LTE approximation remains untested.

While even the next generation of computers may balk at solving the full set of equations, *present* computers are capable of solving reduced sets including about 1000 atomic transitions and 3000 frequencies. Thus, a carefully chosen set of explicitly treated transitions and an integration of the effects of all other transitions now can lead to an essentially complete solution and understanding of the classical model atmosphere problem.

With that goal in mind, I have developed a numerical formalism for the solution of thousands of coupled equations of transfer under any specified constraints in one dimension. In the formalism, the radiation is segregated into a relatively small number (of order 100) of frequency *blocks*, such that all photons in a block experience similar absorption, emission, and scattering probabilities. In § II below I discuss the overall flow of the calculation and qualitatively describe the central algorithm. A condensed version of this description can be found in Anderson (1985). In § III I derive the equations for the numerical solution in detail. In § IV I present some comparisons and in § V I discuss the outlook for future applications. In addition, I have adapted the powerful ODF formalism to nonLTE transfer problems, which allows the treatment of millions of bound-bound transitions. This work will appear in another paper, together with models for an atmosphere in radiative and hydrostatic equilibrium with the solar flux and gravity.

2. A QUALITATIVE DESCRIPTION

The overall structure of the formalism closely follows the algorithm introduced by Feautrier (1964; which makes use of symmetric and antisymmetric averages of the radiation field) and its application to the model atmosphere problem by Mihalas and co-workers (cf. Mihalas 1978, hereafter MSA; chapter 7). For a review of this approach see Mihalas (1984). The Feautrier algorithm casts the transfer equations in second-order differential form with two-point boundary conditions. These and the constraint equations are fully linearized with respect to a set of dependent variables described below. As difference equations they become a tridiagonal block of matrix equations which are solved by forward and back substitution.

The heart of the transfer problem lies in the dependence of the atomic statistics on the radiation through radiation induced transitions and the inverse dependence of the radiation on the atomic statistics through the material opacity and emissivity. It is this interdependence which is responsible for the long-distance scattering of photons. Since many transitions contribute to the opacity and emissivity at each frequency, these transfer coefficients represent a reduced, intermediate data set. In addition, the number of dependent variables against which the transfer equations must be linearized is relatively small. Thus I have chosen to eliminate the atomic rate equations analytically by accumulating the partial derivatives of the transfer coefficients with respect to these variables at the same time as the coefficients themselves are accumulated.

The most significant departure from other, similar, approaches to the model atmosphere problem is the choice of dependent variables. Although the frequency spectrum of the radiation contains vast amounts of information, much of this information is redundant. The atomic rate equations and the fluid momentum and energy balance are not particularly sensitive to the *detailed* distribution of radiation over various energy ranges (e.g. the Lyman continuum, the *cores* of all Lyman lines together, etc.). Instead, state variables (which include the radiative source function) depend on integrals of the radiation over frequency and angle. The truly "independent" radiation fields consist of a small number of *blocks* of the spectrum. The algorithm developed here constructs the set of dependent variables from the electron temperature and pressure and the *total* energy density of radiation in each of these blocks. The code iteratively corrects these variables assuming that the spectral distributions *within* blocks remain unchanged.

The formalism is similar to the *multi-frequency / grey method* introduced by Freeman, *et al.* (1968) for VERA, a radiation-hydrodynamics code, and to the use of frequency averaged mean opacities suggested by Lucy (1964; the Unsold-Lucy temperature correction procedure) and Castor (1972). However, in these formulations there is only one radiation block (the whole spectrum) and the atomic populations have Boltzmann-Saha statistics. Since the distributions within blocks are determined by Λ-iteration, the full scattering problem can not be treated with only one block. With many blocks, photon redistribution within a transition can enter the linearization explicitly through the partial derivatives of the opacity and emissivity in one block with respect to the energy density of radiation in other blocks. Thus, *individual* transitions may require more than one block, but *related* transitions (e.g. several bound-bound transitions with a common lower state) may be grouped together into the same blocks. *The user chooses blocks by hand in advance.* The most important criterion for block membership is that all photons in the block should experience approximately the same physics: they should have similar probabilities for destruction, creation, and scattering with redistribution to another block. An appropriate name for the algorithm is *multi-frequency / multi-grey* (MF/MG).

The MF/MG algorithm is roughly equivalent to Cannon's (1973,1984) frequency quadrature perturbation technique. I will discuss this equivalence in the concluding section. The algorithm is also analogous to the variable Eddington factor formalism (hereafter VEF) developed by Auer and Mihalas (1970) to close the system of angle moment equations. Used together with VEF, it reduces the number of radiation variables from $N \times M \times L$, where N is the number of high-resolution frequency points, M is the number of propagation directions, and L is the the number of depth points in a one-dimensional problem, to $B \times L$, where B is the number of MF/MG blocks. As in the VEF formalism, one must perform a formal solution (i.e. with "known" opacities and emissivities) of the equations of transfer for all $N \times M \times L$ specific intensities in order to determine the Eddington factors and radiation distributions within blocks. However, this calculation is relatively rapid in one dimension, and can be performed once each iteration without much added cost.

To illustrate the reduction of variables achieved by MF/MG, consider the problem posed by a ten level hydrogen atom plus the ion. For simplicity assume each line and each continuum needs ten frequency points for adequate resolution. The total number of frequencies amounts to 550. If all frequencies and level populations were treated explicitly with VEF only, in the manner of Mihalas, Heasley, and Auer (1975), the linearization-correction matrices would be dimensioned 561×561 (assuming no other background continuum contributions or electron donors). Now, in the same spirit, assume each line and continuum needs only two *blocks* to represent its redistribution and escape adequately (Lyman α will need more, but transitions between high-lying levels will need only one each). The continua overlap and will need at most ten independent blocks. With the exception of Lyman α and perhaps Hα, all frequencies in lines of the same Rydberg series may be put into the *same* two blocks. About 30 blocks will be sufficient to treat this problem, giving matrices dimensioned 30×30. Since the linearization-correction effort in the Feautrier formalism scales as the order of the matrix cubed, the cost reduction in this example is a factor of 6539 per iteration. The number of iterations required for convergence is at most about half again as many as required for VEF alone. With MF/MG higher resolution in the spectral profiles and more transitions from the same lower state can be included at (almost) no extra cost. In fact, for typical problems with N=2000 and B=100 the linearization-correction and the formal solution constitute only about 10% of the total cost. Most of the cost is incurred by the calculation of the derivatives of the transport coefficients with respect to the dependent variables.

Two points should be stressed. First, as with VEF, the algorithm is used *only* to correct the total energy density of radiation in each block, under the assumption that distributions within blocks are not sensitive to the dependent variables. Since the distributions are recalculated on each iteration, *the converged solution is that of the detailed spectrum.* Second, the overall solution is *not* by Λ-iteration. Linear scattering sources implied by the atomic rate equations are included by the partial derivatives of those equations with respect to the block energy densities. Approximations appear *in the correction matrices only,* by assuming certain photons behave similarly and can be grouped together. The Λ-iteration used to determine distributions within blocks only sorts out the (small) differences between these "similar" photons. Since these differences do not propagate in the same way as individual photons, the Λ-iteration adequately recovers them.

An iteration proceeds as follows.

(1) With an estimate of the radiation field, solve the atomic rate equations for the state populations and their partial derivatives. Then calculate the opacity and emissivity at each high-resolution frequency. Accumulate source terms and weighted derivatives for the block corrections.

(2) With the transfer coefficients from step (1) perform a formal solution of the equations of transfer for the specific intensities at all frequencies, directions, and depths, and accumulate the radiation stress moments for the block corrections and store the high-resolution distributions.

(3) With the accumulated *present* values of the block quantities solve the linearized equations for the corrections to the block radiation densities and state variables.

(4) With the distributions from (2) recover the *corrected* high-resolution radiation field and return to step (1).

Convergence is greatly improved if one alternates a true Λ-iteration (return to step 1 directly from step 2) with a full correction. Thus a full iteration proceeds

1-2-1-2-3-4-1. The solution converges to a thermal equilibrium of better than one part in 10^5 and all corrections smaller than one part in 10^3 in 6 to 10 full iterations, depending on the original departure.

3. THE NUMERICAL EQUATIONS

In this section I derive the MF/MG algorithm equations. Throughout I assume static plane parallel geometry. Formally it is not difficult to generalize to other geometries.

3.1. *Atomic Statistics and Transfer Coefficients*

The atomic statistics, opacities, and emissivities are calculated from a given radiation field in step (1) of the flow pattern in the previous section. Since the calculation of statistical equilibrium and transfer coefficients in static media is tedious but straightforward, I only refer to the relevant chapters in MSA and continue now with the formal solution of the equations of transfer for each individual frequency.

3.2. *The Formal Solution*

The equation of transfer for the intensity $I(\lambda,\mu,\nu)$ of radiation of frequency ν traveling with zenith angle cosine μ at depth λ where $\lambda = ln(m)$ and m is the column density inward in grams cm^{-2} is

$$\mu \frac{d}{d\lambda} I(\lambda,\mu,\nu) = \kappa(\lambda,\nu) I - \eta(\lambda,\nu) - \frac{m}{\bar{\mu}} N_s \sigma \frac{1}{2} \int_{-1}^{1} I d\mu \,, \tag{1}$$

where κ is the opacity in dimensionless units and η is the emissivity similarly normalized:

$$\kappa(\lambda,\nu) = \frac{m}{\bar{\mu}} \sum_{trans} N_l \left[a_{lu}(\nu) - \frac{N_u}{N_l} a_{ul}(\nu) R_{lu}{}^{*} e^{-h\nu/kT} \right] + \frac{m}{\bar{\mu}} N_s \sigma \tag{2a}$$

$$\eta(\lambda,\nu) = \frac{m}{\bar{\mu}} \sum_{trans} N_u \, a_{ul}(\nu) \frac{2h\nu^3}{c^2} R_{lu}{}^{*} e^{-h\nu/kT} \,. \tag{2b}$$

Here $\bar{\mu}$ is the mean mass of atomic nuclei, N_l is the fraction of atoms in the lower state of a transition, $a_{lu}(\nu) = (\pi e^2/m_e c)(df/d\nu)$ is the transition absorption cross section in cm^2 and $R_{lu}{}^*$ is the Saha-Boltzmann ratio $N_l{}^*/N_u{}^*$. The expression for η assumes isotropic redistribution in angle. The redistribution in frequency is contained in the emissive $df\ /d\nu$. Equation (1) explicitly contains a frequency coherent, isotropic, scattering contribution proportional to a particle fraction N_s (e.g. Rayleigh or Thomson scattering).

Following Feautrier, one defines the symmetric and anti-symmetric averages $P(\lambda,\mu,\nu) \equiv (I^+ + I^-)/2$ and $R(\lambda,\mu,\nu) \equiv (I^+ - I^-)/2$, where the superscripts refer to the sign of the direction cosine μ. By writing equation (1) individually for the two streams and taking the half-sum and half-difference, one derives

$$\mu \frac{d}{d\lambda} R = \kappa P - \eta - \frac{m}{\overline{\mu}} N_s \sigma \int_0^1 P d\mu \ , \tag{3}$$

$$\mu \frac{d}{d\lambda} P = \kappa R \ . \tag{4}$$

Substitution of equation (4) into (3) achieves a second-order equation for P:

$$\mu^2 \frac{d}{d\lambda}\left[\frac{1}{\kappa}\frac{d}{d\lambda}P\right] = \kappa P - \eta - \frac{m}{\overline{\mu}} N_s \sigma \int_0^1 P d\mu \ . \tag{5}$$

This form has the advantage of being the angular equivalent of the moment equation for the radiation energy density I will derive below. The boundary conditions for equation (5) are derived from equation (4). At depth, if one assumes diffusion and $\eta/\kappa = B_\nu$, where B_ν is the Planck function, the condition is simply $P = B_\nu$. Alternatively, if the interior solution is sufficiently converged, at some intermediate boundary the condition is equation (4) with $R \equiv \mathbf{R}$ set from the formal solution of a previous iteration. At the surface, the condition is

$$\frac{\mu}{\kappa}\frac{d}{d\lambda} P = P - I_0^- \ , \tag{6}$$

where I_0^- is the radiation incident from above. I assume that above the upper boundary κ/m is constant and η/κ is linear in optical depth. Then at the upper boundary the optical depth τ_0 is κ_0 and the incident radiation is

$$I_0^- = (S_0 - b\,\mu)(1 - e^{-\tau_0/\mu}) + b\,\tau_0\, e^{-\tau_0/\mu} \tag{7}$$

assuming no external irradiation. Here S_0 and b express the linearity of

$\eta/\kappa = S_0 + b(\tau - \tau_0)$.

One must solve equation (5) with its boundary conditions for P at each point l, m, n on finite grids 1, 2, ...l, ...L in λ; 1, 2, ...m, ...M in μ; and 1, 2, ...n, ...N in ν. Typically $\Delta\lambda$ is of order 0.25 and the μ-quadrature is the positive half of 6-point Gaussian. Setting up the ν-quadrature is not straightforward. One wishes to insure that each depth layer can properly radiate those photons which are escaping from that depth. First, I select out the relatively few strong, resonant, bound-bound transitions. I assign all *other* bound-bound transitions a quadrature of from one to five points. The assigned weights contain a thermal Doppler core with optional wings. For the continua and the *selected* transitions, using an initial model and a very fine ν-quadrature I calculate the monochromatic surface $\lambda_e(\nu)$ such that $\tau(\lambda_e, \nu) = 2/3$. This surface and the next procedure are illustrated in Figure 1.

Figure 1. Frequency quadrature through a resonant transition. λ_e is the depth at which the monochromatic optical depth is 2/3. Δ is the maximum allowed weight. Vertical lines mark quadrature boundaries (see text). Heavy points mark the final frequency quadrature.

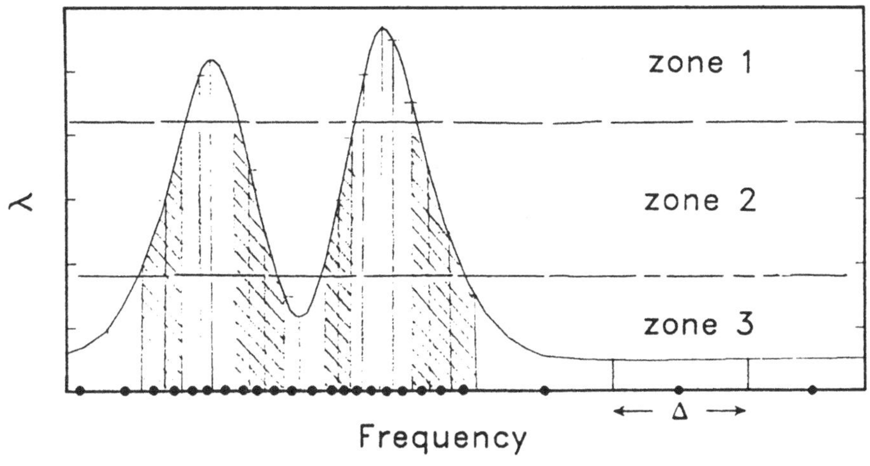

Moving along the surface toward increasing frequency, I place a quadrature *boundary* each time either *a*) λ_e crosses an integer value in the positive direction, *b*) λ_e crosses a half-integer value in the negative direction, *c*) $d\lambda_e/d\nu$

changes sign, d) $\Delta log\,(\nu)$ from the previous boundary exceeds some value, or e) a subordinate transition quadrature is encountered.

Given the quadrature, using ordinary second-order differences the differential equations become the following difference equations:

$$- \mu_m^2 \left[\left\langle \frac{1}{\kappa_n} \right\rangle^{+} \frac{P_{l+1,m,n} - P_{l,m,n}}{\Delta\lambda^{+}} - \left\langle \frac{1}{\kappa_n} \right\rangle^{-} \frac{P_{l,m,n} - P_{l-1,m,n}}{\Delta\lambda^{-}} \right]$$
$$+ \kappa_{l,n} P_{l,m,n} - \frac{m}{\bar{\mu}} N_s \sigma \sum_{m'} \Delta\mu_{m'} P_{l,m',n} = \langle \Delta\lambda \rangle \eta_{l,n} \,, \tag{8a}$$

$$- \mu_m \left\langle \frac{1}{\kappa_n} \right\rangle^{+} \frac{1}{\Delta\lambda^{+}} (P_{2,m,n} - P_{1,m,n}) + \frac{1}{2}(P_{2,m,n} + P_{1,m,n})$$
$$= (S_o - b\,\mu_m)(1 - e^{-\tau_o/\mu_m}) + b\,\tau_o\, e^{-\tau_o/\mu_m} \,, \tag{8b}$$

$$P_{L,m,n} = B_{L,n} \quad or$$
$$\mu_m \left\langle \frac{1}{\kappa_n} \right\rangle^{-} \frac{1}{\Delta\lambda^{-}} (P_{L,m,n} - P_{L-1,m,n}) = \mathbf{R}_{m,n} \,, \tag{8c}$$

with

$$\Delta\lambda^{+} = \lambda_{l+1} - \lambda_l \,, \quad \Delta\lambda^{-} = \lambda_l - \lambda_{l-1}$$

$$\langle \Delta\lambda \rangle = 0.5\,(\Delta\lambda^{+} + \Delta\lambda^{-})$$

$$\left\langle \frac{1}{\kappa_n} \right\rangle^{\pm} = 0.5\,(1/\kappa_{l\pm1,n} + 1/\kappa_{l,n})$$

$$\tau_o = \kappa_{1,n} + \Delta\lambda^{+}(0.375\kappa_{1,n} + 0.125\kappa_{2,n})$$

$$b = (\eta/\kappa_{2,n} - \eta/\kappa_{1,n})/[0.5\Delta\lambda^{+}(\kappa_{1,n} + \kappa_{2,n})]$$

$$S_o = 0.5(\eta/\kappa_{1,n} + \eta/\kappa_{2,n}) \,.$$

The equations form a linear, tridiagonal set of L $M{\times}M$ matrix equations of over depth.

3.3. *The Moment Equations*

The MF/MG algorithm linearizes with respect to and corrects the *block energy densities.* I now derive the equations of transfer for these densities. In the following, J, H, and K are the usual zeroth, first, and second moments of I with respect to μ. The integration of equation (5) over μ leads to the second-order equation

$$\frac{d}{d\lambda}\left[\frac{1}{\kappa}\frac{d}{d\lambda}K(\lambda,\nu)\right] = \frac{1}{2}\int_{-1}^{1}(\kappa I - \eta)\,d\mu \equiv \gamma(\lambda,\nu)\,. \tag{9}$$

Here I have defined a new net source term γ, which the code determines explicitly when calculating the opacity and emissivity. The contributions of frequency coherent scattering in κ and η are not included in its evaluation. When differenced over the depth grid this equation becomes

$$\left\langle\frac{1}{\kappa_n}\right\rangle^{+}\frac{1}{\Delta\lambda^{+}}(K_{l+1,n} - K_{l,n}) - \left\langle\frac{1}{\kappa_n}\right\rangle^{-}\frac{1}{\Delta\lambda^{-}}(K_{l,n} - K_{l-1,n}) = \left\langle\Delta\lambda\right\rangle\gamma_{l,n}\,. \tag{10a}$$

Similarly one has boundary conditions derived from VEF. At the surface,

$$\left\langle\frac{1}{\kappa_n}\right\rangle^{+}\frac{1}{\Delta\lambda^{+}}(K_{2,n} - K_{1,n}) - \Phi_n\frac{1}{2}(J_{2,n} + J_{1,n}) = 0\,, \tag{10b}$$

where

$$\Phi_n \equiv H_n/J_n = \sum_m \mu_m\,\Delta\mu_m\,(R_{2,n} + R_{1,n}) / \sum_m \Delta\mu_m\,(P_{2,n} + P_{1,n})$$

is the Eddington flux factor. At depth, after integrating the choices in the formal solution over μ,

$$J_{L,n} = B_{L,n}\,, \quad or \quad \left\langle\frac{1}{\kappa_n}\right\rangle^{-}\frac{1}{\Delta\lambda^{-}}(K_{L,n} - K_{L-1,n}) = \mathbf{H}_n \tag{10c}$$

where $\mathbf{H}_n \equiv \sum_m \mu_m\,\Delta\mu_m\,\mathbf{R}_{m,n}$. Notice that for the first choice the *total* flux $\mathbf{H} = \sigma T_{eff}^4/4\pi$ never enters the solution. If $\mathbf{H}$ is a constraint it must appear in a separate equation, namely the condition of radiative equilibrium.

We now come to the heart of MF/MG. While the number N of high resolution frequency bins may be large, the photons in each bin n are experiencing interactions similar to those of photons in other bins. The trick is to match up these similar bins and sum over them. I have already mentioned some circumstances under which bins are similar. Here I will list all grouping criteria.

(1) The entire frequency range of bins with opacity dominated by bound-free and free-free transitions may be divided into about fifteen blocks with boundaries at prominent bound-free edges. In principle, the widths of these blocks should not exceed kT/h. However, the photon emission in continuum processes is usually proportional to the Planck function due to the thermalization of free electrons. Therefore, as long as the Planck function is properly integrated over a block, the only *real* change in interactions occurs when the dominant transition (with its collision and radiative rates) changes. Thus even at frequencies much greater than kT/h wide blocks ($\Delta\log(\nu) \approx 0.3$) are permissible.

(2) Bins dominated by strong, *resonant*, bound-bound transitions should not be grouped with bins dominated by other transitions. However, within the transition (including all multiplet components unless one or more interacts differently from the rest) one may collect together all photons with similar mean free paths. These photons interact with the atom at similar places on the transition profile so experience similar probabilities for redistribution. In Figure 1 I have divided the depth into *zones*. These zones typically span a factor of ten in column mass. All bins in which photons leave the atmosphere from a particular zone are grouped together to form a block. For example, the shaded bins in the figure all have λ_e within zone 2.

(3) Bins dominated by individual, *subordinate*, bound-bound transitions are blocked much as those in resonant transitions. However, the destruction channel for these photons is usually via *another* transition with probability of order 0.1 to 1.0. All such transitions out of the same lower state have similar properties. Thus one may collect together all bins which radiate from the same zone *and* have opacity dominated by the same atomic state. In practice this collection includes all transitions in the three Rydberg series with $\Delta L=-1,0,+1$ (except resonant members).

(4) Bins dominated by transitions between highly excited states (with $\Delta E \lesssim kT$) in one *ion* may be grouped together, since the state populations will be mostly proportional to the statistical weights. Exceptions involve particular channels which may significantly alter these populations, for ex-

ample via dielectronic recombination.

While one can imagine clever schemes to assign block membership automatically, at present the assignment is done by hand with the help of trial plots of $\lambda_e(\nu_n)$. Within the code is an array b_n of length N. The n-th element contains the block b (of the list of blocks 1, 2, ...b, ...B) to which n is assigned. Using this array, one simply sums the equations of transfer over the bins in a block, at the same time multiplying by $4\pi\Delta\nu_n/c$ to convert to total energy density. With the following definitions,

$$\rho_l^{\,b} \equiv \frac{4\pi}{c}\sum_{n\in b}\Delta\nu_n\, J_{l,n} = \frac{4\pi}{c}\sum_{n\in b}\Delta\nu_n \sum_m \Delta\mu_m\, P_{l,m,n} \tag{11a}$$

$$k_l^{\,b} \equiv \frac{4\pi}{c}\sum_{n\in b}\Delta\nu_n\, K_{l,n} = \frac{4\pi}{c}\sum_{n\in b}\Delta\nu_n \sum_m \mu_m^2\,\Delta\mu_m\, P_{l,m,n} \tag{11b}$$

$$\gamma_l^{\,b} \equiv \frac{4\pi}{c}\sum_{n\in b}\Delta\nu_n\, \gamma_{l,n} \tag{11c}$$

$$\xi_{l,l'}^{\,b} \equiv \frac{4\pi}{c}\sum_{n\in b}\Delta\nu_n \frac{1}{\kappa_{l,n}} K_{l',n}/k_{l'}^{\,b} \quad , \quad l' = l-1,\, l,\, l+1 \tag{11d}$$

$$\Phi^b \equiv \frac{4\pi}{c}\sum_{n\in b}\Delta\nu_n\, \Phi_n\,(J_{2,n}+J_{1,n})/(\rho_2^{\,b}+\rho_1^{\,b}) \tag{11e}$$

$$\beta_L^{\,b} \equiv \frac{4\pi}{c}\sum_{n\in b}\Delta\nu_n\, B_{L,n} \quad \text{and} \quad h_l^{\,b} \equiv \frac{4\pi}{c}\sum_{n\in b}\Delta\nu_n\, \mathbf{H}_{l,n}\ , \tag{11f}$$

the transfer equations (10) become:

$$\begin{aligned} &\frac{1}{\Delta\lambda^+}(\xi_{l+1,l+1}^{\,b}+\xi_{l,l+1}^{\,b})k_{l+1}^{\,b} + \frac{1}{\Delta\lambda^-}(\xi_{l,l-1}^{\,b}+\xi_{l-1,l-1}^{\,b})k_{l-1}^{\,b} \\ &- [\frac{1}{\Delta\lambda^+}(\xi_{l+1,l}^{\,b}+\xi_{l,l}^{\,b})+\frac{1}{\Delta\lambda^-}(\xi_{l,l}^{\,b}+\xi_{l-1,l}^{\,b})]\,k_l^{\,b} \\ &= 2\langle\Delta\lambda\rangle\gamma_l^{\,b} \end{aligned} \tag{12a}$$

$$(\xi_{2,2}^{\,b}+\xi_{1,2}^{\,b})k_2^{\,b} - (\xi_{2,1}^{\,b}+\xi_{1,1}^{\,b})k_1^{\,b} - \Delta\lambda^+\Phi^b\,(\rho_2^{\,b}+\rho_1^{\,b}) = 0 \tag{12b}$$

$$\rho_L^b = \beta_L^b \ , \quad or$$
$$(\xi^b_{l,l}+\xi^b_{l-1,l})k^b_l - (\xi^b_{l,l-1}+\xi^b_{l-1,l-1})k^b_{l-1} = 2\Delta\lambda^- h_l^b \ . \tag{12c}$$

I have introduced *three inverse opacity means* $\xi^b_{l,l'}$ at each depth l. They appear because the summation over $n \in b$ is done *after* the equations are written in difference form over depth. These means are always well behaved, as $k^b_{l'}$ is always positive. Flux, or $dK/d\lambda$, means are *not* well behaved, as situations do come up where the block flux may go through zero to negative values.

Then, as in VEF, one assumes $k^b_l \equiv f_l^b \rho_l^b$. Equations (12) then become the desired equations of transfer for the block radiation densities.

Once solved (via linearization and correction) one must recover the high resolution bin densities $\rho_{l,n} \equiv 4\pi\Delta\nu_n J_{l,n}/c$ to calculate the opacities and emissivities for the next iteration. During the formal solution step, the code calculates the spectral *distribution*

$$a_{l,n} = \Delta\nu_n J_{l,n} / \sum_{n \in b} \Delta\nu_n J_{l,n} \ . \tag{13}$$

This distribution is assumed unchanged by the linearized correction, so one recovers the corrected $\rho_{l,n}$ by

$$\rho_{l,n} = a_{l,n} \rho_l^{b_n} \ . \tag{14}$$

Finally I provide the equations for the classical constraints of radiative and hydrostatic equilibrium. These equations describe the state of the material particles, which I assume to have Maxwellian velocity distributions. One must choose two variables. The variables which appear directly in the atomic rate equations are the electron pressure and temperature. If one chooses these two as the fundamental dependent variables along with ρ_l^b, then the rate equations are linear in the state population fractions N_j and can be solved directly without iteration (assuming no significant molecular coupling).

Radiative equilibrium takes two forms, corresponding to the boundary conditions on the radiative transfer. At all depths except the deep boundary where $\rho_L^b \equiv \beta_L^b$ the code employs the standard balance of net absorption against net emission:

$$\sum_b \gamma_l^b = 0 \ . \tag{15a}$$

At the deep boundary, I specify the total flux:

$$\sum_b [(\xi^b_{L,L} + \xi^b_{L-1,L}) k^b_L - (\xi^b_{L,L-1} + \xi^b_{L-1,L-1}) k^b_{L-1}] = 2\Delta\lambda^- \frac{\sigma}{c} T^4_{eff} \ . \tag{15b}$$

The hydrostatic equation is:

$$\varsigma_l p_l^e - \varsigma_{l-1} p_{l-1}^e + \sum_b (k_l^b - k_{l-1}^b) = (m_l - m_{l-1}) g \ . \tag{16a}$$

Here $\varsigma_l \equiv (N_e + 1 + q)/N_e$, where N_e is the fractional number of electrons relative to nuclei and $q = 0.5mv^2/kT$ is the fractional contribution of microturbulence, and g is the surface gravity. The hydrostatic boundary condition is

$$\varsigma_1 p_1^e = m_1 g - \sum_b (k_1^b - k_0^b)$$

where k_0^b is the stress moment at $m = 0$. While one does not know k_0^b, one may estimate it several ways. Mihalas, Heasley and Auer (1975) replace the radiation correction term with $(4\pi/c)\sum \Delta\nu_n \, \kappa_n \, \mathbf{H}_n$, which assumes that κ/m and the flux are constant with height above the upper boundary. Alternatively one may assume k^b is linear in optical depth and column mass (equivalent to constant $\mathbf{H}$ and κ/m). Then the boundary condition becomes

$$\varsigma_1 p_1^e + \frac{1}{\Delta\lambda^+} \sum_b (k_2^b - k_1^b) = m_1 g \ , \tag{16b}$$

where the linearity is determined from k_2^b and k_1^b.

3.4. *Linearization and Correction*

Let $\mathbf{\Psi}_l = (T_e, p_e, \rho^1, \rho^2, \ldots \rho^b, \ldots \rho^B)_l$ be the vector of dependent variables at depth l. The linearization of the moment and constraint equations with respect to this vector proceeds in the standard manner. One replaces each dependent variable Ψ^i with $\Psi_o^i(1+\psi^i)$ where Ψ_o^i is the iteration's current value, and deletes all terms of order ψ^2 or higher. The code *applies* the corrections in the form $\Psi^i = \Psi_o^i \, exp\,(\psi^i)$ to avoid negative results. Coefficients are linearized by constructing the sum of partial derivatives. Thus,

$$\Xi \rightarrow \Xi_o + \sum_i \frac{\partial \Xi}{\partial \ln \Psi^i} \psi^i ,$$

where the derivatives are evaluated using the current solution. Below I list those derivatives which must be calculated. For convenience I have dropped subscripts l with the understanding that quantities at one l do not explicitly depend on the variables at another l.

$$\frac{\partial k^b}{\partial \ln \Psi^i} = \delta_{i,b} \, k^b \quad via \text{ VEF} \tag{17a}$$

$$\frac{\partial \xi^b_{l'}}{\partial \ln \Psi^i} = -\frac{4\pi}{c} \sum_{n \in b} \Delta \nu_n \frac{1}{\kappa_n^2} \frac{\partial \kappa_n}{\partial \ln \Psi^i} K_{l',n} / k^b_{l'} \tag{17b}$$

$$\frac{\partial \gamma^b}{\partial \ln \Psi^i} = \frac{4\pi}{c} \sum_{n \in b} \Delta \nu_n \frac{\partial (\kappa_n J_n - \eta_n)}{\partial \ln \Psi^i} \tag{17c}$$

$$\frac{\partial \varsigma}{\partial \ln \Psi^i} = -\frac{(1+q)}{N_e^2} \frac{\partial N_e}{\partial \ln \Psi^i} - \delta_{i,T} \frac{q}{N_e} \tag{17d}$$

Here $\delta_{i,i'}$ is the Kronecker delta. Notice that the weighting factors $(4\pi/c)\Delta\nu_n K_{l',n}/k^b_{l'}$ are assumed independent of Ψ in accordance with MF/MG. *Calculating all $\partial\kappa_n/\partial\ln\Psi^i$'s and $\partial\eta_n/\partial\ln\Psi^i$'s consumes the bulk of the time and primary memory of the entire code.*

The physics of frequency redistribution enters through the derivatives $\partial \Xi^b / \partial \ln \rho^{b'}$, where $b \neq b'$. These derivatives in turn derive entirely from the corresponding derivatives of the atomic population fractions. If one writes the matrix of rate equations for one atomic species as

$$\mathbf{A} \cdot N = B , \tag{18}$$

then

$$\frac{\partial N}{\partial \ln \Psi^i} = -\mathbf{A}^{-1} \cdot \frac{\partial \mathbf{A}}{\partial \ln \Psi^i} \cdot N . \tag{19}$$

Similarly $\partial N_e / \partial \ln \Psi^i$ derives from the charge conservation equation:

$$\frac{\partial N_e}{\partial \ln \Psi^i} = \sum_j Z_j \frac{\partial N_j}{\partial \ln \Psi^i} , \tag{20}$$

where Z_j is the charge number of atomic state j. Since it costs almost no extra time, the code accumulates *all* derivatives of $\mathbf{A}$ and calculates all derivatives of $\boldsymbol{N}$ from equation (19).

Once linearized, one may write the set of equations for the corrections $\boldsymbol{\psi}$ in matrix form (cf. MSA p.233):

$$-\mathbf{A}_l \cdot \boldsymbol{\psi}_{l-1} + \mathbf{B}_l \cdot \boldsymbol{\psi}_l - \mathbf{C}_l \cdot \boldsymbol{\psi}_{l+1} = \boldsymbol{Q}_l \ . \tag{21}$$

The boundary conditions at $l=1$ and $l=L$ drop the $\mathbf{A}$ and $\mathbf{C}$ matrices respectively. This tridiagonal set is solved by Feautrier elimination. For completeness I provide the complete set of matrix elements here. The second superscript of an element refers to the "row" or equation (with $r=$ radiative equilibrium, $h=$ hydrostatic equilibrium, and $b=$ radiative transfer for block b), and the first to the variable ψ^i .

At $l \neq 1$ *or* L :

$$\mathbf{A}_l^{i,r} = 0.0$$

$$\mathbf{A}_l^{i,h} = p_{l-1}^e \frac{\partial \varsigma_{l-1}}{\partial \ln \Psi_{l-1}^i} + \delta_{i,p}\, \varsigma_{l-1} p_{l-1}^e + \delta_{i,b}\, k_{l-1}^b$$

$$\mathbf{A}_l^{i,b} = \frac{1}{\Delta\lambda^-} k_{l-1}^b \frac{\partial \xi_{l-1,l-1}^b}{\partial \ln \Psi_{l-1}^i} - \frac{1}{\Delta\lambda^-} k_l^b \frac{\partial \xi_{l-1,l}^b}{\partial \ln \Psi_{l-1}^i} + \delta_{i,b} \frac{1}{\Delta\lambda^-}(\xi_{l,l-1}^b + \xi_{l-1,l-1}^b) k_{l-1}^b$$

$$\mathbf{B}_l^{i,r} = \sum_b \frac{\partial \gamma_l^b}{\partial \ln \Psi_l^i}$$

$$\mathbf{B}_l^{i,h} = p_l^e \frac{\partial \varsigma_l}{\partial \ln \Psi_l^i} + \delta_{i,p}\, \varsigma_l\, p_l^e + \delta_{i,b}\, k_l^b$$

$$\mathbf{B}_l^{i,b} = \left(\frac{1}{\Delta\lambda^+} + \frac{1}{\Delta\lambda^-}\right) k_l^b \frac{\partial \xi_{l,l}^b}{\partial \ln \Psi_l^i} - \frac{1}{\Delta\lambda^+} k_{l+1}^b \frac{\partial \xi_{l,l+1}^b}{\partial \ln \Psi_l^i} - \frac{1}{\Delta\lambda^-} k_{l-1}^b \frac{\partial \xi_{l,l-1}^b}{\partial \ln \Psi_l^i}$$

$$+ \delta_{i,b}\left[\frac{1}{\Delta\lambda^+}(\xi_{l+1,l}^b + \xi_{l,l}^b) + \frac{1}{\Delta\lambda^-}(\xi_{l,l}^b + \xi_{l-1,l}^b)\right] k_l^b + 2\langle\Delta\lambda\rangle \frac{\partial \gamma_l^b}{\partial \ln \Psi_l^i}$$

$$\mathbf{C}_l^{i,r} = \mathbf{C}_l^{i,h} = 0.0$$

$$\mathbf{C}_l^{i,b} = \frac{1}{\Delta\lambda^+} k_{l+1}^b \frac{\partial \xi_{l+1,l+1}^b}{\partial \ln \Psi_{l+1}^i} - \frac{1}{\Delta\lambda^+} k_l^b \frac{\partial \xi_{l+1,l}^b}{\partial \ln \Psi_{l+1}^i}$$

$$+ \delta_{i,b} \frac{1}{\Delta\lambda^+} (\xi_{l,l+1}^b + \xi_{l+1,l+1}^b) k_{l+1}^b$$

$$Q^r = -\sum_b \gamma_l^b$$

$$Q^h = (m_l - m_{l-1}) g - \varsigma_l p_l^e + \varsigma_{l-1} p_{l-1}^e - \sum_b (k_l^b - k_{l-1}^b)$$

$$Q^b = \frac{1}{\Delta\lambda^+} (\xi_{l+1,l+1}^b + \xi_{l,l+1}^b) k_{l+1}^b + \frac{1}{\Delta\lambda^-} (\xi_{l,l-1}^b + \xi_{l-1,l-1}^b) k_{l-1}^b$$

$$- [\frac{1}{\Delta\lambda^+} (\xi_{l+1,l}^b + \xi_{l,l}^b) + \frac{1}{\Delta\lambda^-} (\xi_{l,l}^b + \xi_{l-1,l}^b)] k_l^b - 2\langle \Delta\lambda \rangle \gamma_l^b$$

At $l = 1$:

$$\mathbf{B}_l^{i,r} = \sum_b \frac{\partial \gamma_l^b}{\partial \ln \Psi_l^i}$$

$$\mathbf{B}_l^{i,h} = p_l^e \frac{\partial \varsigma_l}{\partial \ln \Psi_l^i} + \delta_{i,p} \varsigma_l p_l^e + \delta_{i,b} \frac{1}{\Delta\lambda^+} k_l^b$$

$$\mathbf{B}_l^{i,b} = k_l^b \frac{\partial \xi_{l,l}^b}{\partial \ln \Psi_l^i} - k_{l+1}^b \frac{\partial \xi_{l,l+1}^b}{\partial \ln \Psi_l^i} + \delta_{i,b} [(\xi_{l+1,l}^b + \xi_{l,l}^b) k_l^b + \Delta\lambda^+ \Phi^b \rho_l^b]$$

$$\mathbf{C}_l^{i,r} = 0.0$$

$$\mathbf{C}_l^{i,h} = -\delta_{i,b} \frac{1}{\Delta\lambda^+} k_{l+1}^b$$

$$\mathbf{C}_l^{i,b} = k_{l+1}^b \frac{\partial \xi_{l+1,l+1}^b}{\partial \ln \Psi_{l+1}^i} - k_l^b \frac{\partial \xi_{l+1,l}^b}{\partial \ln \Psi_{l+1}^i}$$

$$+ \delta_{i,b} \left[(\xi_{l,l+1}^b + \xi_{l+1,l+1}^b) k_{l+1}^b - \Delta\lambda^+ \Phi^b \rho_{l+1}^b \right]$$

$$\boldsymbol{Q}^r = -\sum_b \gamma_l^b$$

$$\boldsymbol{Q}^h = m_l\, g - \varsigma_l\, p_l^e - \frac{1}{\Delta\lambda^+} \sum_b (k_{l+1}^b - k_l^b)$$

$$\boldsymbol{Q}^b = (\xi_{l+1,l+1}^b + \xi_{l,l+1}^b) k_{l+1}^b - (\xi_{l+1,l}^b + \xi_{l,l}^b) k_l^b - \Delta\lambda^+ \Phi^b (\rho_{l+1}^b + \rho_l^b)$$

At $l = L$:

$$\mathbf{A}_l^{i,r} = \sum_b k_{l-1}^b \frac{\partial \xi_{l-1,l-1}^b}{\partial \ln \Psi_{l-1}^i} - \sum_b k_l^b \frac{\partial \xi_{l-1,l}^b}{\partial \ln \Psi_{l-1}^i} + \delta_{i,b} (\xi_{l,l-1}^b + \xi_{l-1,l-1}^b) k_{l-1}^b$$

$$or \; = 0.0$$

$$\mathbf{A}_l^{i,h} = p_{l-1}^e \frac{\partial \varsigma_{l-1}}{\partial \ln \Psi_{l-1}^i} + \delta_{i,p}\, \varsigma_{l-1} p_{l-1}^e + \delta_{i,b}\, k_{l-1}^b$$

$$\mathbf{A}_l^{i,b} = 0.0$$

$$or \; = k_{l-1}^b \frac{\partial \xi_{l-1,l-1}^b}{\partial \ln \Psi_{l-1}^i} - k_l^b \frac{\partial \xi_{l-1,l}^b}{\partial \ln \Psi_{l-1}^i} + \delta_{i,b} (\xi_{l,l-1}^b + \xi_{l-1,l-1}^b) k_{l-1}^b$$

$$\mathbf{B}_l^{i,r} = \sum_b k_l^b \frac{\partial \xi_{l,l}^b}{\partial \ln \Psi_l^i} - \sum_b k_{l-1}^b \frac{\partial \xi_{l,l-1}^b}{\partial \ln \Psi_l^i} + \delta_{i,b} (\xi_{l,l}^b + \xi_{l-1,l}^b) k_l^b$$

$$or \; = \sum_b \frac{\partial \gamma_l^b}{\partial \ln \Psi_l^i}$$

$$\mathbf{B}_l^{i,h} = p_l^e \frac{\partial \varsigma_l}{\partial \ln \Psi_l^i} + \delta_{i,p}\, \varsigma_l\, p_l^e + \delta_{i,b}\, k_l^b$$

$$\mathbf{B}_l^{i,b} = \delta_{i,b}\, \rho_l^b$$

$$or \quad = k_l^b \frac{\partial \xi_{l,l}^b}{\partial \ln \Psi_l^i} - k_{l-1}^b \frac{\partial \xi_{l,l-1}^b}{\partial \ln \Psi_l^i} + \delta_{i,b}\,(\xi_{l,l}^b + \xi_{l-1,l}^b) k_l^b$$

$$Q^r = 2\Delta\lambda^{-} \frac{\sigma}{c} T_{eff}^4 - \sum_b (\xi_{l,l}^b + \xi_{l-1,l}^b) k_l^b + (\xi_{l,l-1}^b + \xi_{l-1,l-1}^b) k_{l-1}^b$$

$$or \quad = -\sum_b \gamma_l^b$$

$$Q^h = (m_l - m_{l-1}) g - \varsigma_l\, p_l^e + \varsigma_{l-1} p_{l-1}^e - \sum_b (k_l^b - k_{l-1}^b)$$

$$Q^b = \beta_l^b - \rho_l^b$$

$$or \quad = 2\Delta\lambda^{-} h_l^b - (\xi_{l,l}^b + \xi_{l-1,l}^b) k_l^b + (\xi_{l,l-1}^b + \xi_{l-1,l-1}^b) k_{l-1}^b$$

4. COMPARISONS AND TIMING

Soon after writing the first working version of the code, several tests were run to compare with the work of Mihalas and Auer (1970; hereafter MA V) and Auer and Mihalas (1972; hereafter AM VII). While these comparisons were reported on by Anderson (1982), none of the details have appeared in the literature. Figures 2 and 3 show the results of a computation with conditions as close as reasonably possible to those of MA V, for models in radiative and hydrostatic equilibrium at $T_{eff} = 35000$ K and Log $g = 4.0$. Included were 6 hydrogen lines (Lα,β,γ; Hα,β; Pα) with depth dependent Doppler profiles. Figure 2 shows the electron temperature as a function of depth and Figure 3 shows the hydrogen LTE departure coefficients. The agreement is remarkably and encouragingly close, especially considering that the departure coefficients are very sensitive to the total flux at these effective temperatures. All differences can be accounted for by four computational differences: a) correct free-free Gaunt factors in my calculation, b) slightly different collision rates, c) the quadrature and numerical details of the recombination integration, and d)

Figure 2. Electron temperature vs. column mass. Solid line: Anderson (1982). Dashed line: Mihalas and Auer (1970).

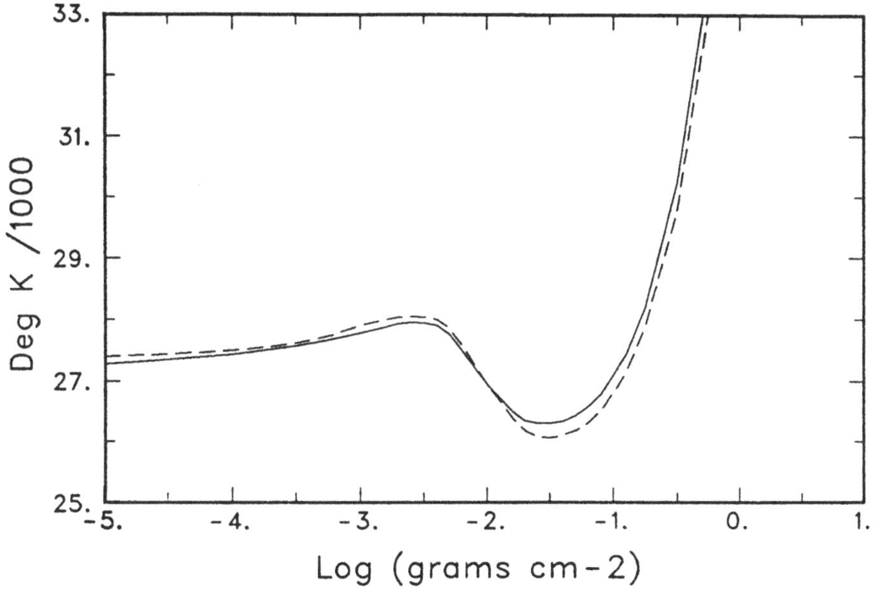

Figure 3. Hydrogen LTE departure coefficients. Number indicates quantum state. Solid line: Anderson (1982). Dashed line: Mihalas and Auer (1970).

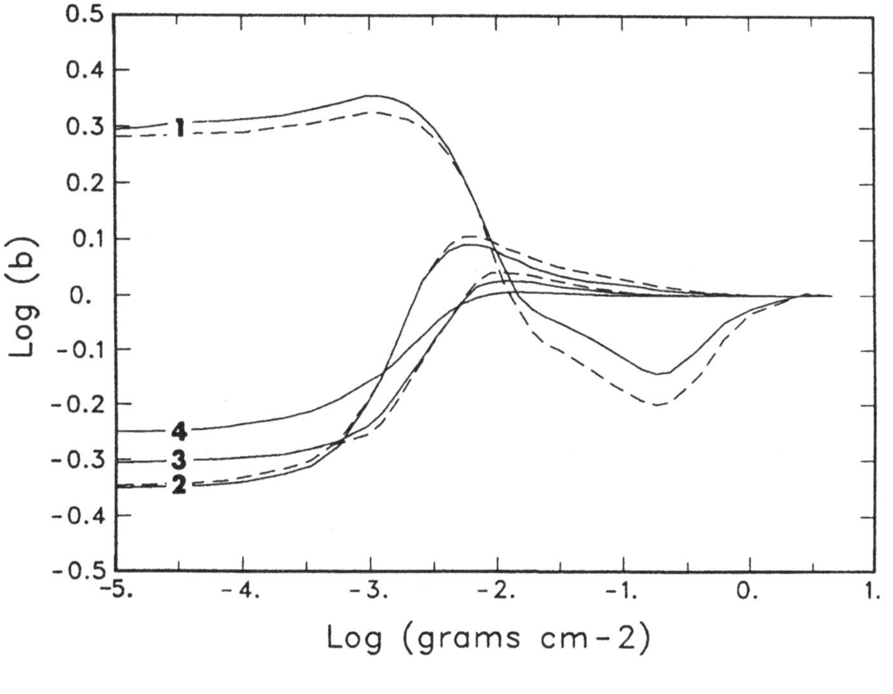

the quadrature and depth dependence of line profiles. The temperature profile is sensitive to the difference between free-free Gaunt factors set to unity (MA V) and set to their correct value at about the one percent level. The hydrogen departure coefficients are most sensitive to the collision rates. I used Johnson's (1972) formulae. The most significant difference is that his collisional ionization rate from the $n = 2$ state is about 1.35 times that used by MA V. The recombination is quite sensitive to the frequency quadrature, particularly for $\Delta\nu \gtrsim kT/h$. One achieves greater accuracy for the rate and the corresponding emissivity if one analytically integrates the exponential term over $\Delta\nu$. These results are encouraging in the sense that they closely match the previous work when the physical approximations are the same. To the best of my knowledge my code is the first completely independent construction after the work of Mihalas and his co-workers; it is good to see their monumental efforts corroborated.

More discouraging is the *insensitivity* of the emergent line profiles to rather radical temperature structure variations. After the MA V match several models of hydrogen-helium atmospheres with increasing numbers of transitions were computed, culminating in one with 24 hydrogen lines, 28 hydrogen and helium continua, and 5$\pm$ Gaussian angle quadrature. The major structural change was to raise the boundary plateau temperature by 4000 K. This increase results from the condition that heating in the upper layers occurs through Lyman continuum absorption followed by radiative recombination. Even if a recombining electron returns to the ground state via radiative transitions, the net result is heating due to the thermal degradation of the Lyman continuum photon caused by the free electron's collisional redistribution. Cooling occurs mainly by bound-bound transitions. Each increase in the number of included transitions allows more channels for electron decay to the ground level. Thus there are more atoms in the ground level acting as sites for Lyman continuum absorption. For the same reason, in each bound-bound transition the lower level population increases relatively more than the upper level population. The resulting decrease in the source function reduces cooling efficiency. Figure 4 shows the emergent flux at Hα, and compares it with that given by AM VII. Hα is slightly deeper in the 24-line model than in AM VII, due to the population shifts just discussed.

Figure 4. Emergent Eddington flux at Hα. Solid line: Anderson (1982). Dashed line: Auer and Mihalas (1972).

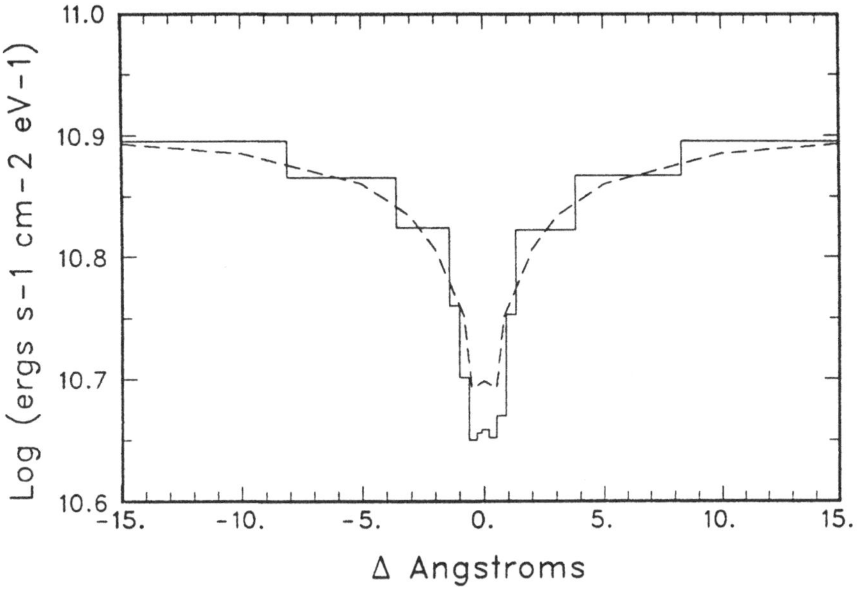

The discouragement increases when one looks at the results from a model with 112 bound-bound transitions, including 57 in CIII and CIV (Anderson 1985). There the CIV resonance transition forces the temperature structure of the outer atmosphere to decrease monotonically, mimicking the LTE solution (heaven forbid!). Even so, the emergent spectrum shows no significant trace of the much reduced boundary temperatures. In hindsight this insensitivity might have been expected, since the populations are controlled by radiation which is thermally created at depths relatively insensitive to the line cooling. Some effort will have to be expended to find adequate diagnostics for the temperature structure of the continuum-thin layers in hot stars. The local temperature has importance through its influence on the pressure scale height and conditions at the critical points in the transsonic flow.

As mentioned in § II, it is the calculation of the transport coefficient derivatives rather than the solution of the linearized equations which determines the cost of a model. The linearization requires the derivative of the opacity and emissivity contribution from each transition at each individual frequency and with respect to each dependent variable (cf. equations 17b and c). Thus the cost increases roughly linearly with the sum over all transitions of the number of frequency points in each transition, times the number of radiation blocks. While it is difficult to compare codes run on different machines in different decades, one may make estimates based on approximate conversion factors. Per iteration, the model in Anderson (1985) with 809 frequencies, 112 bound-bound transitions, 93 bound-free transitions, and 64 radiation blocks, if run on the same machine, would take about ten times the computing time of the complete linearization code written by Mihalas, Heasley, and Auer (1975) with only 94 frequencies, 6 bound-bound and 11 bound-free transitions. Each iteration of the 1985 model consumed approximately 90 seconds of CRAY 1A CPU time. A recently completed model in radiative and hydrostatic equilibrium with the Solar flux and gravity, with 1622 frequencies, 124 explicit bound-bound transitions (20 of which represent about 10^6 bound-bound transitions with cross section distribution functions in nonLTE), 102 bound-free transitions, and 96 radiation blocks consumed about 300 seconds of CRAY 1A time per iteration. The number of explicit transitions is about the same in both models. However, the Solar model has about twice the number of frequencies and 1.5 times the number of dependent variables, the product of which accounts for the factor of three increase in time.

5. CONCLUSIONS

In this paper I present an approach to the solution of multiple equations of transfer, each of which is coupled to the others through non-linear, global constraints. As I have mentioned, the approach is similar to the perturbation techniques of Cannon, even to the method of application using the Feautrier approach outlined by Cannon (1984, p.158f). Further references to Cannon refer to this citation. Equations labeled with a preceding C are his. The linearized form of the moment equation (12) is essentially equivalent to Cannon's coursely gridded second-order equation of transfer, derived in equations (C15-22):

$$\frac{1}{\Delta\tau}\left[\frac{1}{\Delta\tau^+}(\Phi_{j+1}^{(l)}-\Phi_j^{(l)}) - \frac{1}{\Delta\tau^-}(\Phi_j^{(l)}-\Phi_{j-1}^{(l)})\right]$$

$$= \Phi_j^{(l)}(\nu_i^*,\mu_k^*) - (1-\epsilon)\sum_{i=1}^{N_\nu^*}\sum_{k=1}^{N_\mu^*} w_{ik}^*\,\phi(\nu'{}_i^*)\Phi_j^{(l)}(\nu'{}_i^*,\mu'{}_k^*)$$

$$- \mathcal{E}_j^{(l)}(\nu_i^*,\mu_k^*) .$$

Here $\Phi_j^{(l)}$ is the l-th order correction to the Feautrier symmetric mean intensity at depth j and τ is the usual monochromatic optical depth. Cannon describes an already linear problem, where the source function is known to be a linear function of the specific intensity. My block array length B is Cannon's N_ν^*, for the moment equation $N_{\mu'}^* = 1$, and my Q^b in equation (21) is his driving error $\mathcal{E}^{(l)}$. My atomic statistics and formal solution are equivalent to Cannon's calculation of the driving error, equations (C27-31).

The main difference between Cannon's and my approach is that Cannon solves his linear system for corrections to the actual intensities on a specific, course, ν^*,μ^* grid assuming complete redistribution, whereas I solve for corrections to the integrated radiation density within entire blocks. For a linear problem, the difference amounts to a redefinition of the frequency and angle quadrature weights. He Λ-iterates $\mathcal{E}^{(l)}(\nu,\mu)$, which contains corrections resulting from the error incurred by replacing the integral operator

$$\hat{\mathcal{L}} = \frac{1}{2\phi_\nu}\int d\nu' \int d\mu'\, R\,(\nu',\mu';\nu,\mu) \tag{C4}$$

with the course grid, complete redistribution approximation

$$\hat{\mathcal{L}}^* = \frac{1}{2}\sum_{i=1}^{N_\nu^*}\sum_{k=1}^{N_\mu^*} w_{ik}^*\,\phi(\nu'{}_i^*) . \tag{C8}$$

I Λ-iterate the fine grid spectral distribution a_n (equation 13). My choice is dictated by the need to satisfy additional integral constraints such as radiative equilibrium. Partial redistribution in frequency is *not* Λ-iterated in MF/MG, but appears explicitly in the linear derivatives contained in the Feautrier **B** matrices. One could do the same with appropriate weighting factors in Cannon's definition of the approximate operator $\hat{\mathcal{L}}^*$. Finally, the problems I wish to address are non-linear, having opacities and emissivities which are functions of the dependent variables, thus I must iterate the Feautrier matrices. These differences are more in name than substance; therefore the convergence properties demonstrated by Cannon apply to MF/MG as well (as long as one remains within the circle of convergence for the linearization).

The multi-frequency / multi-grey algorithm developed here is ideal for those classes of radiative transfer problems which must include the interaction of photons and atoms through many transitions and under global constraints. The classical stellar atmosphere problem is one example. The implementation of the algorithm in the code I have developed is extremely flexible. The code only "knows" how to solve the atomic rate equations, radiative transfer, and constraint equations. The physics of the individual model atoms is read in in input files and is thus entirely up to the user. At present, the choice of a frequency quadrature and its grouping into blocks is also up to the user. Thus a certain amount of insight is required *a priori*. It is possible to write a set-up program which will make those choices.

As mentioned in § III, the formal generalization to other one dimensional geometries (e.g. spherical) is not difficult. Most of the changes occur in the formal solution for the radiation field, and by the need to include an equation relating the geometrical depth to the column density in the linearization matrix. At the time of writing this article, a spherical version of the program has been written and tested with favorable results using continuum opacities in LTE. Since lines and nonLTE are not directly related to geometry, no problems should occur on their inclusion.

A formulation of the algorithm for moving media has been explored, but it is limited in application to static and low velocity ($\leq$the sound speed) problems. This limitation occurs because the frequency *blocking* remains fixed in both the Lagrangian and Eulerian frames. In reality, of course, the Lagrangian blocks see different photons from the Eulerian blocks, as a function of the angle cosine between the photons' direction of propagation and the flow velocity. This difference effectively "blurs" the *derivatives* used in the correction matrices. I hasten to add that *only* the derivatives are affected. Since the narrowest blocks occurring in the cores of lines need be of order one thermal (or microturbulent) Doppler width, one must so limit the "blurring." However, even with this limitation one may explore the origins of stellar winds in the photospheres of stars, and calculate non-LTE transfer through low amplitude waves and convective motions.

The algorithm is not limited to steady-state problems. However, it does cost time and resources to run. The availability of those resources will limit the time-dependent applications.

While one-dimensional geometry probably remains the limit in the near future, MF/MG, with its clear separation of atomic statistics, the formal solution, and moment corrections, is promising for multi-dimensional geometries as well.

ACKNOWLEDGMENTS

I am pleased to thank Drs. Eugene Avrett, Wolfgang Kalkofen, and Dimitri Mihalas for reading the manuscript and offering their suggestions. I specially thank Dr. Mihalas for his encouragement and advice while developing and encoding this algorithm, and the High Altitude Observatory for its generous support in applications.

REFERENCES

Anderson, L. S. 1982, *Bull. Amer. Astron. Soc.*, **14**, 921.

____________. 1985, *Astrophys. J.*, **298**, 848.

Auer, L. H., and Mihalas, D. 1970, *Mon. Not. R. Astr. Soc.*, **149**, 65.

____________. 1972, *Astrophys. J. Suppl.*, **23**, 193 (AM VII).

Cannon, C. J. 1973, *Astrophys. J.*, **185**, 621.

____________. 1984, *Methods in radiative transfer*, Edited by W. Kalkofen, (Cambridge: Cambridge University Press), p. 157.

Castor, J. I. 1972, *Astrophys. J.*, **178**, 779.

Freeman, B. E., Hauser, L. E., Palmer, J. T., Pickard, S. O., Simmons, G. M., Williston, D. G., and Zerkle, J. E. 1968, 1968, DASA Report No. 2135, Vol. I, (La Jolla: Systems, Science, and Software Inc.).

Feautrier, P. 1964, *Comptes Rendus Acad. Sci. Paris,* **258**, 3189.

Johnson, L. C. 1972, *Astrophys. J.*, **174**, 227.

Kurucz, R. L. 1970, *ATLAS, A Computer Program for Calculating Model Stellar Atmospheres*, Smithsonian Astrophysical Observatory, Special Report No. 309.

____________. 1979, *Astrophys. J. Suppl.*, **40**, 1.

Labs, D. 1951, *Z. f. Astrophysik,* **29**, 199.

Lucy, L. B. 1964, *Proceedings of the First Harvard-Smithsonian Conference on Stellar Atmospheres*, Smithsonian Astrophisical Observatory, Special Report No. 167.

Mihalas, D. 1972, *NCAR Technical Note NCAR-TN/STR-76.*

____________. 1978, *Stellar Atmospheres*, (San Francisco: W. H. Freeman and Co.; MSA).

____________. 1984, *J. Comput. Phys.*, **57**, 1.

Mihalas, D. and Auer, L. H. 1970, *Astrophys. J.*, **160**, 1161 (MA V).

Mihalas, D., Heasley, J., and Auer, L. 1975, *NCAR Technical Note NCAR-TN/STR-104.*

Sneden, C., Johnson, H., and Krupp, B. 1976, *Astrophys. J.*, **204**, 281.

OPERATOR PERTURBATION FOR DIFFERENTIAL EQUATIONS

W. Kalkofen
Harvard-Smithsonian Center for Astrophysics, Cambridge, USA

ABSTRACT: Cannon's suggestion to solve transfer problems by perturbing the exact differential equations about lower-order, approximate equations provided the impetus for the powerful and efficient modern methods in numerical radiative transfer. Yet his original approach, intended for line transfer with partial frequency redistribution, introduced extraneous terms not present in the unperturbed equations; the inconsistencies are present even for isotropic scattering. We discuss a general method of deriving perturbation equations in differential form from the exact differential equations, using integral equations in an intermediary step, and apply the procedure to an isotropic scattering problem where the temperature is given and to a transfer problem where the temperature is determined from the radiative equilibrium constraint.

1. INTRODUCTION

In a seminal paper, Cannon (1973a) suggested that radiative transfer problems for lines with partial redistribution, in which the frequency or direction of scattered photons is correlated with the frequency or direction before the absorption, be solved by perturbing the high-order, exact differential equations about lower-order, approximate equations. His derivation of the perturbation equations introduced additional terms representing formal solutions of the transfer equation for known source function. The terms are not present in the original equations. As a consequence, the hierarchy of perturbation equations is different from the original, and a numerical solution demonstrating the features of the method gives a converged result with a finite error even though the solution should have reproduced the exact result to machine accuracy.

The inconsistency of Cannon's equations is found also in recent descriptions of the method (Cannon 1984, 1985; Auer 1986). We highlight the difficulties of the approach in the simplest problem that exhibits them, namely, monochromatic radiative transfer with isotropic scattering. In section 2 we follow Cannon's derivation for partial redistribution to show that his perturbation equations converge to a solution that is different from that of the exact equations even in the far simpler case of complete redistribution, *i.e.*, when the incident and scattered photons are uncorrelated. In section 3 we give Auer's version of

Cannon's derivation. In section 4 we show how Cannon's perturbation equations for coherent isotropic scattering can be correctly formulated by means of integral equations as an intermediary step; but the method is restricted to complete redistribution; and in section 5 we turn to a transfer problem with a constraint to describe in detail how the same basic approach can be used there and discuss the role played by the boundary conditions. The final section summarizes the findings.

2. CANNON'S PERTURBATION OF DIFFERENTIAL EQUATIONS

We discuss Cannon's seminal idea, which is to solve radiative transfer problems involving large, coupled systems of equations by reducing the order via the perturbation of the exact operator of a problem about a simpler operator. The paper (Cannon 1973a) describing the method presented the theory for coherent scattering with arbitrary phase function and gave the application for a line with completely non-coherent, isotropic scattering. We analyze his equations for the still simpler case of coherent, isotropic scattering.

Consider a monochromatic transfer problem for isotropic scattering. The equation of radiative transfer may be written as

$$\begin{aligned} DI &= I - S \ , \\ S &= \Omega' I + \beta \ , \end{aligned} \tag{2.1}$$

where D is a differential operator whose precise form we need not specify at this point, S is the source function, which was angle and depth-dependent in Cannon's case because he admitted a general phase function, but depends only on depth here because we assume isotropic scattering; and Ω' is the angle integral,

$$\Omega' \sim (1-\epsilon)\,\tfrac{1}{2}\int_{-1}^{1} d\mu \ldots \quad , \tag{2.2}$$

into which we have absorbed the scattering factor, $1-\epsilon$. Recall that our aim is to show difficulties and inconsistencies in Cannon's (1973a) treatment. For this purpose it is immaterial whether D is a first-order difference operator as suggested by equations (2.1) or is the more common second-order difference operator. We will be more specific when we describe the solution of a transfer problem subject to a constraint (*cf.* section 5). — Note also that we follow the common practice to use the terms *differential* operator, *differential* equation, and *integral* even though we usually mean their finite difference analogues.

Recall that if we assume that Ω', which is normalized, and the inhomogeneous term β are depth-independent, the solution in a semi-infinite medium has the exact surface value of $S(0) = \frac{1}{\sqrt{\epsilon}}\beta$. The thermalization length, *i.e.*, the depth where the solution approaches its asymptotic value of β/ϵ, is of the order of $\tau \sim 1/\sqrt{\epsilon}$.

For the numerical representation of the angle integral we discretize the angle space. The "exact" equation uses M division points, hence the integral operator Ω' becomes

$$\Omega' = (1-\epsilon)\sum_{l=1}^{M} W_l \quad , \tag{2.3}$$

and the approximate treatment uses m division points, hence the approximate integral operator is

$$\omega' = (1-\epsilon)\sum_{l=1}^{m} w_l \quad , \tag{2.3'}$$

where m is assumed to be much smaller than M.

Because of the coupling of the intensities in the scattering integral (2.2) the equations (2.1) constitute a system of coupled equations; for the exact problem the system is of order M, and for the approximate treatment there are m coupled equations.

In order to solve the "high-order" system (2.1) of M coupled equations we perturb the exact operator Ω' about the lower-order operator ω',

$$\Omega' = \omega' + (\Omega' - \omega') \quad , \tag{2.4}$$

and expand the intensity in a series,

$$I = \sum_{k=0}^{\infty} i^{(k)} \quad . \tag{2.5}$$

Instead of the differential operator D at the large number of angle points we use an operator at the smaller number m of angle points. To indicate this change in the angle set of the system of equations we designate the corresponding differential operator by d. Introducing the expansion for the intensity and the expression (2.4) for the angle operators into equations (2.1) we obtain a hierarchy of differential equations,

$$\begin{aligned} d\, i^{(k)} &= i^{(k)} - s^{(k)} \quad , \\ s^{(k)} &= \omega' i^{(k)} + e^{(k)} \quad , \end{aligned} \tag{2.6}$$

with the source term given by

$$e^{(k)} = \begin{cases} \beta & , \quad k = 0 \ , \\ (\Omega' - \omega')i^{(k-1)} & , \quad k > 0 \ . \end{cases} \tag{2.7}$$

Note that the perturbation equations are to be solved at the reduced number m of angle division points but that the source terms $e^{(k)}$ require integrals over the intensity at the larger set of M points.

For the solution of the equations we proceed as follows (*i.e.*, following Cannon 1973a): For $e^{(0)} = \beta$ we solve the coupled equations (2.6) at the m division points, resulting in the intensity $i^{(0)}$. The source term $s^{(1)}$ for the intensity $i^{(1)}$ in first order requires the intensity, designated $\hat{\imath}^{(0)}$, at the larger set of M angle points. In order to compute $\hat{\imath}^{(0)}$ we use the solution for $i^{(0)}$ just obtained to determine an explicit expression for the source function,

$$s^{(0)} = \omega' i^{(0)} + \beta \ , \tag{2.8}$$

and solve the scalar differential equations at the larger set of M angle divisions,

$$D\hat{\imath}^{(0)} = \hat{\imath}^{(0)} - s^{(0)} \ , \tag{2.9}$$

one angle point at a time. Since these equations are uncoupled the computation is very fast. The solution may be written formally as

$$\begin{aligned} \hat{\imath}^{(0)} &= (1 - D)^{-1} s^{(0)} \ , \\ &= \Lambda s^{(0)} \ , \end{aligned} \tag{2.10}$$

defining the Λ-operator as the inverse of the difference operator $(1 - D)$. The solution allows us to evaluate the source term contribution, $\Omega'\hat{\imath} = \Omega'\Lambda s^{(0)}$, from which we obtain the source function for the intensity in first order,

$$s^{(1)} = \omega' i^{(1)} + \Omega'\Lambda s^{(0)} - \omega' i^{(0)} \ . \tag{2.11}$$

We can now solve the coupled system of equations (2.6) for $i^{(1)}$. To continue this process, we use the solution for the intensity $i^{(1)}$ to determine the source function $s^{(1)}$ with explicitly given intensity and solve the M uncoupled equations for the intensity $\hat{\imath}^{(1)}$ at the larger set of angle points. This yields the source function for the intensity contribution in second order,

$$s^{(2)} = \omega' i^{(2)} + \Omega'\Lambda s^{(1)} - \omega' i^{(1)} \ , \tag{2.12}$$

and so on. The important point to note is that the source function $s^{(1)}$ contains a single application of the Λ-operator at the larger angle set, $s^{(2)}$ contains two such applications, *etc.*

It is now interesting to ask whether the perturbation series in the intensities $i^{(n)}$ is equivalent to the original equation for the intensity I. We therefore add the hierarchy of differential equations (2.6) and source terms (2.7). This results in

$$dI = I - \tilde{S} \quad ,$$

where (2.13)

$$\tilde{S} = \Omega'(\hat{\imath}^{(0)} + \hat{\imath}^{(1)} + \cdots) + \beta \quad ,$$

assuming the contributions $i^{(n)}$ and $\hat{\imath}^{(n)}$ approach zero with increasing value of n.

The comparison with equation (2.1) shows that there are two differences between the original differential equation and the summed perturbation set: (1) the differential operator d of the perturbation set differs from the operator D, and (2) the source function $\tilde{S}$ differs from S, containing multiple additional operations with the integral operator Λ. Thus the perturbation problem is different from the original problem. Therefore convergence of the equations cannot guarantee convergence to the correct solution and, in general, the answers will be different.

To demonstrate the properties of his method Cannon (1973a) solved a transfer problem for a Doppler-broadened line with completely noncoherent scattering in a finite, constant-property atmosphere of the total optical thickness $\tau = 10$ and with the scattering parameter $\epsilon = 10^{-4}$. Thus the medium was effectively thin, *i.e.*, $\tau \ll 1/\epsilon$. The "exact" equation, with which he measured the accuracy and convergence of the perturbation solution, was solved for $M = 3$ angle points and the approximate equation for $m = 1$.

The converged solution did not agree with the reference solution; at the midpoint of the slab, $\tau = 5$, it had an error of 7×10^{-4}, an indication that the series converges to a solution different from that of equations (2.1). The obvious inconsistency of the two sets of equations explains why Cannon's solution has a finite error even though the perturbation equations should have reproduced the exact solution to machine accuracy.

That the error is small in the present case of an effectively thin slab is no consolation since other problems, for thicker media for example or for partial redistribution, for which the method was designed, may yield much larger errors and, what is more insidious, they may go undetected while leaving the impression that the problem has been solved to high accuracy.

3. AUER'S APPROACH TO PARTIAL REDISTRIBUTION

We return to the problem of partial redistribution and present Auer's (1986) description of Cannon's operator perturbation treatment. Even though Auer's equations purport to represent Cannon's equations, they differ from them in interesting ways. However, what is true for Cannon's equations is also true for Auer's: the perturbed and the unperturbed systems describe different problems. Nevertheless, Auer's approach may be useful in some applications and therefore merits another look at this particular approach to partial redistribution.

One of the ways in which Auer's equations depart from Cannon's is in Auer's explicit use of the intensity mean along a ray (*i.e.*, the Feautrier variable) rather than the specific intensity. This implies a restriction to scattering problems with forward-backward symmetry. Another difference is in his use of partial sums for the intensity mean (analogous to the partial sums 4.6 for the source function). Thus, in each step of the iteration he solves the full problem. This is safer than using the expansion terms $i^{(k)}$ (*cf.* equation 2.5) since it prevents a possible drift of the iterative solution. And a third difference is that Auer performs an additional formal solution within each iteration cycle, making the perturbation equations even more inconsistent with the original equations. But that need not be a serious flaw; in practical applications, additional formal solutions may in fact be useful. And finally, even though Cannon perturbs the exact redistribution operator about the isotropic operator in his demonstration problem, his original equations admit a general redistribution function as an approximate operator; Auer uses only the isotropic operator. But since the isotropic operator does not reduce the order of the system of equations, the simplification is insignificant. Because there is additional power in an approximate anisotropic redistribution function, we return to Cannon's original equations here.

As in Cannon's original, the basic strategy of the perturbation approach is to reduce the order of the system of equations to be solved. This means that they are written for two different angle spaces: one, with M points, is very dense; the other, with the smaller number of m points, is very sparse. We use a compact notation in which angle points in the dense set are designated by n and

n', and in the sparse set by $\tilde{n}$ and $\tilde{n}'$. The primed variables n' and $\tilde{n}'$ are used in sums over the primed index, with summation implied. We suppress the depth variable.

Instead of the complete redistribution equations (2.1), the prototype transfer problem for partial redistribution is described by the transfer equation

$$DJ(n) = J(n) - S(n) \quad , \tag{3.1}$$

with the source function

$$S(n) = \mathcal{R}(n, n')J(n') + \beta \quad , \tag{3.2}$$

where D is the second-order difference operator of the Feautrier equation, R is the redistribution operator (which contains the factor $\underline{1-\epsilon}$) describing the scattering of photons from the beam travelling along the ray specified by the direction index n' into the direction specified by n. — It is immaterial for this discussion whether R depends only on angle as implied in this description of the physics contained in R, or whether R depends also on frequency (and perhaps polarization), as it does in most interesting partial redistribution problems. — The thermal emission term β in typical situations depends only on depth, but it could also depend on angle and frequency (and polarization) without affecting the form of the equation.

If these equations, which are of order M, were to be solved directly they would be combined into the single equation

$$[(1 - D) - \mathcal{R}]J = \beta \tag{3.3}$$

and solved for the intensity $J(n)$ (for the numerical solution, see references following equations 5.5). In the perturbation approach, these equations are rewritten for the reduced angle space of m points. The transfer equation is then given by

$$(1 - d)J^{(k)}(\tilde{n}) = S^{(k)}(\tilde{n}) + \Delta S^{(k)}(\tilde{n}) \quad , \tag{3.4}$$

where the intensity $J^{(k)}$ is the partial sum of an expansion (*cf.* the analogous definition 5.12) and the source function is

$$S^{(k)}(\tilde{n}) = r(\tilde{n}, \tilde{n}')J^{(k)}(\tilde{n}') + \beta \quad , \tag{3.5}$$

with the approximate redistribution operator r; the correction term $\Delta S^{(k)}(\tilde{n})$ is discussed below. We note parenthetically that Auer describes the problem with the isotropic angle operator ω,

$$S^{(k)} = \omega(\tilde{n}')J^{(k)}(\tilde{n}') + \beta \quad , \tag{3.6}$$

resulting in a source function that depends on depth only; since the use of the operator ω rather than of r does not lower the order of the system of equations but reduces the numerical work only minimally, the diminished generality of ω makes it preferable to employ the more general operator r.

For the numerical solution, the transfer and source function equations are combined into the single equation

$$[(1-d)-r]J^{(k)} = \beta + \Delta S^{(k)} \quad , \tag{3.7}$$

which is solved for the intensity $J^{(k)}(\tilde{n})$ in the reduced angle space of m division points.

The next step follows Cannon's procedure, to estimate an improved source function $\hat{S}^{(k)}(n)$ in the space of M angle points using the intensity with only m points,

$$\hat{S}^{(k)}(n) = \mathcal{R}(n,\tilde{n}')J^{(k)}(\tilde{n}') + \beta \quad , \tag{3.8}$$

and to determine an improved intensity $\hat{J}^{(k)}(n)$ in the denser space by means of formal solutions of M separate transfer equations for given source function,

$$(1-D)\hat{J}^{(k)}(n) = \hat{S}^{(k)}(n) \quad , \tag{3.9}$$

resulting in $\hat{J}^{(k)}(n)$.

Now Auer determines a further improved source function $\hat{\hat{S}}^{(k)}(n)$,

$$\hat{\hat{S}}^{(k)}(n) = \mathcal{R}(n,n')\hat{J}^{(k)}(n') + \beta \quad , \tag{3.10}$$

and a mean intensity $\hat{\hat{J}}^{(k)}(n)$ from a formal solution,

$$(1-D)\hat{\hat{J}}^{(k)}(n) = \hat{\hat{S}}^{(k)}(n) \quad . \tag{3.11}$$

This last step is not present in Cannon's approach. It is a useful addition to the method, however, since, at least for one part of the iteration cycle, an intensity in the dense space of M angle points is used to compute another intensity in the same space, which is the essence of partial redistribution (*cf.* equations 3.1 and 3.2).

The lagged error term, $\Delta S^{k+1}(\tilde{n})$ (*cf.* eq. 3.4), can now be computed,

$$\Delta S^{k+1}(\tilde{n}) = \mathcal{R}(\tilde{n},n')\hat{\hat{J}}^{(k)}(n') - r(\tilde{n},n')\hat{\hat{J}}^{(k)}(n') \quad , \tag{3.12}$$

using for both terms on the right-hand side the intensity in the space of M points.

We ask whether the perturbation equations have the same physical content as the original equations. If it is assumed that the iterations have converged to $\hat{\hat{J}}$, the equations (3.4 – 3.12) for the mean intensity can be combined into the single equation

$$[1 - \Lambda\mathcal{R}\Lambda\mathcal{R}\lambda(\mathcal{R} - r)]\hat{\hat{J}} = \Lambda(1 + \mathcal{R}\Lambda + \mathcal{R}\Lambda\mathcal{R}\lambda)\beta \quad , \tag{3.13}$$

where we have used the definitions $[(1-d) - r]^{-1} = \lambda$ and $(1-D)^{-1} = \Lambda$. The perturbation equation should match the exact equation (3.3) which may be written in a similar form,

$$(1 - \Lambda\mathcal{R})J = \Lambda\beta \quad . \tag{3.14}$$

It is clear that the perturbation equations converge to a solution $\hat{\hat{J}}(n)$ that is different from the solution $J(n)$ of the exact equations, except in the special case $r = R$ [where $\lambda = (1 - \Lambda\mathcal{R})^{-1}\Lambda = \Lambda(1 - \mathcal{R}\Lambda)^{-1}$]. So what could be the merit of the approach?

The important question is whether the Λ-iterations (3.9) and (3.11) converge rapidly enough to yield an acceptable solution, either with the procedure as outlined here or with a modification that adds further formal solutions, perhaps only to the very last iteration cycle. One point that needs to be stressed is that the difficulties encountered in the solution of scattering problems by means of Λ-iterations need not defeat the present approach (for a graphic depiction of the failure of Λ-iteration, *cf.* Auer 1984, p265). Recall that the reason for the slow convergence of the conventional Λ-iteration is that the Λ-operator describes only the short-range behavior of the transfer, approximately extending over a distance of the order of a photon mean-free-path in the center of a line, whereas the influence of the integral operator $(1 - \Lambda\mathcal{R})^{-1}\Lambda$ extends over a distance of the order of a thermalization length. The well-known difficulties of Λ-iteration need not be a problem here, however. With an appropriate choice of the approximate operator r, the long-range depth behavior of the scattering kernel may be properly accounted for in the approximate solution. Then the corrections merely have to redistribute photons over beams (*i.e.*, direction and frequency), and not also over depth. This depends on the properties of the approximate operator. Therefore the choice of r is important and the flexibility afforded by an anisotropic redistribution function may be critical.

The Cannon-Auer approach to partial redistribution clearly merits further study. The equations given by Auer suggest, especially if further Λ-iterations are inserted to calculate the intensity from a source function that is (at least approximately) consistent with this intensity, that a problem be split into two parts: a calculation of the transfer of photons in depth, and a calculation of the transfer in angle-frequency space. The latter is a process that may proceed much faster than transfer in depth since a relatively small number of scattering events should suffice to spread photons over the few (in a typical problem) Doppler widths covering the full bandwidth of a line. Therefore, one would naively expect that diffusion in frequency is not as obstinate a problem as diffusion in depth is in conventional Λ-iteration. If this expectation is borne out by numerical experiment the approach, which is like a complete redistribution calculation in depth alternating with a calculation of the photon distribution in frequency, could be a convenient way to solve partial redistribution problems and constitute an alternative to Scharmer's (1983) approach.

It is interesting to note that Scharmer's (1983) solution of partial redistribution problems can be viewed in just this manner: as a calculation of a function that depends only on depth, making this essentially a complete redistribution problem, and a calculation of the distribution of photons in angle and frequency at a given depth. His formulation is based on integral equations and is written in terms of the expansion terms that make up the partial sums of the source function. Unlike the Cannon-Auer equations, Scharmer's perturbation equations do not contain any inconsistencies. However, it is essential in this approach that the approximate redistribution operator be isotropic. Thus, widely differing mean-free-paths along different directions, as can occur in some polarization problems, might not be easy to treat with this approach. This is one of the reasons why alternative methods are worth studying.

4. FORMULATION VIA INTEGRAL EQUATIONS

We consider again equations (2.1) for isotropic scattering. To derive a low-order system of differential equations we transform the exact system of differential equations into the equivalent integral equation, perturb this integral equation using an approximate integral operator, and transform back to differential equations. The low-order system of differential equations that we will obtain is the one sought by Cannon, but with the restriction to complete redistribution.

This hierarchy of perturbation equations converges to the solution of the exact equations if a convergence criterion is satisfied.

The first step in the derivation of the integral equation is to solve the scalar (*i.e.*, uncoupled) differential equations (2.1) formally for the intensity, which is given by the so-called formal solution,

$$I = (1 - D)^{-1} S \quad . \tag{4.1}$$

If we needed the Λ-operator we could obtain it by inverting the bidiagonal operator $(1-D)$ if the transfer equation is a first-order differential equation for the specific intensity, or by inverting the tridiagonal operator $(1 - D)$ if the transfer equation is in the more common second-order form, *i.e.*, the Feautrier equation. But our interest is not in solving the transfer problem as an integral equation. The use of integral equations is merely a device for deriving the requisite differential equations of the perturbation problem.

We insert the formal solution into the expression (2.1) for the source function to obtain the integral equation,

$$S = \Omega'(1 - D)^{-1} S + \beta \quad , \tag{4.2}$$

which we write in a more compact form as

$$\mathcal{L} S = \beta \quad , \tag{4.3}$$

with the "exact" integral operator $\mathcal{L}$ given by

$$\mathcal{L} = 1 - \Omega'(1 - D)^{-1} \quad . \tag{4.4}$$

In order to solve the integral equation by means of operator perturbation we follow Cannon's basic prescription by expressing the exact operator in terms of an approximate operator plus a correction,

$$\mathcal{L} = L + (\mathcal{L} - L) \quad . \tag{4.5}$$

At the same time we expand the source function in a series,

$$S^{(n)} = \sum_{k=0}^{n} s^{(k)} \ , \qquad S = \lim_{n \to \infty} S^{(n)} \quad . \tag{4.6}$$

Putting this expansion and the operator expression (4.5) into equation (4.3) yields the hierarchy of integral equations for the expansion terms s,

$$L \, s^{(n+1)} = \mathcal{E}^{(n)} \quad , \tag{4.7}$$

with the driving term $\mathcal{E}$ given in terms of the partial sums S by

$$\mathcal{E}^{(n)} = \beta - \mathcal{L}S^{(n)} \quad . \tag{4.8}$$

Note that the inhomogeneous vector $\mathcal{E}^{(n)}$ in the $(n+1)^{th}$-order equation is the error made by the (known) n^{th}-order source function in the conservation equation (4.3). For the approximate operator we use

$$L \;=\; 1 - \omega'(1-d)^{-1} \quad , \tag{4.9}$$

i.e., a representation in terms of the corresponding operators of the approximate differential equations we are seeking.

We evaluate the term $\mathcal{L}S^{(n)} \;=\; S^{(n)} - \Omega'(1-D)^{-1}S^{(n)}$ by separating the calculation into several parts: We define an intensity, written with half-integral index to emphasize that it was obtained by means of a Λ−operation,

$$I^{(n+\frac{1}{2})} \;=\; (1-D)^{-1}S^{(n)} \tag{4.10}$$

with appropriate boundary conditions, the corresponding mean intensity,

$$J^{(n+\frac{1}{2})} \;=\; \Omega' I^{(n+\frac{1}{2})} \quad . \tag{4.11}$$

and the source function

$$S^{(n+\frac{1}{2})} \;=\; J^{(n+\frac{1}{2})} + \beta \quad . \tag{4.12}$$

Thus the error term is equal to the difference between the nth-order source function and the source function obtained from it by applying the integrated Λ-operator,

$$\mathcal{E}^{(n)} \;=\; S^{(n+\frac{1}{2})} - S^{(n)} \quad . \tag{4.13}$$

Note that this equation would be used directly (with $\mathcal{E}$ set equal to zero) if the transfer problem were solved by means of a Λ-iteration.

In order to derive the differential equations of the perturbed problem we exploit the analogy between the integral and differential equation statements of the exact problem. The coupled system of differential equations (2.1) that corresponds to the integral equation (4.3) may be written as

$$\begin{aligned} [(1-D) - \Omega']I \;&= \beta \;, \\ S \;&= \Omega' I + \beta \quad . \end{aligned} \tag{4.14}$$

Therefore the differential equations corresponding to the perturbation equations (4.7) in integral form are given by

$$\begin{aligned}[(1-d)-\omega']i^{(n+1)} &= \mathcal{E}^{(n)} , \\ s^{(n+1)} &= \omega' i^{(n+1)} + \mathcal{E}^{(n)} \quad .\end{aligned} \tag{4.15}$$

These equations might be solved with the starting solution $s^{(0)} = 0$, $S^{(0)} = 0$, $\mathcal{E}^{(0)} = \beta$ for the intensities $i^{(n)}$ (starting with $n = 1$), from which the source function corrections $s^{(n)}$, the partial sums $S^{(n)}$ and, hence, the error terms $\mathcal{E}^{(n+1)}$ are computed. The treatment of the boundary conditions in the perturbed equations and the numerical solution will be discussed in detail in the following section.

To check whether the system of perturbed differential equations agrees with the original equations we use their integral equation equivalents. Thus, summing over the series (4.7) and (4.8) we obtain for the partial sum $S^{(n+1)}$:

$$S^{(n+1)} = \sum_{l=0}^{n}(1 - L^{-1}\mathcal{L})^l L^{-1}\beta \quad , \tag{4.17}$$

where we have assumed that the starting solution is $S^{(0)} = 0$.

Since the infinite sum is equal to $\mathcal{L}^{-1}L$, this series converges to the solution of the exact integral equation (4.3), *i.e.*,

$$S = \mathcal{L}^{-1}\beta \quad , \tag{4.18}$$

provided the eigenvalues of the operator $(1 - L^{-1}\mathcal{L})$ are all smaller than unity in absolute value. The equations converge also when the starting solution differs from zero (*cf.* Kalkofen 1984). Thus the perturbed differential equations (4.15) are equivalent to the original differential equations (2.1), and they represent the sought reduction in the order of the system of differential equations.

5. TRANSFER PROBLEM WITH CONSTRAINT

For the further discussion of the operator perturbation approach in differential equations and for the treatment of boundary conditions we add the constraint of radiative equilibrium to the previous equations. The prescription given in the preceding section, where we constructed an integral operator acting on the given inhomogeneous term, namely the thermal emission term, does not work here since that term must be determined together with the radiation field.

The procedure we will use instead is to turn the boundary condition into a term on which a suitable integral operator can act. In practical applications, this boundary term is usually stated in the form of a given net radiative flux. For ease of exposition we will assume, however, that the boundary intensity is given explicitly. The extension of the equations to the usual case is straightforward and will not be discussed further except for the numerical results, which are for prescribed net flux.

We consider a monochromatic transfer problem with isotropic scattering where the structure of the atmosphere is not known but must be calculated from the condition of monochromatic radiative equilibrium. Our aim is again to determine the integral equation form of the transfer problem, perturb the integral equation, and use the analogy between integral and differential equation statements to derive the differential equations corresponding to the perturbed integral equation. For this purpose it is useful to express the exact differential equations in several forms, which we write as Feautrier equations, *i.e.*, as second-order differential equations for the even part of the specific intensity along a ray (*cf.* Feautrier 1964; Mihalas 1978, p152; Kalkofen, this volume). Thus D is a second-order difference operator except at the boundaries, where it is a first-order operator. The transfer equation for this intensity mean, the source function, and the radiative equilibrium constraint, respectively, are given by

$$\begin{aligned} DJ &= J - S \;, \\ S &= (1-\epsilon)\Omega J + \epsilon B \;, \\ S &= \Omega J \;, \end{aligned} \tag{5.1}$$

where J is the intensity mean along a ray defined by the angle cosine μ, S is the source function, and B is the Planck function; the integral operator Ω now represents the angle integral

$$\Omega \sim \tfrac{1}{2}\int_{-1}^{1} d\mu ...$$

without the scattering factor. The discrete operators Ω and ω are defined analogously to the previous primed operators (2.3) and (2.3′), except for the factor $(1-\epsilon)$, which is now omitted from the definition.

Because of the radiative equilibrium constraint, $S = \Omega J$, which implies the equality of the source terms,

$$S = \Omega J = B \quad , \tag{5.2}$$

the equations (5.1) can be simplified to

$$\begin{aligned}(1-D)J &= B\ ,\\ \Omega J &= B\ .\end{aligned}\tag{5.3}$$

Thus for a grey atmosphere the equations *with* scattering are identical to the equations *without* scattering, except for the definition of the optical depth, which now contains both the true absorption and the scattering coefficients. These equations are to be solved subject to boundary conditions $I_{bc}^{\mp}$ at the outer and inner boundaries, respectively. Stated in detail these equations are

$$\begin{aligned}(1-d/d\tau)J &= I_{bc}^{-}\ ,\\ (1-d^2/d\tau^2)J-B &= 0\quad ,\\ (1+d/d\tau)J &= I_{bc}^{+}\ ,\\ \Omega J-B &= 0\quad ,\end{aligned}\tag{5.4}$$

which are second-order in the interior of the atmosphere and first-order at the boundaries. With M angle points, these differential equations are written as M separate tridiagonal equations, coupled by the radiative equilibrium constraint. For the numerical solution they are put into a single compact system,

$$\begin{pmatrix}(1-D) & -1\\ \Omega & -1\end{pmatrix}\begin{pmatrix}J\\ B\end{pmatrix}=\begin{pmatrix}I_{bc}^{-}\\ 0\end{pmatrix}_{\tau_1}\ ,\ \begin{pmatrix}0\\ 0\end{pmatrix}_{\tau_i}\ ,\ \begin{pmatrix}I_{bc}^{+}\\ 0\end{pmatrix}_{\tau_N}\quad ,\tag{5.5}$$

where the first M rows are transfer equations and the $(M+1)^{th}$ row is the energy condition. In this set, for N depth points τ_i, we have defined N vectors in which the first M components are the specific intensities J at a given depth and the $(M+1)^{th}$ component is the unknown Planck function at the same depth. The right-hand sides at the outer boundary, an inner point, and the inner boundary, respectively, are vectors ordered in a similar way. This block tridiagonal system of order N, with the matrix elements being themselves matrices of order $M+1$, is solved in the usual manner by Gaussian elimination (*cf.* Cuny 1967; Mihalas 1978, p156; Auer 1984, section 4.3). (Note that the matrix on the left-hand side does not take proper account of the absence of the B term from the transfer equation (5.4) at the two boundaries; at the boundaries, the column vector 1 in the upper right-hand corner should be interpreted as $1-\delta_{i,1}-\delta_{i,N}$.)

An alternative formulation of the transfer problem, one that we need for deriving the integral equation statement of the problem, is obtained by separating the radiation field into two components, one being due to sources in the interior of the medium and the other due to the radiation incident at the boundaries. Then the total intensity is given by

$$J = J_{int} + [I_{bc}^{-} g^{-}(\tau) + I_{bc}^{+} g^{+}(\tau)] \quad , \tag{5.6}$$

where the functions $g^{\pm}$ describe the exponential decay of the incoming intensities with distance from the respective boundaries along inclined rays. The transfer problem for the intensity due to internal sources is described by the transfer equation at the upper boundary, in the interior, and at the lower boundary, respectively, and by the energy conservation equation,

$$\begin{aligned} (1 - d/d\tau) J_{int} &= 0 \, , \\ (1 - d^2/d\tau^2) J_{int} - B &= 0 \, , \\ (1 + d/d\tau) J_{int} &= 0 \, , \\ \Omega J_{int} - B &= -G \, , \end{aligned} \tag{5.7}$$

where G is the angle integral of the external radiation penetrating to depth τ,

$$G(\tau) = \Omega [I_{bc}^{-} g^{-}(\tau) + I_{bc}^{+} g^{+}(\tau)] \quad . \tag{5.8}$$

The corresponding compact form of the coupled equations is

$$\begin{pmatrix} (1-D) & -1 \\ \Omega & -1 \end{pmatrix} \begin{pmatrix} J_{int} \\ B \end{pmatrix} = \begin{pmatrix} 0 \\ -G \end{pmatrix} \quad , \tag{5.9}$$

Note that the boundary conditions on these equations are zero (the parenthetic comments following equation 5.5 concerning the coupling matrix apply here as well).

From the differential equations (5.7) for the internal intensity, written analogously to equations (5.3), we obtain the integral equation statement of the problem by first writing the specific intensity as the formal integral over the source function (*cf.* equation 4.1) and then evaluating the mean intensity using the integral operator Ω. From the energy equation (the last equation in 5.7) we then obtain the integral equation

$$\begin{aligned} \mathcal{L} B &= G \quad , \\ \mathcal{L} &= 1 - \Omega (1 - D)^{-1} \quad , \end{aligned} \tag{5.10}$$

with the matrix operator $\mathcal{L}$, whose order is identical to the number of depth points, N. The order of the system of differential equations is no longer apparent, having contributed only to the number of terms that are added together in the calculation of the individual matrix elements of $\mathcal{L}$.

To solve the integral equation by means of operator perturbation we perturb the exact matrix operator $\mathcal{L}$ about the "simpler" operator L, given by

$$L = 1 - \omega(1-d)^{-1} \quad . \tag{5.11}$$

As an integral operator, L is no simpler than $\mathcal{L}$, both having the same order, N. Only their differential equation representations are different.

The expression $\mathcal{L} = L + (\mathcal{L} - L)$ is introduced into the exact integral equation together with the expansion

$$B^{(n)} = \sum_{k=0}^{n} b^{(k)} \tag{5.12}$$

for the Planck function. The resulting hierarchy of integral equations for the expansion terms b is

$$L\, b^{(n+1)} = \mathcal{E}^{(n)} \quad , \tag{5.13}$$

where the driving term is given in terms of the partial sums B by

$$\mathcal{E}^{(n)} = G - \mathcal{L}B^{(n)} \quad , \tag{5.14}$$

which is again the error in a conservation statement, namely equation (5.10). To evaluate this error term (*cf.* section 4) we have two options. We can either solve the differential equations

$$(1-D)J_{int}^{(n+\frac{1}{2})} = B^{(n)} \tag{5.15}$$

with zero boundary conditions for the *internal* specific intensities and determine from them the total mean intensity

$$\bar{J}^{(n+\frac{1}{2})} \equiv \Omega J^{(n+\frac{1}{2})} = \Omega J_{int}^{(n+\frac{1}{2})} + G \quad , \tag{5.16}$$

or we can solve the transfer equations directly for the *total* intensities by taking the prescribed boundary conditions for the total intensities into account, and then determine the total mean intensity. Either way the error term becomes

$$\mathcal{E}^{(n)} = \bar{J}^{(n+\frac{1}{2})} - B^{(n)} \quad . \tag{5.17}$$

Thus the driving term of the correction equation is given by the difference between the nth-order Planck function and its Λ-iterated value.

TABLE 1
Maximal Relative Corrections and Errors for a Grey Atmosphere

Iteration	1	2	3	4	5	6	7
Correction	1006	0.016	6.3^{-4}	2.5^{-5}	1.1^{-6}	4.4^{-8}	1.9^{-9}
Error	-0.017	-6.6^{-4}	-2.6^{-5}	-1.1^{-6}	-4.6^{-8}	-1.9^{-9}	-1.4^{-10}

In order to calculate the corrections $b^{(n)}$ we do not solve the integral equation (5.13) directly but transform it first into the equivalent system of differential equation, exploiting the analogy between the exact integral equation (5.10) and the corresponding system of differential equations (5.9) for the *internal* intensity. Thus,

$$\begin{pmatrix} (1-d) & -1 \\ \omega & -1 \end{pmatrix} \begin{pmatrix} j^{(n+1)} \\ b^{(n+1)} \end{pmatrix} = \begin{pmatrix} 0 \\ -\mathcal{E}^{(n)} \end{pmatrix} , \qquad (5.18)$$

which we solve for zero incident intensities, except in lowest order ($n = 1$), where it may be advantageous to use the form for the *total* intensity, with the prescribed boundary conditions. Then the right-hand side of equation (5.18) is the same as that of equation (5.5), except for the lower order of the column vectors. Note that the error term $\mathcal{E}^{(0)}$ is zero in that case. Again, for the matrix at the boundaries, see the parenthetic remarks following equation (5.5).

Our actual procedure for solving the transfer problem is as follows. After having solved the equations for the corrections $b^{(n+1)}$ and $j^{(n+1)}$ we determine the partial sum $B^{(n+1)}$, but we do not use the intensity corrections $j^{(n+1)}$. Instead, we follow the prescription defined by equations (5.13) and (5.14) and solve the individual transfer equations (5.15) for the components of the total intensity and evaluate the corresponding error term before solving again the low-order coupled system (5.18) for the intensity and Planck function corrections.

The solution of the transfer problem is complete when the Planck function is known to the required accuracy. Since, for a well-chosen approximate operators L, convergence is very rapid, the accuracy can often be estimated from the magnitude of the corrections b/B (*cf.* Table 1). From the solution $B(\tau)$, the specific mean intensities J and in particular the emergent intensities $I = J(0)$ can then be calculated by means of simple quadratures, most efficiently by using the Feautrier equation.

Table 1 lists the maximal relative corrections and the maximal relative errors in the iterative solution of the equations for a grey atmosphere with constant net flux, $\frac{1}{2}\int_{-1}^{1} I(\mu)\mu d\mu$ = const. The lower boundary condition in this problem is therefore different from that of equations (5.5): Instead of the explicitly given incident intensities $I_{bc}^{+}(\mu)$, the boundary intensities at τ_N must be determined together with the solution. Consequently, the intensities at the lower boundary are coupled not only via the flux derivative constraint but also via the flux constraint (in the form of the boundary condition), which modifies the matrix acting on J and B in equation (5.5) at the point τ_N. The radiation incident at $\tau = 0$ is assumed to be zero. The exact solution of the problem is computed with $M = 4$ angle points per hemisphere (so-called eight-stream approximation), the approximate solution with $m = 2$. The starting solution was given by the exact solution reduced by a factor of 10^3 throughout. Since a non-zero initial estimate of the solution was available here, the form (5.18) for the *internal* intensity was used also in the first iteration, $n = 1$. Note that convergence, though linear, is sufficiently rapid in this demonstration problem that the error in one iteration is approximately equal to the correction in the subsequent iteration.

It is interesting to note that while the exact equations (5.5) and (5.9) for the total and the internal intensities, respectively, are entirely equivalent as *differential* equations (as can easily be seen for a single angle point, for example, by solving them analytically), they are not in general equivalent as *difference* equations and, therefore, numerically. The reason is that whereas the total intensity is nearly a *linear* function of optical depth (for a single angle point it is strictly linear), the intensity due to internal sources alone depends *exponentially* on depth. This behavior is exactly matched, of course, by the intensity incident on the boundaries, which has the compensating exponential behavior. Unless the depth gridding is matched to the depth dependence of the internal intensity, the exponential behavior is extremely poorly represented by the second-order difference equation with the usual logarithmic depth spacing commonly used in semi-infinite media. Only when this behavior is taken into account in the choice of the depth points will the numerical solution of the equations for the internal intensity yield the same result as that for the total intensity. The preferred procedure for calculating the error term (5.17) is therefore to solve equations (5.15) *with* boundary conditions, *i.e.*, for the *total* intensity. The corresponding equation (5.16) is then without the additional term G representing the boundary intensity.

6. CONCLUSIONS

We have shown that Cannon's operator perturbation equations for partial redistribution problems contain extraneous formal solutions not present in the exact equations, and that Auer's equations describing Cannon' method contain further formal solutions. These additional terms can become the source of serious — and undetected as well as unsuspected — error, even when the perturbation equations are solved to high accuracy.

We argue that, in spite of the inherent shortcoming of this perturbation method, the Cannon-Auer approach may prove useful for solving some partial redistribution problems, perhaps with a modification adding further Lambda iterations. This will be feasible if diffusion of photons in frequency and angle is sufficiently rapid to make the complete redistribution source function a good starting solution. To check this suggestion, numerical experiments should be performed in which the exact solutions of the partial redistribution equations are compared with perturbation solutions. The aim would be to determine the limits of such an approach for typical cases of practical interest.

The inconsistency between the exact and the perturbed *differential* equations for partial redistribution problems is found in Cannon (1973a, 1976, 1984, 1985; Cannon *et al.* 1975) and Auer (1986). Cannon's perturbed *integral equations* for *complete redistribution* (Cannon 1973b, 1984, 1985; also Cram & Lopert (1976) when restricted to complete redistribution) do not contain extra terms and are correct. Likewise, Scharmer's (1983) perturbation equations for partial redistribution, as well as those for complete redistribution (Scharmer 1981 *etc.*) and its many derivatives by other authors, are not affected by this criticism.

For the case of complete redistribution we give a general prescription for deriving perturbation equations in differential form. The procedure uses integral equations in an intermediary step: First, the exact differential equations are cast into the equivalent integral equation form. Then this exact integral equation is written as a perturbation series of integral equations in a Cannon-type expansion, using an approximate integral operator that corresponds to a simple differential operator. Finally, the perturbed integral equations are transformed back into the equivalent set of differential equations. The result is a set of approximate differential equations whose order is much lower than that of the exact differential equations. The exact differential operator is used only in the formal solution of scalar (*i.e.*, uncoupled) Feautrier equations for calculating the error in a conserved quantity. This error becomes the driving term in the low-order set of correction equations. We sketch the method for the case of monochromatic, isotropic scattering and describe it in detail for a radiative equilibrium problem, for which we also give a numerical solution showing the properties of the method.

ACKNOWLEDGMENTS

I am grateful to Larry Auer and Dimitri Mihalas for helpful comments on the manuscript and stimulating discussions of the subject matter of this paper.

REFERENCES

Auer, L. H. 1984, *Methods in Radiative Transfer*, W. Kalkofen ed., Cambridge University Press, Cambridge, 237.

———— 1986 *J. Quant. Spectrosc. Rad. Transfer*, submitted.

Cannon, C. J. 1973a, *J. Quant. Spectrosc. Rad. Transfer*, **13**, 627.

———— 1973b, *Astrophys. J.*, **185**, 621.

———— 1976, *Astron. Astroph.*, **52**, 337.

———— 1984, *Methods in Radiative Transfer*, W. Kalkofen ed., Cambridge University Press, Cambridge, 157.

———— 1985, *The Transfer of Spectral Line Radiation*, Cambridge University Press, Cambridge.

Cannon, C. J., Lopert, P. B. & Magnan, C. 1975, *Astron. Astroph.*, **42**, 347.

Cram, L. E. & Lopert, P. B. 1976, *J. Quant. Spectrosc. Rad. Transfer*, **16**, 347.

Cuny, Y. 1967, *Ann. d'Astroph.*, **30**, 143.

Feautrier, P. 1964, *Compt. Rend. Acad. Sci. Paris*, **258**, 3189.

Kalkofen, W. 1984, *Methods in Radiative Transfer*, W. Kalkofen ed., Cambridge University Press, Cambridge, 427.

Mihalas, D. 1978, *Stellar Atmospheres*, Second Edition, W. H. Freeman & Co., San Francisco.

Scharmer, R. 1981, *Astrophys. J.*, **249**, 720.

———— 1983, *Astron. Astroph.*, **117**, 83.

A GENTLE INTRODUCTION TO POLARIZED RADIATIVE TRANSFER

David E. Rees

Department of Applied Mathematics
University of Sydney, NSW, 2006, Australia
and
High Altitude Observatory
National Center for Atmospheric Research [1]
Boulder, Colorado, USA.

ABSTRACT: The equations of polarized radiative transfer, especially those relevant to the formation of magnetically split spectral lines, are probably unfamiliar to most researchers in astrophysics. This chapter is an introductory tour of the field. We discuss the representation and measurement of polarized light in terms of the Stokes parameters. Possible sources of error because of ambiguities about sign conventions are highlighted by an analysis of the LTE transfer equations for the Stokes parameters of a normal Zeeman triplet. A simple derivation of the analytic solution of these equations is given for a Milne-Eddington model atmosphere. The discussion is complemented by a brief review of developments in non-LTE polarized transfer. Finally we summarize the papers on polarization in the latter part of this monograph.

1. INTRODUCTION

Intensity is a concept with which researchers in astrophysical radiative transfer feel comfortable. We all have a good idea of a photon's energy (or frequency or wavelength) and direction. But polarization has a phantom quality.

The language of polarization can be very confusing. Sign conventions, notably with regard to the 'handedness' of circular polarization, have been treated in a cavalier fashion in the astrophysical literature. In his survey of this problem Clarke (1974) waxed Biblical with the quotation (St. Matthew, Ch.6,v.3), "*let not thy left hand know what thy right hand doeth*" , because so many authors do not explain how to interpret their conventions.

This is not a trivial issue. Elusive sign errors lurk in many a polarization transfer computer code. This author can attest to long hours of searching

[1] The National Center for Atmospheric Research is sponsored by the National Science Foundation.

for such 'bugs' . A frustrating aspect of this field is that it is so easy to forget *one's own* sign conventions! Careful documentation is essential, since a brief absence from working at the coalface of polarization transfer invariably leads to gnawing doubts about at least one negative sign.

Polarized light is usually characterized by the Stokes parameters I,Q, U and V. Here I denotes intensity, while Q and U are connected with linear polarization, and V with circular polarization. A formal mathematical definition can be given for the Stokes parameters, but perhaps more immediate insight is gained by relating them to a set of intensity measurements made with different optical devices. This is called an operational definition.

Many excellent books discuss the representation and measurement of polarized light (e.g. Born & Wolf 1959; Shurcliff 1962; Clarke & Grainger 1971; Robson 1974). Even so, before we embark on the intricacies of the transfer equations it seems appropriate here to reiterate the basics of the subject. A quantum mechanical treatment of polarization transfer requires an understanding of density matrix formalism (Fano 1957). However one can survive quite well in the field while embedded in the paradigm of classical electromagnetic theory. This is the viewpoint adopted in Section 2 where sign conventions, as well as the mathematical and operational definitions of the Stokes parameters are summarized.

The landmark paper in the theory of spectral line formation in the presence of a magnetic field is Unno's (1956) heuristic derivation of the LTE transfer equations for a normal Zeeman triplet. This theory was generalized by Rachkovsky (1962a) to include magneto-optical (birefringence) effects, and later by Beckers (1969a,b) to treat arbitrary Zeeman multiplets. The literature spawned by Unno's paper contains many ambiguities, mainly associated with sign conventions. Staude (1971b), and more recently Martin & Wickramasinghe (1981), drew attention to differences between various authors' formulations of the transfer equations. To resolve these ambiguities Landi degl'Innocenti & Landi degl'Innocenti (1972) initiated the Florentine Renaissance of polarization transfer by deriving the LTE equations quantum mechanically. After a careful intercomparison of different formulations the brothers Landi disconcertingly concluded: "We think that experiments are useful in order to clarify this argument, with the aim of acquiring a well established theory to interpret the observational results."

Whose transfer equations, then, are correct?

In Section 3 we present the LTE transfer equations for a Zeeman triplet. Various sign traps are highlighted. Also, to justify the 'correctness' of the equations we show that they are consistent with well established laboratory observations. This 'surgical' treatment of the transfer equations has been an invaluable tool for validating computer programs. Of course, a complete check requires dedication from the reader who must rederive the equations from first principles!

Approximate analytic solutions of the LTE transfer equations for a Milne-Eddington atmosphere (source function linear with optical depth) and constant magnetic vector (Unno 1956; Rachkovsky 1962b,1967) play a central role in the interpretation of Stokes parameter profiles (e.g. Lites & Skumanich 1985). As Staude (1971a) noted, derivations of these solutions tend to be unnecessarily complicated. The elementary steps leading to an analytic solution are detailed in Section 4.

Given the focus of this monograph it may seem inappropriate to dwell on LTE equations and analytic solutions. It should be recognized, however, that by far the majority of practical applications of polarization transfer theory to data analysis of Zeeman split spectral lines has relied on such simple models. Considerable progress has been made in formulating the non-LTE transfer theory for polarized radiation, but the development of viable numerical tools is in its infancy. In Section 5 we survey the field with a 'broad brush' to provide an historical setting for the chapters on polarization that follow. These contributions are summarized in Section 6 and we conclude in Section 7 with an overview of future prospects in the field.

2. REPRESENTATION OF POLARIZED LIGHT

The discussion in this section draws heavily on the book by Born & Wolf. In classical electromagnetic theory the properties of light may be described in terms of the electric wave vector **E** in the plane perpendicular to the direction of propagation. Consider a quasi-monochromatic wave defined as the superposition of many randomly timed and statistically independent wave trains which have a mean frequency ν, say, and corresponding mean wavelength λ. Let the resultant wave have a spectral width $\Delta\nu \ll \nu$. For such a wave propagating in the positive z direction let

$$E_x(t) = a_x(t)\ e^{i(\tilde{\phi}_x(t)-2\pi\nu t+2\pi z/\lambda)} \quad ; \quad E_y(t) = a_y(t)\ e^{i(\tilde{\phi}_y(t)-2\pi\nu t+2\pi z/\lambda)} \tag{1}$$

be complex analytic representations of the mutually orthogonal components of **E**. The axes x, y and z form a right-handed co-ordinate system. The amplitudes a_x, a_y and the phases $\tilde{\phi}_x$, $\tilde{\phi}_y$ vary slowly with time, i.e. are approximately constant over any interval that is short compared with the coherence time, $(\Delta\nu)^{-1}$, of the wave. For the particular case of a strictly monochromatic (i.e completely polarized) wave the amplitudes and the phases are constants.

At a given point z the tip of the vector **E** traces out a path that, for a completely polarized wave, is in general an ellipse. It is very important to establish a convention for describing the sense of rotation of **E**. We adopt the 'traditional' convention commonly used in optics. Thus a *clockwise* rotation as seen by an observer *receiving* the radiation is called *right-handed*, and a *counterclockwise* rotation is called *left-handed*. The sense of rotation is governed by the phase difference

$$\delta = \tilde{\phi}_x - \tilde{\phi}_y \tag{2}$$

such that $\sin\delta > 0$ for right-handed and $\sin\delta < 0$ for left-handed.

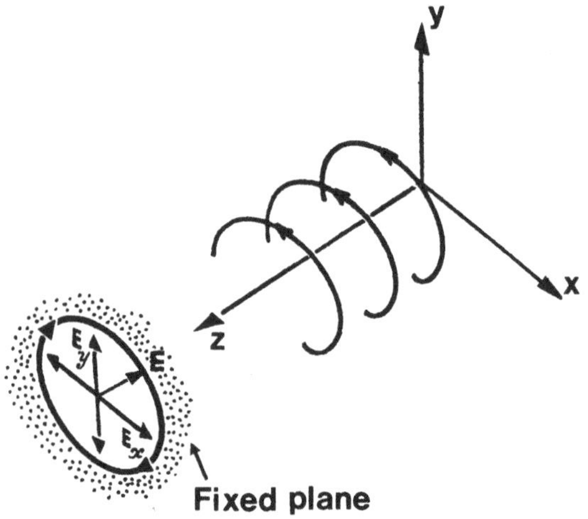

Fig. 1. For *right-hand* circularly polarized light the snapshot picture shown here is a *right-handed* circular helix traced by the electric vector at some instant t. As the helix propagates *without rotation* along the z-axis its intersection with the fixed plane executes a *clockwise* circular path.

It is possible to define right- and left-handed in a way that is independent of the observer's viewpoint. This is the so-called *snapshot picture* (e.g. Clarke & Grainger). Einstein notwithstanding, consider the wave defined by Equations (1) at a particular instant, $t = 0$ say. The tip of the electric vector viewed as a function of z traces out an elliptical helix which will have the sense of a right- (or left-) handed screw depending on whether $\sin\delta > 0$ (or < 0). As the wave propagates this helical pattern advances along the z axis *without rotation.* Thus as the helix traverses a fixed plane perpendicular to z, the vector $\mathbf{E}$ in that plane rotates in a clockwise (or counterclockwise) sense if the helix is right- (or left-) handed. The snapshot picture for right-hand circularly polarized light is illustrated in Figure 1.

The amplitudes and phases of $\mathbf{E}$ are not directly observable. However the properties of the polarization ellipse can be inferred by passing the light through appropriate optical devices and measuring the transmitted intensities. The following experiment designed for this purpose leads naturally to the definition of the Stokes parameters.

Suppose the light passes through a compensator which subjects E_y to a phase retardance ϵ relative to E_x and this is is followed by a polarizer which transmits linearly polarized light oriented at an angle θ counterclockwise to the x-axis. Then in the θ direction the component of the electric vector of the transmitted light is

$$E(t;\theta,\epsilon) = E_x \cos\theta + E_y e^{i\epsilon} \sin\theta. \tag{3}$$

Note that the retardance enters as a positive exponent in $e^{i\epsilon}$ because of the negative frequency exponent in $e^{-2\pi i\nu t}$ in the definition of E_x and E_y.

The corresponding intensity (a real quantity) is, omitting a constant of proportionality,

$$I_{trans}(\theta,\epsilon) = < E(t;\theta,\epsilon) E^*(t;\theta,\epsilon) >, \tag{4}$$

where $*$ denotes complex conjugate and $< ... >$ denotes a time average over the observation period which is $\gg \nu^{-1}$, the natural period of the wave.

Imagine six transmitted intensity measurements with the following angle settings:

$\epsilon = 0$ (no compensator)

$\theta = 0, \frac{1}{4}\pi, \frac{1}{2}\pi$, and $\frac{3}{4}\pi$,

so the device transmits only linearly polarized light at these respective orientations; and

$\epsilon = \frac{1}{2}\pi$ (e.g. compensator is a quarter-wave plate)

$\theta = \frac{1}{4}\pi$ and $\frac{3}{4}\pi$,

so the device transmits respectively right and left circularly polarized light.

It is helful to visualize these six configurations as filters transparent only to the respective polarizations represented symbolically by $\ominus, \varobar, \oslash, \obslash, \circlearrowright$, and $\circlearrowleft$, and to denote the transmitted intensities by

$$
\begin{aligned}
I_{\ominus} &= I_{trans}(0,0) & I_{\varobar} &= I_{trans}(\tfrac{1}{2}\pi,0) \\
I_{\oslash} &= I_{trans}(\tfrac{1}{4}\pi,0) & I_{\obslash} &= I_{trans}(\tfrac{3}{4}\pi,0) \\
I_{\circlearrowright} &= I_{trans}(\tfrac{1}{4}\pi,\tfrac{1}{2}\pi) & I_{\circlearrowleft} &= I_{trans}(\tfrac{3}{4}\pi,\tfrac{1}{2}\pi).
\end{aligned}
\tag{5}
$$

The Stokes parameters associated with the incident wave $\mathbf{E}$ are then defined as combinations of these measured intensities:

$$
\begin{aligned}
I &= I_{\ominus} + I_{\varobar} \\
Q &= I_{\ominus} - I_{\varobar} \\
U &= I_{\oslash} - I_{\obslash} \\
V &= I_{\circlearrowright} - I_{\circlearrowleft}.
\end{aligned}
\tag{6}
$$

Using Equations (1)-(6) one may translate this operational definition into a formal mathematical definition involving the amplitudes and phases of $\mathbf{E}$:

$$
\begin{aligned}
I &= < a_x^2 > \; + \; < a_y^2 > \\
Q &= < a_x^2 > \; - \; < a_y^2 > \\
U &= \; 2 < a_x a_y \cos\delta > \\
V &= \; 2 < a_x a_y \sin\delta > .
\end{aligned}
\tag{7}
$$

It is evident from Equations (6) and (7) that $V > 0$ implies an excess of *right-handed (clockwise)* polarization.

For convenient reference the sign conventions defined in this section are summarized in Table 1.

TABLE 1

Sign Conventions

left-handed	right-handed
counterclockwise	clockwise
$\sin\delta < 0$	$\sin\delta > 0$
$V < 0$	$V > 0$

3 . CHECKING THE TRANSFER EQUATIONS

In terms of the Stokes vector $\mathbf{I} = (I, Q, U, V)^\dagger$, ($\dagger$ means transpose), the vector transfer equation for polarized light is

$$\frac{d\mathbf{I}}{dz} = -\mathbf{K}\,(\mathbf{I} - \mathbf{S}). \tag{8}$$

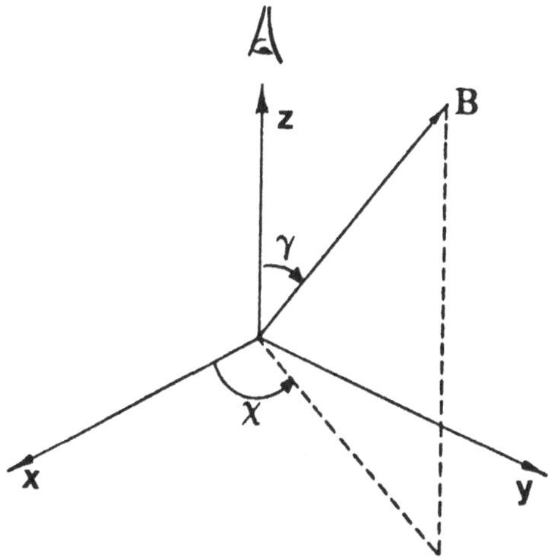

Fig. 2. Reference frame for definition of Stokes vector **I** and magnetic vector **B**.

The Stokes parameters are defined relative to a fixed frame xyz with the z-axis towards the observer (Figure 2). **K** is the absorption matrix and **S** is the source vector. For a Zeeman split spectral line the absorption matrix depends in a rather complex way on the magnetic field vector $\mathbf{B}$, specified here by its magnitude B, inclination γ to the line of sight, and azimuth χ relative to the x-axis (Figure 2). The aim of this section is to try to unravel some of this complexity and thereby provide some tools for computer code verification. Except for minor differences in notation the theory presented below follows the formulation by Landi degl'Innocenti (1976) and Landolfi & Landi degl'Innocenti (1982).

If continuum polarization is neglected and both line and continuum processes are governed by LTE then,

$$\mathbf{S} = B_\nu(T_e)\,\mathbf{I}_o \tag{9}$$

where $\mathbf{I}_o = (1,0,0,0)^\dagger$ and $B_\nu(T_e)$ is the Planck function at the local electron temperature T_e, and

$$\mathbf{K} = \kappa_c \mathbf{1} + \kappa_o \mathbf{\Phi}, \tag{10}$$

where κ_c is the continuum opacity, $\mathbf{1}$ is the unit 4×4-matrix and κ_o is the line centre opacity for zero damping and zero magnetic field.

The line absorption matrix $\mathbf{\Phi}$ is

$$\mathbf{\Phi} = \begin{pmatrix} \phi_I & \phi_Q & \phi_U & \phi_V \\ \phi_Q & \phi_I & \phi'_V & -\phi'_U \\ \phi_U & -\phi'_V & \phi_I & \phi'_Q \\ \phi_V & \phi'_U & -\phi'_Q & \phi_I \end{pmatrix}, \tag{11}$$

where

$$\begin{aligned}
\phi_I &= \tfrac{1}{2}\phi_p \sin^2\gamma + \tfrac{1}{4}(\phi_r + \phi_b)(1+\cos^2\gamma) \\
\phi_Q &= \tfrac{1}{2}(\phi_p - \tfrac{1}{2}(\phi_r + \phi_b))\sin^2\gamma\cos 2\chi \\
\phi_U &= \tfrac{1}{2}(\phi_p - \tfrac{1}{2}(\phi_r + \phi_b))\sin^2\gamma\sin 2\chi \\
\phi_V &= \tfrac{1}{2}(\phi_r - \phi_b)\cos\gamma \\
\phi'_Q &= \tfrac{1}{2}(\phi'_p - \tfrac{1}{2}(\phi'_r + \phi'_b))\sin^2\gamma\cos 2\chi \\
\phi'_U &= \tfrac{1}{2}(\phi'_p - \tfrac{1}{2}(\phi'_r + \phi'_b))\sin^2\gamma\sin 2\chi \\
\phi'_V &= \tfrac{1}{2}(\phi'_r - \phi'_b)\cos\gamma.
\end{aligned} \tag{12}$$

$\phi_{p,b,r}$ are absorption profiles and $\phi'_{p,b,r}$ are anomalous dispersion profiles. Following Skumanich (1987) we introduce here the prime notation to signify that an anomalous dispersion profile has a shape similar (but not equal!) to the negative of the derivative of the corresponding absorption profile. The indices refer to the respective components of a normal Zeeman triplet (p = unshifted π component; b, r = blue, red shifted σ components). Let m_{upper} and m_{lower} be the magnetic quantum numbers of Zeeman sublevels in the upper and lower energy levels of the line forming transition and

$$\Delta m = m_{upper} - m_{lower}. \tag{13}$$

Then

$$\Delta m = \begin{cases} +1 & \equiv \quad b \\ 0 & \equiv \quad p \\ -1 & \equiv \quad r \end{cases} \tag{14}$$

Note that the alternative convention $\Delta m = m_{lower} - m_{upper}$ is quite common, so care must be taken in coding to avoid swapping the sense of splitting of the σ components.

The profiles are

$$\begin{aligned} \phi_p &= H(a, v + v_{los}) & \phi'_p &= 2F(a, v + v_{los}) \\ \phi_b &= H(a, v + v_B + v_{los}) & \phi'_b &= 2F(a, v + v_B + v_{los}), \\ \phi_r &= H(a, v - v_B + v_{los}) & \phi'_r &= 2F(a, v - v_B + v_{los}) \end{aligned} \tag{15}$$

where $H(a, v)$ and $F(a, v)$ are the Voigt and Faraday-Võigt functions:

$$H(a, v) = \frac{a}{\pi} \int_{-\infty}^{\infty} \frac{e^{-y^2}}{(v - y)^2 + a^2} \, dy \tag{16}$$

and

$$F(a, v) = \frac{1}{2\pi} \int_{-\infty}^{\infty} \frac{(v - y)e^{-y^2}}{(v - y)^2 + a^2} \, dy. \tag{17}$$

Here a is the line damping, v is the *wavelength* measured from laboratory line centre, v_B is the *wavelength* Zeeman splitting and v_{los} is the *wavelength* Doppler shift induced by the macroscopic velocity field component (positive *towards* the observer). Each of a, v, v_B and v_{los} is measured in units of the Doppler width. One of the most accurate and efficient methods for computing $H(a, v)$ and $F(a, v)$ is Humlicek's (1982) rational approximation algorithm for the complex probability function.

The reason for emphasising wavelength in the previous paragraph is to draw attention to the risk of confusing the symbols with *frequency* measurements.

If one wishes to work in frequency rather than wavelength space the following transformations must be strictly adhered to in Equations (15):

$$v, v_B, v_{los}(\text{wavelength}) \longrightarrow -v, -v_B, -v_{los}(\text{frequency}). \tag{18}$$

In the absence of Zeeman splitting an error in this transformation is easy to trace because it manifests itself in an obviously incorrect asymmetry in the computed I profile. The situation is more complex when $v_B \neq 0$. $H(a,v)$ is a symmetric function and $F(a,v)$ is an asymmetric function of v, i.e.

$$F(a,-v) = -F(a,v) \quad \text{and} \quad F(a,v) > 0 \quad \text{if} \quad v > 0. \tag{19}$$

An error in the wavelength-frequency transformation, or an error in the definition of Δm can cause non-trivial changes in Equations (15) that may have serious consequences for data analysis. A notorious example of this is interpretation of sunspot Stokes profile asymmetries and net circular polarization (for a thorough discussion of this problem see Skumanich 1987).

Most of the confusion in the literature about the correct form of the transfer equations has been associated with the expressions for ϕ_V and ϕ'_V. This confusion is related to how one defines the Stokes parameter V. We shall not catalogue and compare all the forms and notations that have been published (see Martin & Wickramasinghe 1981). Rather we shall verify the signs of ϕ_V and ϕ'_V in Equations (12) by showing that they are consistent with standard laboratory experiments, the results of which are of course independent of any particular formulation of the transfer equations. Throughout the rest of this section we assume $v_{los} = 0$.

Longitudinal Zeeman Effect for an Emission Line

To test ϕ_V we set $\gamma = 0$ and neglect line absorption, anomalous dispersion and both continuum absorption and emission. Then Equations (8)-(12) reduce to

$$\begin{aligned} \frac{dI}{dz} &= \kappa_o \phi_I B_\nu(T_e) \\ \frac{dQ}{dz} &= 0 \\ \frac{dU}{dz} &= 0 \\ \frac{dV}{dz} &= \kappa_o \phi_V B_\nu(T_e) \end{aligned} \tag{20}$$

with

$$\begin{aligned}\phi_I &= \tfrac{1}{2}(H(a, v - v_B) + H(a, v + v_B)) \\ \phi_V &= \tfrac{1}{2}(H(a, v - v_B) - H(a, v + v_B)).\end{aligned} \tag{21}$$

Consider a thin layer between z and $z + dz$ with no radiation incident at z. Suppose that the Zeeman components are completely separated and choose a wavelength v near the *red* σ component. The Stokes parameters emerging at $z + dz$ are approximately

$$\begin{aligned}I(z + dz) &= V(z + dz) = \tfrac{1}{2}\kappa_o H(a, v - v_B) B_\nu(T_e)\, dz \\ Q(z + dz) &= U(z + dz) = 0.\end{aligned} \tag{22}$$

This *positive* value of V implies that the *red* component radiation is *right-hand (clockwise)* circularly polarized (see Table 1), a result in agreement with laboratory observations of the longitudinal Zeeman effect in emission (e.g. Jenkins & White 1976, p. 683 Figure 32C(a)).

The Faraday Effect

To test ϕ'_V we again set $\gamma = 0$ but this time neglect absorption and emission in both the line and the continuum, so that only Faraday rotation terms remain. Then Equations (8)-(12) become

$$\begin{aligned}\frac{dI}{dz} &= 0 \\ \frac{dQ}{dz} &= -\kappa_o \phi'_V U \\ \frac{dU}{dz} &= +\kappa_o \phi'_V Q \\ \frac{dV}{dz} &= 0\end{aligned} \tag{23}$$

with

$$\phi'_V = F(a, v - v_B) - F(a, v + v_B). \tag{24}$$

One can think of the layer dz as a circular retarder with retardance angle

$$\delta_V = \kappa_o \phi'_V dz. \tag{25}$$

Now

$$\delta_V > 0 \quad \text{if} \quad v > v_B \quad \text{or} \quad v < -v_B, \tag{26}$$

i.e. the layer acts as a *left-hand (counterclockwise)* circular retarder in these wavelength ranges (see Jenkins & White p. 687 Figure 32F(c)).

For small dz we have $\cos\delta_V \approx 1$ and $\sin\delta_V \approx \delta_V$, so one can use Equations (23) to relate the incident and the transmitted Stokes vectors $\mathbf{I}(z)$ and $\mathbf{I}(z+dz)$ approximately thus,

$$\mathbf{I}(z+dz) = \mathbf{M}\,\mathbf{I}(z), \tag{27}$$

where $\mathbf{M}$ is the Mueller matrix for the layer,

$$\mathbf{M} = \begin{pmatrix} 1 & 0 & 0 & 0 \\ 0 & \cos\delta_V & -\sin\delta_V & 0 \\ 0 & \sin\delta_V & \cos\delta_V & 0 \\ 0 & 0 & 0 & 1 \end{pmatrix}. \tag{28}$$

This accords with the expression for the Mueller matrix of a left-handed circular retarder in Shurcliff (p.170).

A similar test can be applied to ϕ'_Q and ϕ'_U by considering the Voigt effect in a transverse magnetic field, but the details will be omitted here as the signs of these terms seem well established in the literature.

4. ANALYTIC SOLUTION OF THE TRANSFER EQUATIONS

The long algebraic expressions for the analytic solutions of the Stokes transfer equations, particularly for the case of a Milne-Eddington model atmosphere, frequently appear in the literature (see Landolfi & Landi degl'Innocenti for a clear presentation). Perhaps because of their apparent complexity, these solutions have become enshrined in mystery at least in the unwritten folklore of the subject. The purpose of this section is to expose their essential simplicity.

Consider a plane-parallel model atmosphere. Suppose that the line of sight z axis is at an angle $\cos^{-1}\mu$ to the outward normal. Let τ be the optical depth in the continuum so that $d\tau/\mu = -\kappa_c dz$. In the Milne-Eddington model the source function is a linear function of τ,

$$B_\nu(T_e) = B_0 + B_1\tau, \tag{29}$$

and the opacity ratio $\eta_o = \kappa_o/\kappa_c$, the line damping and the Doppler width are constants. If we assume also that the magnetic vector and the line of sight velocity are a constants, then $\mathbf{\Phi}$ is a constant matrix.

From Equations (8) and (10) we have

$$\mu \frac{d\mathbf{I}}{d\tau} = (\mathbf{1} + \eta_o \mathbf{\Phi})(\mathbf{I} - \mathbf{S}), \tag{30}$$

which has the formal solution for the emergent Stokes vector at $\tau = 0$,

$$\mathbf{I}(0, \mu) = \int_0^\infty e^{-(\mathbf{1}+\eta_o \mathbf{\Phi})\tau/\mu} (\mathbf{1} + \eta_o \mathbf{\Phi}) \, \mathbf{S} \, d\tau/\mu. \tag{31}$$

The integrand can be written so compactly in terms of the matrix exponential because η_o and $\mathbf{\Phi}$ are constants. Integrating by parts we obtain

$$\mathbf{I}(0, \mu) = \mathbf{S}(0) + \int_0^\infty e^{-(\mathbf{1}+\eta_o \mathbf{\Phi})\tau/\mu} \frac{d\mathbf{S}}{d\tau} \, d\tau, \tag{32}$$

where, from Equations (9) and (29),

$$\mathbf{S}(0) = B_0 \, \mathbf{I}_o, \tag{33}$$

and

$$\frac{d\mathbf{S}}{d\tau} = B_1 \, \mathbf{I}_o. \tag{34}$$

Thus the integral in Equation (32) can be evaluated analytically to give

$$\mathbf{I}(0, \mu) = B_0 \, \mathbf{I}_o + \mu B_1 \, (\mathbf{1} + \eta_o \mathbf{\Phi})^{-1} \, \mathbf{I}_o, \tag{35}$$

where the matrix inverse can be obtained by Cramer's rule (note that only the first column of the inverse is needed).

Let

$$\eta_{I,Q,U,V} = \eta_o \, \phi_{I,Q,U,V} \tag{36}$$

and

$$\rho_{Q,U,V} = \eta_o \, \phi'_{Q,U,V}. \tag{37}$$

Then the emergent Stokes parameters are

$$\begin{aligned}
I &= B_0 + \mu B_1 \left[(1 + \eta_I)((1 + \eta_I)^2 + \rho_Q^2 + \rho_U^2 + \rho_V^2) \right] / \Delta \\
Q &= -\mu B_1 \left[(1 + \eta_I)^2 \, \eta_Q + (1 + \eta_I)(\eta_V \rho_U - \eta_U \rho_V) + \rho_Q W \right] / \Delta \\
U &= -\mu B_1 \left[(1 + \eta_I)^2 \, \eta_U + (1 + \eta_I)(\eta_Q \rho_V - \eta_V \rho_Q) + \rho_U W \right] / \Delta \\
V &= -\mu B_1 \left[(1 + \eta_I)^2 \, \eta_V + \rho_V W \right] / \Delta
\end{aligned} \tag{38}$$

where

$$W = \eta_Q \rho_Q + \eta_U \rho_U + \eta_V \rho_V \tag{39}$$

and

$$\Delta = (1+\eta_I)^2[(1+\eta_I)^2 - \eta_Q^2 - \eta_U^2 - \eta_V^2 + \rho_Q^2 + \rho_U^2 + \rho_V^2] - W^2 \tag{40}$$

is the determinant of the matrix $(\mathbf{1} + \eta_o \mathbf{\Phi})$.

A generalization of this analytic solution suitable for modelling strong non-LTE lines such as Mg I *b* in the low chromosphere of the Sun is presented in Lites *et al.* (1987).

5. NON-LTE POLARIZED RADIATIVE TRANSFER

The earlier chapters in this monograph have focused mainly on the development of efficient numerical methods to solve the non-LTE multilevel atom line formation problem in stellar atmospheres, neglecting the effects of polarization. Extension of these methods to include polarization can be a major step. In general polarized line formation involves four coupled transfer equations for the Stokes parameters and often demands attention to the details of subtle quantum mechanical effects. As discussed in Section 3, even in the LTE case, the subject is an algebraic minefield which has to be negotiated with great care when one writes a computer code. In this section we briefly review developments in non-LTE polarization transfer in spectral lines in the context of solar and stellar atmospheres. Thus we omit reference to a large body of polarization transfer literature relevant to the study of planetary atmospheres (see Chandrasekhar 1950; van de Hulst 1980). Specific details of numerical methods are not discussed. Rather the purpose is to provide a broad historical perspective to the chapters that follow.

Much of the impetus in the study of polarized line transfer theory has come from the need to interpret solar polarimetric data. The Huntsville Workshop (Hagyard 1985) provides a good introductory summary of current observational and theoretical activity in this area (see also Stenflo 1978a, 1986 and Rees 1982). The emphasis in this section reflects somewhat the author's interest in solar spectroscopy. However applications in other astrophysical contexts should not be overlooked. For example, recently there have been important advances in the following areas: continuum and line polarization transfer in magnetic white dwarfs (Martin & Wickramasinghe 1979a,b, 1981, 1982; Nagendra & Peraiah 1984, 1985a,b);

cyclotron line formation in accreting magnetized neutron stars (Meszaros 1984; Meszaros & Nagel 1985); maser line formation in molecular clouds (Goldreich & Kylafis 1982; Western & Watson 1983; Deguchi & Watson 1985).

It is interesting to note that solar observations of non-magnetic line polarization played an important role in the early formulation of scattering redistribution theory, a subject central to present-day non-LTE line formation studies. Indeed the roots of the subject can be found "in a pit rather more than 3 metres deep" where Redman (1941) attempted to measure the linear polarization in the Ca I 4227 Å resonance line near the solar limb. The unexpectedly low polarization detected led Zanstra (1941) to revise the classical theory of scattering polarization to include depolarization and frequency redistribution by collisions.

Unno (1956), Rachkovsky (1962a,b) and Beckers' (1969a,b) formulations of the LTE transfer equations for Zeeman split spectral lines were based on the classical oscillator picture of absorption and emission. Rachkovsky (1963) proposed the first non-LTE line formation model for a normal Zeeman triplet assuming complete redistribution in frequency during scattering. With occasional exceptions, such as Charvin's (1965) quantum mechanical analysis of scattering polarization in coronal emission lines, the classical model for spectral line polarization prevailed during the 1960's (see review by Stenflo 1971).

Since then there has been significant progress towards a quantum electrodynamical theory starting with Landi degl'Innocenti & Landi degl'Innocenti's (1972) rederivation of the Unno-Rachkovsky LTE equations. Some of the key papers relevant to the quantum theory of scattering redistribution and polarization transfer in resonance lines are: House (1970a,b,1971), Lamb & ter Haar (1971), Omont *et al.* (1972, 1973), Stenflo (1976,1978b), Ballagh & Cooper (1977), Cooper *et al.* (1982) and Streater *et al.* (1987). Important contributions to non-LTE transfer under the Zeeman effect were made by Sidlichovsky (1974), Dolginov & Pavlov (1974), House & Steinitz (1975), Landi degl'Innocenti & Landi degl'Innocenti (1974), Landi degl'Innocenti *et al.* (1976), Bommier & Sahal-Brechot (1978), and Mathys (1982). Recently Landi degl'Innocenti (1983a,b,1984) has embarked on a grand synthesis of quantum line formation theory with a view to encompassing and extending all previous work. Here the algebraic minefield reaches formidable proportions and only deeply committed transfer theorists dare enter!

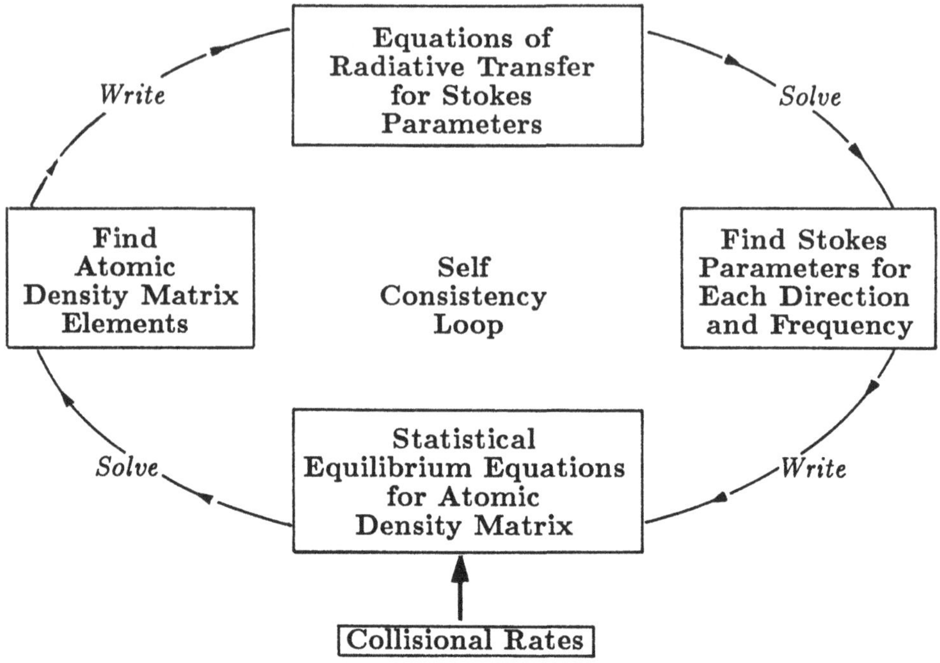

Fig. 3. Flow chart of numerical solution of the general non-LTE line formation problem including polarization (adapted from Landi degl'Innocenti 1983b).

The central numerical task in non-LTE formation of optically thick spectral lines is to obtain a self-consistent solution of the equations of transfer and the equations of statistical equilibrium. In unpolarized transfer problems statistical equilibrium refers to the steady state population densities of the energy levels of the atomic model. In polarized transfer one is concerned with the populations of the Zeeman sublevels, so that, in principle, even a simple two-level atom may require a multi-sublevel treatment. A further complication may arise if the magnetic splitting is so small so that there is significant overlap or interference between the broadened sublevels. Then one must adopt a density matrix representation of the atom, associating diagonal elements with sublevel populations and off-diagonal elements with the interferences. Landi degl'Innocenti (1983b) summarizes the overall problem graphically in the self-consistency loop shown in Figure 3. He rates the numerical closure of the loop as "one of the most challenging tasks in the future of solar polarimetry ".

The full non-LTE 'catastrophe' pictured in Figure 3 is indeed a daunting numerical problem. However, under appropriate physical conditions it does reduce to a reasonably tractable form. Consider first the case where magnetic splitting is large enough so that the sublevel interferences can be neglected. If the collisional transition rates between the sublevels are high then the sublevel populations within any given atomic level will be essentially equal. Then one need only treat the level populations in the equations of statistical equilibrium. This is called the *complete level depolarization* or *dealignment* approximation. Data on inter-sublevel collision rates are available for only a few atomic configurations, but crude estimates (Lamb 1970) indicate that this is a viable approximation for many spectral lines formed in the photosphere and low chromosphere of the Sun.

A numerical solution of the non-LTE polarization transfer problem assuming complete dealignment was obtained by Rees (1969) for Rachkovsky's (1963) two level atomic model, neglecting magneto-optical effects. The major conclusion of that calculation was that the level populations are very close to those obtained by ignoring magnetic splitting. This implies that one could safely adopt a *field-free* approximation, using populations calculated with an existing non-magnetic non-LTE code and then simply apply a formal numerical integration to the Stokes transfer equations accounting for the Zeeman effect. Later investigations of the two level problem including magneto-optics by Domke (1970), Domke & Staude (1973a,b) and Landi degl'Innocenti (1978) confirmed this conclusion. All these papers assumed a plane-parallel model atmosphere. Stenholm & Stenflo (1978) studied a multi-dimensional transfer problem for a two-level Fe I atom in a cylindrical magnetic flux tube and again found that the field-free approximation gave accurate results. The *tour de force* in this context is the work by Auer *et al.* (1977) who generalized the complete linearization method (Auer & Heasley 1976) to solve both the Stokes transfer and the statistical equilibrium equations consistently for a multi-level model of the Ca II ion. They found that the field-free approximation leads typically to errors of less than 10% in the intensity profiles and less than 3% in the rates of linear and circular polarization.

Another physical situation which is amenable to relatively easy numerical solution is the limiting case where there is no magnetic splitting, but one takes account of the interferences between the sublevels. Polarization then arises from the scattering of the *anisotropic* radiation field. An extensive survey of non-magnetic linear polarization in the solar spectrum covering the wavelength range

3165 Å to 9950 Å has been published by Stenflo *et al.* (1983a,b). They find evidence of a wide variety of quantum interference effects within and between atomic levels. Dumont *et al.* (1973) did the first detailed transfer analysis of this phenomenon to try to explain the profiles of linear polarization in the solar Ca I 4227 Å resonance line. They introduced non-LTE effects in the line core by means of empirically determined line source functions and assumed coherent scattering for the line wings. Two-level atom model calculations relevant to the polarization in resonance line Doppler cores were carried out by Stenflo & Stenholm (1976) and Rees (1978) assuming complete redistribution in frequency during scattering. Dumont *et al.* (1977) introduced partial redistribution effects for the case of purely Doppler redistribution in the line core (a generalization of Hummer's (1962) type I redistribution). Auer *et al.* (1980) studied quantum interference between the upper levels of the Ca II H and K lines, solving the transfer in the line wings assuming coherent scattering. Partial redistribution calculations coupling the wings and the line core have been done by Rees & Saliba (1982), McKenna (1984a,b,1985), Saliba (1985) and Faurobert (1987a,b).

The most difficult numerical problem is the intermediate case where the magnetic field is weak and interferences between sublevels are important. This is the regime of the Hanle effect. Stenflo (1978b) proposed a method for incorporating the Hanle effect but as yet no solutions to his formulation of the problem have been attempted. House & Cohen (1969) obtained Monte-Carlo solutions for a heuristic model of Hanle effect scattering and transfer for a two-level atom in a prominence model. But the the most ambitious effort so far to 'close the loop' in Figure 3 is the multilevel study by Landi degl'Innocenti *et al.* (1986) of the Hanle effect in prominences.

6. SUMMARY OF LATER CHAPTERS

Generally the absorption matrix **K** and the source vector **S** are functions of position z, specified either (a) *a priori* (for lines formed in LTE or under conditions where the field-free approximation is reasonable) or (b) as part of an iteration procedure (in the non-LTE self-consistency loop). In each case one must integrate the vector transfer equation (8) numerically.

For problems of type (a) the most commonly used method has been the Runge-Kutta algorithm (Beckers 1969a; Staude 1970; Katz 1972; Wittmann

1974; Landi degl'Innocenti 1976; van Ballegooijen 1985). This can give high accuracy, but usually at the expense of excessive computational time, especially for strong lines (typically with line to continuum opacity ratio $\eta_o > 10^3$). Hardorp *et al.* (1976) achieved about a tenfold increase in speed over the Runge-Kutta method using Gohring's (1971) iteration scheme which involves diagonalizing the absorption matrix and treating the extra terms that arise in the transformed equations as perturbations. However, Martin & Wickramasinghe (1979b) found that in certain cases the convergence of this perturbation method is very slow and sometimes even gives incorrect results, though they were not able to discover why. To avoid the shortcomings of both methods they developed an efficient integration scheme by dividing the atmosphere into layers along a given line of sight and finding an approximate analytic solution to the transfer equations in each layer.

None of the above methods would appear to be easily adapted to tackle problems of type (b), i.e. coupled transfer and statistical equilibrium equations. In the chapters by Landi degl'Innocenti, van Ballegooijen and Rees and Murphy several alternative procedures are discussed which address this issue.

A new method applicable to both types of problem is described by Landi degl'Innocenti (Chapter 11) who shows that the formal solution of Equation (8) can be written as

$$\mathbf{I}(z) = \int_{z_o}^{z} \mathbf{O}(z,z')\,\mathbf{K}(z')\,\mathbf{S}(z')\,dz' + \mathbf{O}(z,z_o)\,\mathbf{I}(z_o), \tag{41}$$

where $\mathbf{O}(z,z')$ is a matrix attenuation operator. Given $\mathbf{O}(z,z')$ the integral in Equation (41) can be evaluated by standard quadrature. Landi degl'Innocenti's main contribution here is the development of an efficient way of calculating the attenuation operator.

The integration path is divided into subintervals and within each subinterval $\mathbf{K}$ is approximated by a constant matrix (e.g. its value at the midpoint). For constant $\mathbf{K}_i$ in the subinterval (z_i, z_{i-1}), say, we have

$$\mathbf{O}(z_i, z_{i-1}) = e^{-(z_i - z_{i-1})\,\mathbf{K}_i}. \tag{42}$$

The matrix exponential is usually defined as an infinite series expansion in powers of $\mathbf{K}$. However by exploiting some beautiful algebraic properties of the absorption matrix one may express the matrix exponential as a sum of only *three* matrices (Landi degl'Inocenti & Landi degl'Innocenti 1985). Landi degl'Innocenti outlines the principles of an iteration scheme for solving the full non-LTE problem (Figure 3) with this formal integration of the transfer equation as a basis.

Van Ballegooijen's paper (Chapter 12) represents a breakthrough in the numerical analysis of non-LTE line transfer in the presence of a strong magnetic field. He adopts an equivalent-two-level-atom approach to the multilevel problem, generalizing the integral equation method to include the Zeeman effect. His basic assumptions are complete level depolarization, no interference between Zeeman sublevels, complete redistribution in frequency on scattering and a plane-parallel atmosphere.

In this model it is worth noting that, despite the fact that the transfer equation (8) is in *vector* form, one need only solve for a *scalar* line source function S_L for each transition. For brevity we consider here a strictly two-level atom so that

$$(1+\epsilon)S_L = \bar{J} + \epsilon B_\nu(T_e), \tag{43}$$

where ϵ is the ratio of collisional and spontaneous de-excitation rates from the upper to the lower level of the transition. $\bar{J}$ is a generalized mean integrated intensity with contributions from all Stokes parameters at all wavelengths and angles. As in unpolarized non-LTE transfer, one can relate $\bar{J}$ and S_L by a generalized *scalar* lambda operator thus,

$$\bar{J} = \Lambda_{\mathbf{B}}\, S_L, \tag{44}$$

where we use the subscript $\mathbf{B}$ to signify that the operator depends on the magnetic vector. From Equations (43) and (44) we obtain an integral equation for S_L,

$$[(1+\epsilon) - \Lambda_{\mathbf{B}}]\, S_L = \epsilon B_\nu(T_e). \tag{45}$$

To solve this integral equation one must devise a numerical representation of the lambda operator on an appropriate spatial grid $\{z_i\}$. Van Ballegooijen does this by recasting the transfer equation (8) in terms of the 2×2 *complex* coherency matrix (Born & Wolf pp. 545 and 554),

$$\frac{1}{2}\begin{pmatrix} I+Q & U+iV \\ U-iV & I-Q \end{pmatrix}. \tag{46}$$

He then obtains a formal solution for the coherency matrix in a manner somewhat similar to Landi degl'Innocenti's method for the Stokes vector. Finally he constructs the lambda operator using linear and quadratic approximations to S_L between the grid points. Equation (45) is thereby replaced by a system of linear equations for the unknown source function values $S_L(z_i)$ at the grid points and this is solved by a standard matrix method.

The importance of van Ballegooijen's method lies partly in the fact that it offers the opportunity of detailed assessment of the range of applicability of the field-free method. To the author's knowledge there is *no* computer code in current usage that specifically calculates the effect of Zeeman splitting on the non-LTE multilevel atom populations in a semi-infinite model atmosphere. Despite the convergence difficulties which can plague the equivalent-two-level-atom approach (e.g. Mihalas 1978 p. 380) the method should be very productive.

Both Landi degl'Innocenti and van Ballegooijen's methods can be classed as *one way* integration schemes in the sense that at each point on a given ray path the Stokes vectors (or coherency matrices) for opposite directions along the ray are computed *separately*. Feautrier (1964) converted the scalar first-order transfer equation for the intensity to a second-order equation for the *average* of inward and outward directed intensities at each point of a ray path. This second-order system, which is solved by finite-differences, has been the foundation of a large number of computer codes with a reputation for efficiency, accuracy and robustness (see review by Mihalas 1985). The first application of the Feautrier method in the present context was by Auer *et al.* (1977), who recognized that the absorption matrix and the source vector possess certain symmetries which permit a second-order conversion. Auer *et al.* neglected magneto-optical effects. These were included in a Feautrier-type code by Landman & Finn (1979) but they gave no details of their method.

Rees and Murphy (Chapter 10) show how to apply the Feautrier method in this general case. The second-order conversion is not so obvious because, as noted in Section 3, the magneto-optical terms are asymmetric functions of wavelength. The method leads to a familiar block tri-diagonal system of linear equations for the *average* Stokes vector. This system is solved recursively to determine the emergent Stokes parameters and their associated contribution functions. These contribution functions are useful in determining the depth of formation of

Zeeman split lines.

Also a new method of one-way formal integration of Equation (8) is discussed. Here advantage is taken of the fact that the diagonal elements of the absorption matrix are all equal (see Equation (11)). The vector transfer equation is solved using a numerical representation of the lambda operator defined on the optical depth scale related to this diagonal element. The method is not dependent on any symmetry properties of **K** and **S**. Again Stokes contribution functions are obtained in a straightforward manner.

Finally Rees and Murphy consider what could be regarded as the simplest two-level atom non-LTE polarization transfer problem. This is the formation of non-magnetic resonance scattering polarization. In a plane-parallel atmosphere symmetries in the radiation field imply that $U = V = 0$ so one must solve two coupled equations for I and Q, i.e. the radiation exhibits only linear polarization. This problem offers perhaps the easiest conceptual transition between unpolarized and polarized non-LTE transfer.

A hierarchy of approximations for the scattering redistribution matrix (the generalization of the usual redistribution function) is described, retaining the angular dependence in the scattering phase matrix essential to the production of polarization. The polarization is treated as a perturbation and the coupled non-LTE transfer equations are solved by a Scharmer(1981,1983)-type operator perturbation method.

7. CONCLUSION

In this review of polarized radiative transfer in spectral lines I have frequently alluded to the algebraic complexity of the theory and have deliberately tried to spare the reader the grimmer details. Transfer theorists venturing into this area must learn the language of polarization and be prepared to assimilate a host of new physical concepts not encountered in standard non-LTE theory. Writing computer codes to solve the transfer equations for the Stokes parameters often requires dogged detective work hunting sign errors.

The subject is challenging, the physical processes fascinating. But theoretical interests aside, the problem of polarized transfer, especially in magnetically split spectral lines, is assuming increasing observational importance. For example there are several solar instrument projects in train to develop polarimeters which will provide profiles of the Stokes parameters at high spectral resolution (see review by Harvey 1985). Parallel development of polarization transfer diagnostic methods is imperative to provide rapid, reliable data analysis packages for inferring the vector magnetic field distribution in the solar atmosphere.

One point seems clear. The scale of the numerical problem in non-LTE polarized line transfer (see Figure 3) strongly suggests that one try a perturbation technique. Indeed the form of Equation (45) contains a hint of how one might proceed. Introducing another lambda operator $\tilde{\Lambda}$ we could solve Equation (45) iteratively thus,

$$[(1+\epsilon) - \tilde{\Lambda}]\, S_L^{(n+1)} = \epsilon B_\nu(T_e) + (\Lambda_{\mathbf{B}} - \tilde{\Lambda})\, S_L^{(n)}, \tag{47}$$

where n and $n+1$ indicate successive iterates. Experience with the field-free method suggests that $\tilde{\Lambda}$ could be the lambda operator neglecting Zeeman splitting, or better still, one of the approximate operators described elsewhere in this monograph. The term involving $\Lambda_{\mathbf{B}}$ on the RHS of Equation (47) would require rapid and accurate formal integration of the vector transfer equation (8) (or its 2×2 coherency matrix equivalent) for many wavelengths and directions.

In principle the steps that would lead ultimately to a magnetic generalization of the Scharmer & Carlsson (1985) multilevel transfer code do not seem insurmountable, though they surely present an algebraic 'Annapurna' (Herzog 1986).

ACKNOWLEDGMENTS

The author is indebted to Andy Skumanich for constructive criticism and help in detecting sign errors, and apologizes for any such errors that might remain in the manuscript! He thanks Bruce Lites and Graham Murphy for helpful suggestions. The patience of Wolfgang Kalkofen during the long gestation of this chapter has been very much appreciated. This work was supported by ARGS Grant B8415543 and by funds from the High Altitude Observatory Visiting Scientist Program.

REFERENCES

Auer, L.H. & Heasley, J.N. 1976, *Astrophys. J.*, **205**, 165.

Auer, L.H., Heasley, J.N. & House, L.L. 1977, *Astrophys. J.*, **216**, 531.

Auer, L.H., Rees, D.E. & Stenflo, J.O. 1980, *Astron. Astrophys.*, **88**, 302.

Ballagh, R.J. & Cooper, J. 1977, *Astrophys. J.*, **213**, 479.

Beckers, J.M. 1969a, *Solar Phys.*, **9**, 372.

__________ 1969b, *Solar Phys.*, **10**, 262.

Bommier, V. & Sahal-Brechot, S. 1978, *Astron. Astrophys.*, **69**, 57.

Born, M. & Wolf, E. 1959, *Principles of Optics*, Pergamon Press, Oxford.

Charvin, P. 1965, *Annales d'Astrophys.*, **28**, 877.

Chandrasekhar, S. 1950, *Radiative Transfer*, Oxford Uni. Press.

Clarke, D. 1974, *Applied Optics*, **13**, 222.

Clarke, D. & Grainger, J.F. 1971, *Polarized Light and Optical Measurement*, Pergamon Press, Oxford.

Cooper, J., Ballagh, R.J., Burnett, K. & Hummer, D.G. 1982, *Astrophys. J.*, **260**, 299.

Deguchi, S. & Watson, L.R. 1985, *Astrophys. J.*, **289**, 621.

Dolginov, A.Z. & Pavlov, G.G. 1974, *Soviet Astron.*, **17**, 485.

Domke, H. 1970, *Astrofizika*, **5**, 525.

Domke, H. & Staude, J. 1973a, *Solar Phys.*, **31**, 279.

__________ 1973b, *Solar Phys.*, **31**, 291.

Dumont, S., Omont, A. & Pecker, J.-C. 1973, *Solar Phys.*, **28**, 271.

Dumont, S., Omont, A., Pecker, J.-C. & Rees, D.E. 1977, *Astron. Astrophys.*, **54**, 675.

Fano, U. 1957, *Rev. Mod. Phys.*, **29**, 74.

Faurobert, M. 1987a, *Astron. Astrophys.*, in press.

__________ 1987b, *Astron. Astrophys.*, in press.

Feautrier, P. 1964, *C. R. Acad. Sci. Paris*, **258**, 3189.

Gohring, R. 1971, In *Solar Magnetic Fields, IAU Symp.*, **43**, Ed. R. Howard, p. 162 Dordrecht: Reidel.

Goldreich, P. & Kylafis, N.D. 1982, *Astrophys. J.*, **253**, 606.

Hagyard, M.J.(Editor) 1985, *Measurements of Solar Vector Magnetic Fields* (NASA-CP2374).

Hardorp, J., Shore, S.N. & Wittmann, A. 1976, In *Physics of A_p Stars*, p. 419, Universitssternwarte Wien, Vienna.

Harvey, J.W. 1985, In *Measurements of Solar Vector Magnetic Fields* (NASA-CP2374), Ed. M.J. Hagyard, p. 109.

Herzog, M. 1986, *Annapurna: Conquest of the First 8,000-metre Peak*, Triad/Paladin, London.

House, L.L. 1970a, *J. Quant. Spectrosc. Rad. Transfer*, **10**, 909.

__________ 1970b, *J. Quant. Spectrosc. Rad. Transfer*, **10**, 1171.

__________ 1971, *J. Quant. Spectrosc. Rad. Transfer*, **11**, 367.

House, L.L. & Cohen, L.C. 1969, *Astrophys. J.*, **157**, 261.

House, L.L. & Steinitz, R. 1975, *Astrophys. J.*, **195**, 235.

Hummer, D.G. 1962, *Monthly Notices Roy. Astron. Soc.*, **125**,21.

Humlicek, J. 1982, *J. Quant. Spectrosc. Rad. Transfer*, **27**, 437.

Jenkins, F.A. & White, H.E. 1976, *Fundamentals of Optics* (4th Edition), McGraw-Hill, New York.

Katz, J.M. 1972, *Solar Phys.*, **24**, 28.

Lamb, F.K. 1970, *Solar Phys.*, **12**, 186.

Lamb, F.K. & ter Haar, D. 1971, *Phys. Repts.*, **2C**, 253.

Landi degl'Innocenti, E. 1976, *Astron. Astrophys. Suppl.*, **25**, 379.

__________ 1978, *Astron. Astrophys.*, **66**, 119.

__________ 1983a, *Solar Phys.*, **85**, 3.

__________ 1983b, *Solar Phys.*, **85**, 33.

__________ 1984, *Solar Phys.*, **91**, 1.

Landi degl'Innocenti, E., Bommier, V. & Sahal-Brechot, S. 1986, *Astron. Astrophys.*, in press.

Landi degl'Innocenti, E. & Landi degl'Innocenti, M. 1972, *Solar Phys*, **27**, 319.

__________ 1974, *Il Nuovo Cimento*, **27B**, 134.

__________ 1985, *Solar Phys.*, **97**, 239.

Landi degl'Innocenti, M., Landolfi, M. & Landi degl'Innocenti, E. 1976, *Il Nuovo Cimento*, **35B**, 117.

Landman, D.A. and Finn, G.D. 1979, *Solar Phys.*, **63**, 221.

Landolfi, M. & Landi degl'Innocenti, E. 1982, *Solar Phys.*, **78**, 355.

Lites, B.W. & Skumanich, A. 1985, NCAR Technical Note NCAR/TN-248, Boulder, CO.

Lites, B.W., Skumanich, A., Rees, D.E. & Murphy, G.A. 1987, *Astrophys. J.*, submitted.

Martin, B. & Wickramasinghe, D.T. 1979a, *Proc. Astron. Soc. Australia*, **3**, 351.

__________ 1979b *Monthly Notices Roy. Astron. Soc.*, **189**, 883.

__________ 1981, *Monthly Notices Roy. Astron.Soc.*, **196**, 23.

__________ 1982, *Monthly Notices Roy. Astron.Soc.*, **200**, 993.

Mathys, G. 1982, *Astron. Astrophys.*, **108**, 213.

McKenna, S.J. 1984a, *Astrophys. Space Sci.*, **106**, 283.

__________ 1984b, *Astrophys. Space Sci.*, **107**, 61.

__________ 1985, *Astrophys. Space Sci.*, **108**, 31.

Meszaros, P. 1984, *Space Science Rev.*, **38**,325.

Meszaros, P. & Nagel, W. 1985, *Astrophys. J.*, **298**, 147.

Mihalas, D. 1978, *Stellar Atmospheres* (2nd Edition), Freeman, San Francisco.

__________ 1985, *J. Comput. Phys.*, **57**, 1.

Nagendra, K.N. & Peraiah, A. 1984, *Astrophys. Space Sci.*, **104**, 61.

__________ 1985a, *Monthly Notices Roy. Astron.Soc.*, **214**, 203.

__________ 1985b, *Astrophys. Space Sci.*, **117**, 121.

Omont, A., Smith, E.W., & Cooper, J. 1972, *Astrophys. J.*, **175**, 185.

__________ 1973, *Astrophys. J.*, **182**, 283.

Rachkovsky, D.N. 1962a, *Izv. Krymsk. Astrofiz. Obs.*, **27**, 148.

__________ 1962b, *Izv. Krymsk. Astrofiz. Obs.*, **28**, 259.

__________ 1963, *Izv. Krymsk. Astrofiz. Obs.*, **30**, 267.

__________ 1967, *Izv. Krymsk. Astrofiz. Obs.*, **37**, 56.

Redman, R.O. 1941, *Monthly Notices Roy. Astron. Soc.*, **101**, 266.

Rees, D.E. 1969, *Solar Phys.*, **10**, 268.

__________ 1978, *Publ. Astron. Soc. Japan*, **30**, 405.

__________ 1982, *Proc. Astron. Soc. Australia*, **4**, 335.

Rees, D.E. & Saliba, G.J. 1982, *Astron. Astrophys.*, **115**, 1.

Robson, B.A. 1974, *The Theory of Polarization Phenomena*, Clarendon Press, Oxford.

Saliba, G.J. 1985, *Solar Phys.*, **98**, 1.

Scharmer, G.B. 1981, *Astrophys. J.*, **249**, 720.

__________ 1983, *Astron. Astrophys.*, **117**, 83.

Scharmer, G.B. and Carlsson, M. 1985, *J. Comput. Phys.*, **59**, 56.

Shurcliff, W.A. 1962, *Polarized Light*, Harvard Uni. Press, Cambridge, Mass.

Sidlichovsky, M. 1974, *Bull. Astron. Inst. Czech.*, **25**, 198.

Skumanich, A. 1987, *Astrophys. J.*, submitted.

Staude, J. 1970, *Solar Phys.*,**15**, 102.

__________ 1971a, *Solar Phys.*, **18**, 21.

__________ 1971b, *Solar Phys.*, **18**, 24.

Stenflo, J.O. 1971, In *Solar Magnetic Fields, IAU Symp.*, **43**, Ed. R. Howard, p. 101 Dordrecht: Reidel.

__________ 1976, *Astron. Astrophys.*, **46**, 61.

__________ 1978a, *Rep. Prog. Phys.*, **41**, 865.

__________ 1978b, *Astron. Astrophys.*, **66**, 241.

__________ 1986, *Solar Phys.*, **100**, 189.

Stenflo, J.O. & Stenholm, L.G. 1976, *Astron. Astrophys.*, **46**, 69.

Stenflo, J.O., Twerenbold, D. & Harvey, J.W. 1983a, *Astron. Astrophys. Suppl.*, **52**, 161.

Stenflo, J.O., Twerenbold, D., Harvey, J.W. & Brault, J.W. 1983b, *Astron. Astrophys. Suppl.*, **54**, 505.

Stenholm, L.G. & Stenflo, J.O. 1978, *Astron. Astrophys.*, **67**, 33.

Streater, A., Cooper, J. & Rees, D.E. 1987, in preparation.

Unno, W. 1956, *Publ. Astron. Soc. Japan*, **8**, 108.

van Ballegooijen, A.A. 1985, In *Measurements of Solar Vector Magnetic Fields* (NASA-CP2374), Ed. M.J. Hagyard, p. 322.

van de Hulst, H.C. 1980, *Multiple Light Scattering* Vol.2, Academic Press, New York.

Western, L.R. & Watson, W.D. 1983, *Astrophys. J.*, **268**, 849.

Wittmann, A. 1974, *Solar Phys.*, **35**, 11.

Zanstra, H. 1941, *Monthly Notices Roy. Astron. Soc.*, **101**, 273.

NON-LTE POLARIZED RADIATIVE TRANSFER IN SPECTRAL LINES

David E. Rees and Graham A. Murphy

Department of Applied Mathematics
University of Sydney, NSW, 2006, Australia
and
High Altitude Observatory
National Center for Atmospheric Research [1]
Boulder, Colorado, USA.

ABSTRACT: Numerical methods applicable to two kinds of non-LTE spectral line formation with polarization are discussed. First we describe two techniques for formal integration of the Stokes vector transfer equation of a Zeeman split line: one is a generalization of the Feautrier (1964) finite-difference method; the other involves the lambda operator associated with the diagonal element of the absorption matrix. Each method allows efficient computation of the Stokes parameter contribution functions. Then we consider the problem of resonance line polarization in the absence of magnetic fields. An iterative solution method is outlined. This is an adaptation of Scharmer's (1981, 1983) approximate lambda operator method, the polarization being treated as a perturbation.

1. INTRODUCTION

Substantial progress has been made in the quantum mechanical formulation of the non-LTE theory of line formation in the presence of a magnetic field, so that now one can account for many polarization effects (for a review see Chapter 9, henceforth referred to as GI, i.e. "Gentle Introduction"). Also there have been significant advances in high precision instrumentation for polarimetry in solar and stellar spectra (Harvey 1985, Borra *et al.* 1982, Coyne & McLean 1982). These recent developments have stimulated a growing interest in numerical solutions of the equations of polarized radiative transfer and statistical equilibrium.

The present chapter is devoted to numerical methods applied to two types of non-LTE polarized radiative transfer problem.

(*i*) Firstly we consider the problem of computing the Stokes parameters I, Q, U, V

[1] The National Center for Atmospheric Research is sponsored by the National Science Foundation.

of polarized light (see definitions in GI Section 2) in a non-LTE line split by a strong magnetic field. We shall call this the *Zeeman line transfer* problem. Strictly speaking, to solve this problem one should close the self-consistency loop which couples the transfer equations and statistical equilibrium equations(see Figure 3 of GI) as done by Auer *et al.* (1977) (henceforth referred to as AHH) and van Ballegooijen (Chapter 12). Under conditions appropriate to line formation in the solar photosphere and low chromosphere the atomic level populations are very well approximated by the values computed when Zeeman splitting and polarization are neglected (e.g. Rees 1969 and AHH). This *field-free* approximation allows one to specify *a priori* the line opacities and source functions of the various transitions in the atomic model, thereby reducing the computation of the Stokes parameters in each transition to formal integration of the transfer equations. This is the approach adopted by the High Altitude Observatory - University of Sydney group developing data analysis software for the Advanced Stokes Polarimeter under construction in Boulder. We use Carlsson's (1986) non-LTE code to generate the approximate line opacities and source functions. Details of this procedure and applications to solar polarimetric data analysis are given in Rees *et al.* (1987a) (henceforth referred to as RMD) and Lites *et al.* (1987a, b).

Two formal integration methods are outlined below. One, which extends the work of AHH by including magneto-optical (birefringence) effects, is a generalization of Feautrier's (1964) technique. The first-order differential equation of transfer for the Stokes vector $\mathbf{I} = (I, Q, U, V)^{\dagger}$, († means transpose), is converted to a second-order equation for the *average* of the inward and outward directed Stokes vectors at each point along the line-of-sight. The second-order equation is approximated by finite-differences leading to a block tri-diagonal system which is solved recursively to derive the emergent Stokes parameters. This recursive solution can also be used for efficient evaluation of the line-of-sight variations of the contribution functions for the Stokes parameters.

Staude (1969) devised a Neumann series method for formal solution of the Stokes vector transfer equation taking advantage of the fact that the diagonal elements of the absorption matrix are all the same. This solution involves the lambda operator defined on the optical depth scale associated with this diagonal element. Because it converged slowly Staude (1970) later abandoned this method in favour of Runge-Kutta integration. We describe a new method which is an adaptation of Staude's original idea, but which permits a direct, *non-iterative*, evaluation of the emergent Stokes parameters and their contribution functions.

(*ii*) Secondly we focus on the problem of non-LTE formation of resonance line polarization in the absence of magnetic fields. Here polarization arises from scattering of the *anisotropic* radiation field. The discussion is limited to a two-level atom in a plane-parallel atmosphere in which the symmetries in the radiation field imply that $U = V = 0$. The resultant coupled transfer equations for I and Q represent perhaps the simplest extension of non-LTE transfer theory to allow for polarization. As $|Q/I| \ll 1$ we treat the linear polarization Q as a perturbation. The coupled transfer equations are then solved by Scharmer's (1981, 1983) approximate operator technique.

The rest of the chapter is structured as follows. In Section 2.1 the theory of Zeeman line transfer is summarized. For simplicity we consider a normal Zeeman triplet; van Ballegooijen treats the arbitrary multiplet case. We discuss the generalized Feautrier method in Section 2.2, and the diagonal element lambda operator method in Section 2.3.

In Section 3.1 a simple model for the formation of resonance line polarization is presented along with a hierarchy of approximations for the scattering redistribution matrix. The generalization of the Scharmer operator perturbation method is detailed in Section 3.2 for the case of complete redistribution (CRD) following the treatment by Scharmer and Nordlund (1982). The notation in Sections 3.1 and 3.2 differs slightly from previous Sections. It is chosen to be more in accord with that used in unpolarized non-LTE analyses.

In the conclusion (Section 4) we summarize the main points and speculate on the possibility of developing an operator perturbation method to tackle the entire range of line polarization phenomena.

2.1 ZEEMAN LINE TRANSFER – THEORY

The transfer equation for the Stokes vector is

$$\frac{d}{dz}\mathbf{I} = -\mathbf{K}\,\mathbf{I} + \mathbf{j} = -\mathbf{K}\,(\mathbf{I} - \mathbf{S})\,, \tag{1}$$

where z is the position along the line-of-sight towards the observer in the right-handed reference frame xyz illustrated in GI Figure 2. For a Zeeman-split line the total(line plus continuum) absorption matrix $\mathbf{K}$, the total emission vector $\mathbf{j}$ and the total source vector $\mathbf{S}$ depend on the magnetic field vector $\mathbf{B}$ ($\mathbf{B}$ has magnitude B, inclination γ to the z-axis and azimuth χ relative to the x-axis).

The formal solution of Equation (1) can be obtained using the Volterra matrix calculus (Gantmacher 1959). Details of this approach can be found in Section 2 of Chapter 11 by Landi degl'Innocenti (henceforth referred to as LD). Stokes vectors at z and z_o are related by the equation

$$\mathbf{I}(z) = \int_{z_o}^{z} \mathbf{O}(z,z')\,\mathbf{j}(z')\,dz' + \mathbf{O}(z,z_o)\,\mathbf{I}(z_o)\,, \tag{2}$$

where $\mathbf{O}(z,z')$ is a matrix attenuation or evolution operator. Note that our z is equivalent to s in LD.

When the absorption matrix $\mathbf{K}$ is *constant* the evolution operator is the matrix exponential,

$$\mathbf{O}(z,z') = e^{-\mathbf{K}|z'-z|}\,. \tag{3}$$

When $\mathbf{K}$ is not constant LD shows how to approximate the operator in terms of matrix exponentials associated with subintervals of the integration path. LD also provides an elegant method for evaluating the matrix exponential. In Sections 2.2 and 2.3 we present methods which are algebraically less sophisticated but are nevertheless computationally efficient and quite easy to code. In particular we are interested in evaluating the Stokes vector $\mathbf{I}(0)$ emerging at the surface of a semi-infinite atmosphere. For convenience we choose the origin $z = 0$ so that it coincides with the surface. Then

$$\mathbf{I}(0) = \int_{-\infty}^{0} \tilde{\mathbf{C}}(z')\,dz', \tag{4}$$

where

$$\begin{aligned}\tilde{\mathbf{C}}(z) &= (C_I(z), C_Q(z), C_U(z), C_V(z))^{\dagger} \\ &= \mathbf{O}(0,z)\,\mathbf{j}(z)\end{aligned} \tag{5}$$

is the contribution vector whose elements are the contribution functions of the Stokes parameters.

In scalar radiative transfer the intensity contribution function C_I is frequently used to estimate the height of formation of the line. The polarization contribution functions $C_{Q,U,V}$ are important for generalized height of formation analyses, particularly where one wishes to infer magnetic field gradients from simultaneous observations of lines with different strengths. Their interpretation is difficult because, unlike C_I, they can change sign along the line-of-sight (see van Ballegooijen 1985). More work needs to be done to clarify their meaning.

The physical model adopted here for non-LTE line formation is the same as van Ballegooijen's (Chapter 12):
(*i*) the magnetic field is strong enough that there are no quantum interferences between Zeeman sublevels;
(*ii*) collision rates are high so atomic polarization can be neglected (i.e. within each atomic level the Zeeman sublevel populations are equal);
(*iii*) there is complete redistribution in frequency, angle and polarization on scattering;
(*iv*) stimulated emission is treated as negative absorption.

Then, just as in standard non-LTE transfer in the absence of the Zeeman effect, one can define a line centre opacity κ_o (for zero damping and corrected for stimulated emission) and a scalar line source function S_L, each of which depend on the *total* populations of the upper and lower levels of the particular line transition of interest. In addition we neglect polarization in the background continuum and characterize continuum processes by an opacity κ_c and source function B_ν, the Planck function at the local electron temperature.

With these approximations one can write

$$\mathbf{K} = \kappa_c \mathbf{1} + \kappa_o \mathbf{\Phi}\,, \tag{6}$$

and

$$\mathbf{j} = \kappa_c B_\nu \mathbf{I}_o + \kappa_o S_L \mathbf{\Phi} \mathbf{I}_o\,, \tag{7}$$

where $\mathbf{1}$ is the unit 4×4 matrix, $\mathbf{I}_o = (1,0,0,0)^\dagger$ and $\mathbf{\Phi}$ is the line absorption matrix. The form of this matrix is discussed in Section 3 of GI, but for convenient reference the main points will be repeated here. We have

$$\mathbf{\Phi} = \begin{pmatrix} \phi_I & \phi_Q & \phi_U & \phi_V \\ \phi_Q & \phi_I & \phi'_V & -\phi'_U \\ \phi_U & -\phi'_V & \phi_I & \phi'_Q \\ \phi_V & \phi'_U & -\phi'_Q & \phi_I \end{pmatrix}, \tag{8}$$

where

$$\begin{aligned}
\phi_I &= \tfrac{1}{2}\phi_p \sin^2\gamma + \tfrac{1}{4}(\phi_b + \phi_r)(1 + \cos^2\gamma) \\
\phi_Q &= \tfrac{1}{2}(\phi_p - \tfrac{1}{2}(\phi_b + \phi_r)) \sin^2\gamma \cos 2\chi \\
\phi_U &= \tfrac{1}{2}(\phi_p - \tfrac{1}{2}(\phi_b + \phi_r)) \sin^2\gamma \sin 2\chi \\
\phi_V &= \tfrac{1}{2}(\phi_r - \phi_b) \cos\gamma \\
\phi'_Q &= \tfrac{1}{2}(\phi'_p - \tfrac{1}{2}(\phi'_b + \phi'_r)) \sin^2\gamma \cos 2\chi \\
\phi'_U &= \tfrac{1}{2}(\phi'_p - \tfrac{1}{2}(\phi'_b + \phi'_r)) \sin^2\gamma \sin 2\chi \\
\phi'_V &= \tfrac{1}{2}(\phi'_r - \phi'_b) \cos\gamma.
\end{aligned} \tag{9}$$

Here $\phi_{p,b,r}$ and $\phi'_{p,b,r}$ are the absorption and anomalous dispersion profiles. For an arbitrary Zeeman multiplet (i.e. anomalous splitting) they are weighted sums of Voigt and Faraday-Voigt functions associated with the various shifted Zeeman components. However, to demonstrate the principles of the formal numerical solution of the transfer equation, it is sufficient to consider only the normal Zeeman triplet (for accounts of the general case see Landi degl'Innocenti 1976, RMD and Chapter 12). In this case the subscripts p, b and r denote respectively the unshifted π component, and the blue and red shifted σ components so that

$$\begin{aligned}
\phi_p &= H(a, v + v_{los}) & \phi'_p &= 2F(a, v + v_{los}) \\
\phi_b &= H(a, v + v_B + v_{los}) & \phi'_b &= 2F(a, v + v_B + v_{los}) \\
\phi_r &= H(a, v - v_B + v_{los}) & \phi'_r &= 2F(a, v - v_B + v_{los})
\end{aligned} \tag{10}$$

where $H(a,v)$ and $F(a,v)$ are the Voigt and Faraday-Voigt functions:

$$H(a,v) = \frac{a}{\pi} \int_{-\infty}^{\infty} \frac{e^{-y^2}}{(v-y)^2 + a^2} \, dy \tag{11}$$

and

$$F(a,v) = \frac{1}{2\pi} \int_{-\infty}^{\infty} \frac{(v-y)e^{-y^2}}{(v-y)^2 + a^2} \, dy, \tag{12}$$

which are related to the real and imaginary parts of the complex probability function (see e.g. Humlicek's (1982) method of computing this function). All parameters are expressed here in units of the *wavelength* Doppler width $\Delta\lambda_D$ (see GI Section 3 concerning sign conventions and the transformation between *wavelength* and *frequency* units), i.e.,

$$a = \Gamma\lambda_o^2 / 4\pi c \Delta\lambda_D, \tag{13}$$

where Γ is the line damping, λ_o is the laboratory line centre wavelength and c is the velocity of light;

$$v = (\lambda - \lambda_o)/\Delta\lambda_D, \tag{14}$$

where λ is the wavelength in the line;

$$v_B = ge\lambda_o^2 B/4\pi mc^2 \Delta\lambda_D, \tag{15}$$

is the absolute shift of each σ component of a line with Landé g-factor (e and m denote electron charge and mass); and

$$v_{los} = \lambda_o \mathbf{v} \cdot \mathbf{n}/c\Delta\lambda_D \tag{16}$$

is the Doppler shift induced by the macroscopic velocity field ($\mathbf{v}$ is the velocity vector and $\mathbf{n}$ is the unit vector along the z-axis towards the observer).

Note that

$$\lim_{B\to 0} \mathbf{\Phi} = \phi \mathbf{1}, \tag{17}$$

where

$$\phi = H(a, v + v_{los}) . \tag{18}$$

Thus, in the absence of Zeeman splitting, this non-LTE theory reduces to the usual complete redistribution model involving only the scalar transfer equation for I, since $Q = U = V = 0$.

2.2 ZEEMAN LINE TRANSFER – FEAUTRIER METHOD

The transformation of the differential equation of transfer for the Stokes vector from first-order to second-order was done first by AHH omitting the magneto-optical (birefringence) terms, i.e. setting $\phi'_Q = \phi'_U = \phi'_V = 0$. A similar transformation is also possible when these terms are included. To show how this is achieved it is necessary to make explicit the wavelength and directional dependence of the Stokes vector, the absorption matrix and the emission vector. In the interests of brevity, their dependence on position will be left implicit.

Let $\mathbf{I}(+v, +\mathbf{n})$ be the Stokes vector of radiation at wavelength $+v$ propagating in the direction $+\mathbf{n}$ out of the atmosphere and let $\mathbf{I}(-v, -\mathbf{n})$ be its counterpart on the opposite side of the line at the wavelength $-v$ propagating in the opposite direction $-\mathbf{n}$. Here it is understood that $\mathbf{I}(-v, -\mathbf{n})$ is defined with respect to the *right-handed* reference frame $x\tilde{y}\tilde{z}$ where $\tilde{y} = -y$ and $\tilde{z} = -z$ with the $\tilde{z}$-axis parallel to $-\mathbf{n}$. In this frame $\mathbf{B}$ has an inclination $\pi - \gamma$ to the $\tilde{z}$-axis and an azimuth $-\chi$ relative to the x-axis (some minor mental gymnastics applied to GI Figure 2 will confirm this).

From Equation (1) we have

$$+\frac{d}{dz}\mathbf{I}(+v,+\mathbf{n}) = -\mathbf{K}(+v,+\mathbf{n})\mathbf{I}(+v,+\mathbf{n}) + \mathbf{j}(+v,+\mathbf{n}) \tag{19}$$

and

$$-\frac{d}{dz}\mathbf{I}(-v,-\mathbf{n}) = -\mathbf{K}(-v,-\mathbf{n})\mathbf{I}(-v,-\mathbf{n}) + \mathbf{j}(-v,-\mathbf{n}) \tag{20}$$

Then from Equations (10), (11), (12) and (16), noting that

$$H(a,-v) = H(a,+v) \qquad F(a,-v) = -F(a,+v), \tag{21}$$

we obtain

$$\begin{aligned} \phi_p(-v,-\mathbf{n}) &= +\phi_p(+v,+\mathbf{n}) & \phi'_p(-v,-\mathbf{n}) &= -\phi'_p(+v,+\mathbf{n}) \\ \phi_b(-v,-\mathbf{n}) &= +\phi_r(+v,+\mathbf{n}) & \phi'_b(-v,-\mathbf{n}) &= -\phi'_r(+v,+\mathbf{n}) \\ \phi_r(-v,-\mathbf{n}) &= +\phi_b(+v,+\mathbf{n}) & \phi'_r(-v,-\mathbf{n}) &= -\phi'_b(+v,+\mathbf{n}). \end{aligned} \tag{22}$$

Using these in Equations (9) along with the substitutions $\pi - \gamma$ for γ and $-\chi$ for χ when considering the direction $-\mathbf{n}$ we find that

$$\begin{aligned} \phi_I(-v,-\mathbf{n}) &= +\phi_I(+v,+\mathbf{n}) & & \\ \phi_Q(-v,-\mathbf{n}) &= +\phi_Q(+v,+\mathbf{n}) & \phi'_Q(-v,-\mathbf{n}) &= -\phi'_Q(+v,+\mathbf{n}) \\ \phi_U(-v,-\mathbf{n}) &= -\phi_U(+v,+\mathbf{n}) & \phi'_U(-v,-\mathbf{n}) &= +\phi'_U(+v,+\mathbf{n}) \\ \phi_V(-v,-\mathbf{n}) &= +\phi_V(+v,+\mathbf{n}) & \phi'_V(-v,-\mathbf{n}) &= -\phi'_V(+v,+\mathbf{n}). \end{aligned} \tag{23}$$

Introducing a *modified* Stokes vector $\tilde{\mathbf{I}}(-v,-\mathbf{n})$ for the inward radiation such that

$$\tilde{\mathbf{I}} = (I, Q, -U, V)^\dagger, \tag{24}$$

i.e. the Stokes parameter U is replaced $-U$, and using Equations (6), (7), (8) and (23) one can recast Equation (20) as

$$-\frac{d}{dz}\tilde{\mathbf{I}}(-v,-\mathbf{n}) = -\mathbf{K}(+v,+\mathbf{n})\tilde{\mathbf{I}}(-v,-\mathbf{n}) + \mathbf{j}(+v,+\mathbf{n}). \tag{25}$$

Now we revert to the compact notation

$$\begin{aligned} \mathbf{K} &= \mathbf{K}(+v, +\mathbf{n}) \,, \\ \mathrm{j} &= \mathrm{j}(+v, +\mathbf{n}) \,, \end{aligned} \tag{26}$$

and define generalized Feautrier vectors

$$\begin{aligned} \mathbf{J} &= \left(\mathbf{I}(+v, +\mathbf{n}) + \tilde{\mathbf{I}}(-v, -\mathbf{n})\right)/2 \,, \\ \mathbf{H} &= \left(\mathbf{I}(+v, +\mathbf{n}) - \tilde{\mathbf{I}}(-v, -\mathbf{n})\right)/2 \,, \end{aligned} \tag{27}$$

so that Equations (19) and (25) can be replaced by the equivalent pair,

$$\frac{d}{dz}\mathbf{J} = -\mathbf{K}\,\mathbf{H} \tag{28}$$

$$\frac{d}{dz}\mathbf{H} = -\mathbf{K}\,\mathbf{J} + \mathrm{j}. \tag{29}$$

Eliminating $\mathbf{H}$ we obtain a second-order equation for $\mathbf{J}$,

$$\frac{d}{dz}\left(\mathbf{K}^{-1}\frac{d}{dz}\mathbf{J}\right) = \mathbf{K}\,\mathbf{J} - \mathrm{j}. \tag{30}$$

In practice integration is restricted to a finite range $0 = z_1 \geq z \geq z_N$ where $|z_N|$ is effectively "infinite", i.e at a depth where the radiation field is thermalized. At the surface z_1 it is assumed there is no radiation directed into the atmosphere, i.e $\tilde{\mathbf{I}}(-v, -\mathbf{n}) = \mathbf{0}$. Equation (28) then gives a first-order boundary condition at z_1,

$$\mathbf{K}^{-1}\frac{d}{dz}\mathbf{J} = -\mathbf{J}\,. \tag{31}$$

The first-order boundary condition at z_N is

$$\mathbf{K}^{-1}\frac{d}{dz}\mathbf{J} = -\mathbf{I}(+v, +\mathbf{n}) + \mathbf{J}\,, \tag{32}$$

and at this depth one can use the asymptotic approximation for the Stokes vector (cf. AHH)

$$\mathbf{I}(+v, +\mathbf{n}) \sim B_\nu\,\mathbf{I}_\mathrm{o} - \mathbf{K}^{-1}\frac{d}{dz}B_\nu\,\mathbf{I}_\mathrm{o}\,. \tag{33}$$

We approximate Equations (30)-(33) by finite-differences on a grid z_k $(k = 1, ..., N)$. In the following a subscript k indicates that the function is evaluated at z_k. Let

$$\begin{aligned} \delta_k &= |z_{k+1} - z_k|\,, \\ \mathbf{\Delta}_k &= \tfrac{1}{2}\left(\mathbf{K}_{k+1}^{-1} + \mathbf{K}_k^{-1}\right)/\delta_k\,; \quad k = 1, .., N-1\,. \end{aligned} \tag{34}$$

Then the appropriate difference equations are (cf. AHH's Equations (21) - (25); note that the **B** in the following equations is not the magnetic field vector)

$$\begin{aligned} \mathbf{B}_1\mathbf{J}_1 - \mathbf{C}_1\mathbf{J}_2 &= \mathbf{L}_1\,, \\ -\mathbf{A}_k\mathbf{J}_{k-1} + \mathbf{B}_k\mathbf{J}_k - \mathbf{C}_k\mathbf{J}_{k+1} &= \mathbf{L}_k\,; \quad k = 2,...,N-1 \\ -\mathbf{A}_N\mathbf{J}_{N-1} + \mathbf{B}_N\mathbf{J}_N &= \mathbf{L}_N\,, \end{aligned} \tag{35}$$

where

$$\begin{aligned} &\mathbf{B}_1 = \mathbf{\Delta}_1 + \mathbf{1} + \tfrac{1}{2}\delta_1\mathbf{K}_1\,, \qquad \mathbf{C}_1 = \mathbf{\Delta}_1\,, \qquad \mathbf{L}_1 = \tfrac{1}{2}\delta_1\mathbf{j}_1\,, \\ &\mathbf{A}_k = 2\mathbf{\Delta}_{k-1}/\left(\delta_k + \delta_{k-1}\right)\,, \qquad \mathbf{C}_k = 2\mathbf{\Delta}_k/\left(\delta_k + \delta_{k-1}\right)\,, \\ &\mathbf{B}_k = \mathbf{A}_k + \mathbf{C}_k + \mathbf{K}_k\,, \qquad \mathbf{L}_k = \mathbf{j}_k\,; \quad k = 2,...N-1\,, \\ &\mathbf{A}_N = \mathbf{\Delta}_{N-1} - \tfrac{1}{2}\mathbf{1}\,, \qquad \mathbf{B}_N = \mathbf{\Delta}_{N-1} + \tfrac{1}{2}\mathbf{1}\,, \\ &\mathbf{L}_N = \tfrac{1}{2}\mathbf{I}_o(B_{\nu N-1} + B_{\nu N}) + \mathbf{\Delta}_{N-1}\mathbf{I}_o\left(B_{\nu N} - B_{\nu N-1}\right). \end{aligned} \tag{36}$$

Equations (35) form a block tri-diagonal system that can be solved by Gaussian elimination giving the recursion formulae,

$$\mathbf{J}_N = \mathbf{r}_N\,, \qquad \mathbf{J}_k = \mathbf{r}_k + \mathbf{D}_k\,\mathbf{J}_{k+1}\,; \quad k = 1,...,N-1\,, \tag{37}$$

where

$$\begin{aligned} &\mathbf{r}_1 = \mathbf{B}_1{}^{-1}\mathbf{L}_1\,, \qquad \mathbf{D}_1 = \mathbf{B}_1{}^{-1}\mathbf{C}_1\,, \\ &\mathbf{r}_k = \left(\mathbf{B}_k - \mathbf{A}_k\mathbf{D}_{k-1}\right)^{-1}\left(\mathbf{L}_k + \mathbf{A}_k\mathbf{r}_{k-1}\right)\,, \\ &\mathbf{D}_k = \left(\mathbf{B}_k - \mathbf{A}_k\mathbf{D}_{k-1}\right)^{-1}\mathbf{C}_k\,; \quad k = 2,...,N-1\,. \end{aligned} \tag{38}$$

The emergent Stokes vector at the surface is simply

$$\mathbf{I}(0) = 2\,\mathbf{J}_1 \tag{39}$$

since there is no inwardly directed radiation at z_1.

The connection between this solution and the formal integral in Equation (4) is not obvious. In fact a very simple approximate relationship exists which permits rapid evaluation of the contribution vector. To second-order accuracy in the differences δ_k, the evolution operator can be approximated thus

$$\mathbf{O}(z_k, z_{k+1}) = \mathbf{D}_k\,. \tag{40}$$

The proof of this can be established by induction but is omitted here as it is rather tedious (see RMD for details). Using Equations (5) and (40) together with the composition law for the evolution operator (LD Equation (14)),

$$\mathbf{O}(z, z') = \mathbf{O}(z, z'')\,\mathbf{O}(z'', z')\,, \tag{41}$$

we obtain the approximation

$$\tilde{\mathbf{C}}_k = \left(\prod_{i=0}^{k-1} \mathbf{D}_i \right) \mathbf{j}_k \, ; \quad \mathbf{D}_0 = \mathbf{1} \tag{42}$$

for the contribution vector.

This method has been tested thoroughly against analytic solutions and other numerical schemes. For details of these tests and a discussion about the optimal choice of the depth grid see RMD.

2.3 ZEEMAN LINE TRANSFER – LAMBDA OPERATOR METHOD

The Feautrier solution is a *two-way* integration scheme in the sense that one solves for the *average* of the Stokes vectors of radiation directed outwards and inwards along the line-of-sight. In this Section we outline a *one-way* integration method which does not demand any special insight into the symmetries of the absorption matrix **K**.

From Equations (6) and (8) we see that diagonal elements of **K** are all equal to $\kappa_I = \kappa_c + \kappa_o \phi_I$. We define a line-of-sight optical depth τ associated with this element such that

$$d\tau = -\kappa_I \, dz = -(\kappa_c + \kappa_o \phi_I) \, dz \, , \tag{43}$$

a *modified* absorption matrix with zeros on its diagonal,

$$\tilde{\mathbf{K}} = \mathbf{K}/\kappa_I - \mathbf{1} \, , \tag{44}$$

and a *modified* total source vector,

$$\tilde{\mathbf{S}} = \mathbf{j}/\kappa_I \, , \tag{45}$$

(recall from Equation (1) that the actual total source vector is $\mathbf{S} = \mathbf{K}^{-1}\mathbf{j}$).

Equation (1) can be rewritten as

$$\frac{d}{d\tau}\mathbf{I} - \mathbf{I} = \tilde{\mathbf{K}}\mathbf{I} - \tilde{\mathbf{S}} \, , \tag{46}$$

the formal solution of which involves the lambda-transform with respect to τ,

$$\mathbf{I}(\tau) = \int_{\tau}^{\infty} e^{-(\tau' - \tau)} \left(\tilde{\mathbf{S}}(\tau') - \tilde{\mathbf{K}}(\tau') \, \mathbf{I}(\tau') \right) d\tau' \, . \tag{47}$$

This reduces *exactly* to the formal integral for the intensity $I(\tau)$ when the magnetic field vector $\mathbf{B} = \mathbf{0}$, and hence $\tilde{\mathbf{K}} = \mathbf{0}$.

Equation (47) is in a form suitable for solution by lambda iteration, starting with an initial estimate $\mathbf{I}(\tau) = \mathbf{0}$, say. Staude (1969, 1970) tried this procedure but found that the Neumann series generated by successive iterations converged too slowly for the method to be practicable. However a direct, non-iterative, numerical solution is possible. The idea for the method now described was motivated by Kalkofen's (1974) discussion of numerical representations of the lambda integral operator.

We introduce an optical depth grid τ_k $(k = 1, ..., N)$ corresponding to the geometric grid z_k and let

$$\begin{aligned} \tilde{\delta}_k &= \tau_{k+1} - \tau_k = \tfrac{1}{2}(\kappa_{I\,k+1} + \kappa_{I\,k})\,|z_{k+1} - z_k|\,, \\ E_k &= e^{-\tilde{\delta}_k}\,; \quad k = 1, ..., N-1\,. \end{aligned} \tag{48}$$

On the subinterval (τ_k, τ_{k+1}) we have from Equation (46)

$$\mathbf{I}_k = E_k\,\mathbf{I}_{k+1} + \int_{\tau_k}^{\tau_{k+1}} e^{-(\tau'-\tau_k)} \left(\tilde{\mathbf{S}}(\tau') - \tilde{\mathbf{K}}(\tau')\mathbf{I}(\tau')\right) d\tau'\,. \tag{49}$$

The next step is a "bootstrapping" device which provides an explicit relation between $\mathbf{I}_k$ and $\mathbf{I}_{k+1}$. On this subinterval we approximate $(\tilde{\mathbf{S}} - \tilde{\mathbf{K}}\,\mathbf{I})$ by a *linear* function of optical depth, i.e.

$$\tilde{\mathbf{S}}(\tau) - \tilde{\mathbf{K}}(\tau)\,\mathbf{I}(\tau) = (\tilde{\mathbf{S}} - \tilde{\mathbf{K}}\,\mathbf{I})_k + \left((\tilde{\mathbf{S}} - \tilde{\mathbf{K}}\,\mathbf{I})_{k+1} - (\tilde{\mathbf{S}} - \tilde{\mathbf{K}}\,\mathbf{I})_k\right)(\tau - \tau_k)/\tilde{\delta}_k\,. \tag{50}$$

Then Equation (49) can be integrated *analytically* and after some straightforward algebra one obtains

$$\mathbf{I}_k = \mathbf{P}_k + \mathbf{Q}_k\,\mathbf{I}_{k+1}\,, \tag{51}$$

where

$$\begin{aligned} \mathbf{P}_k &= \left(\mathbf{1} + (F_k - G_k)\tilde{\mathbf{K}}_k\right)^{-1} ((F_k - G_k)\tilde{\mathbf{S}}_k + G_k\tilde{\mathbf{S}}_{k+1})\,, \\ \mathbf{Q}_k &= \left(\mathbf{1} + (F_k - G_k)\tilde{\mathbf{K}}_k\right)^{-1} (E_k\mathbf{1} - G_k\tilde{\mathbf{K}}_{k+1})\,, \\ F_k &= 1 - E_k\,, \\ G_k &= (1 - (1 + \tilde{\delta}_k)E_k)/\tilde{\delta}_k\,. \end{aligned} \tag{52}$$

At τ_N we approximate the Stokes vector by

$$\mathbf{I}_N = B_{\nu N}\,\mathbf{I}_o\,, \tag{53}$$

though an improved asymptotic formula like Equation (33) could also be used. Equation (51) can then be applied recursively to derive the emergent Stokes vector at the surface

$$\mathbf{I}(0) = \mathbf{I}_1\,. \tag{54}$$

Analogously to Equation (2) one can write the formal solution in terms of optical depth on the subinterval (τ_k, τ_{k+1}) as

$$\mathbf{I}(\tau_k) = \int_{\tau_k}^{\tau_{k+1}} \mathbf{O}(\tau_k, \tau')\,\tilde{\mathbf{S}}(\tau')\,d\tau' + \mathbf{O}(\tau_k, \tau_{k+1})\,\mathbf{I}(\tau_{k+1})\,, \tag{55}$$

Comparing this with Equation (51) we obtain another approximation for the evolution operator,

$$\mathbf{O}(\tau_k, \tau_{k+1}) = \mathbf{Q}_k\,. \tag{56}$$

Applying the composition law in Equation (41) we again can approximate the contribution vector by a product,

$$\tilde{\mathbf{C}}_k = \left(\prod_{i=0}^{k-1} \mathbf{Q}_i\right) \mathbf{j}_k\,; \qquad \mathbf{Q}_0 = \mathbf{1}\,. \tag{57}$$

For details about the accuracy of this method compared with the Feautrier method see RMD.

3.1 RESONANCE POLARIZATION – THEORY

The quantum theory of scattering and transfer of polarized radiation in resonance lines is algebraically very complex (e.g. Landi degl'Innocenti 1984 and Streater *et al.* 1987). To develop numerical solutions it is helpful to translate the quantum formalism into the familiar notation of unpolarized non-LTE computations. Most numerical work has focused on models where this translation is relatively easy. The rest of this chapter is devoted to a model problem in this category: the formation of resonance scattering polarization in a two-level model atom in the absence of a magnetic field.

We assume a static plane-parallel atmosphere in which the radiation field exhibits azimuthal symmetry about the normal to the surface. This implies that $U = V = 0$ so the polarization state of the radiation can be specified by just the two Stokes parameters I and Q. In a model representing the solar atmosphere, for example, Q is *negative* if the radiation has a excess of linear polarization *tangential* to the solar limb.

Central to resonance scattering theory is the redistribution matrix which accounts for correlations in frequency, angle and polarization between absorbed and emitted photons. The canonical model, which produces maximum polarization, is based on the following assumptions: the line transition corresponds to a classical electric dipole; in the atom's rest frame the lower level has zero width, while the upper level is naturally broadened and scattering in this frame is coherent in frequency. These assumptions lead to a vector transfer equation involving a redistribution matrix $\mathbf{R}_{\text{II}}$ which is a simple generalization of Hummer's (1962, 1969) well-known scalar redistribution function R_{II}. To facilitate comparison with his work we change notation slightly from that used in previous Sections.

The transfer equation for the *two component* Stokes vector $\mathbf{I} = (I, Q)^{\dagger}$ is (Dumont *et al.* 1977 and Rees and Saliba 1982)

$$\mu \frac{d}{d\tau} \mathbf{I}(x, \mu) = (\beta + \varphi(x))\,(\mathbf{I}(x, \mu) - \mathbf{S}(x, \mu))\,. \tag{58}$$

The angle $\theta = \cos^{-1}\mu$ specifies the direction of propagation relative to the outward normal; τ is now the *mean* optical depth in the line; x is the *frequency* measured in units of the Doppler width ; and β is the ratio of continuum to integrated line opacity. The *normalized* line absorption profile is represented as a Voigt function,

$$\varphi(x) = H(a, x)/\sqrt{\pi}\,. \tag{59}$$

Note that here we deal with a *scalar* absorption coefficient rather than an absorption matrix.

The *two component* total source vector is

$$\mathbf{S}(x, \mu) = (\varphi(x)\mathbf{S}_L(x, \mu) + \beta \mathbf{S}_c)/(\beta + \varphi(x)) \tag{60}$$

where we assume an unpolarized continuum specified by the continuum source vector $\mathbf{S}_c = (B_\nu, 0)^{\dagger}$. The line source vector $\mathbf{S}_L(x, \mu)$ can be written in a form analogous to Hummer's (1969) expression for the scalar line source function,

$$(1+\epsilon)\mathbf{S}_L(x,\mu) = \frac{4\pi}{\varphi(x)} \int_{-\infty}^{\infty} dx' \int_{4\pi} d\Omega' \, \mathbf{R}_{\mathrm{II}}(x',\mathbf{n}';x,\mathbf{n})\mathbf{I}(x',\mu') + \epsilon \mathbf{S}_c \,, \qquad (61)$$

where ϵ is the ratio of collisional and spontaneous de-excitation rates from the upper to the lower level of the transition; x' and x are the frequencies of absorbed and emitted photons in a scattering event, and the unit vectors $\mathbf{n}$ and $\mathbf{n}'$ are their respective directions; and $d\Omega'$ is an element of solid angle about $\mathbf{n}'$. Replacing the phase function $g(\mathbf{n}';\mathbf{n})$ in Hummer's redistribution function by the dipole scattering phase matrix $\mathbf{P}(\mathbf{n}';\mathbf{n})/4\pi$, we obtain the redistribution matrix

$$\mathbf{R}_{\mathrm{II}}(x',\mathbf{n}';x,\mathbf{n}) = \frac{1}{4\pi}\mathbf{P}(\mathbf{n}';\mathbf{n})\frac{1}{4\pi^2|\sin\gamma|}\exp\left[-\left(\frac{x'-x}{2}\right)^2 \csc^2\frac{\gamma}{2}\right] H\left(a\sec\frac{\gamma}{2}, \left(\frac{x+x'}{2}\right)\sec\frac{\gamma}{2}\right), \qquad (62)$$

where $\cos\gamma = \mathbf{n}\cdot\mathbf{n}'$. Using an alternative system of Stokes parameters, $I_l = (I+Q)/2$ and $I_r = (I-Q)/2$ (note that in a solar model the subscripts l and r refer respectively to axes perpendicular and tangential to the solar limb) Chandrasekhar (1950) derived the following expression for the phase matrix (see his Equations (216) and (217)),

$$\tilde{\mathbf{P}}(\mathbf{n}';\mathbf{n}) = \frac{3}{2}\begin{pmatrix} (l,l)^2 & (r,l)^2 \\ (l,r)^2 & (r,r)^2 \end{pmatrix}, \qquad (63)$$

where

$$\begin{aligned} (l,l) &= (1-\mu^2)^{\frac{1}{2}}(1-\mu'^2)^{\frac{1}{2}} + \mu'\mu\cos\Delta \,, \\ (r,l) &= \mu\sin\Delta \,, \\ (l,r) &= -\mu'\sin\Delta \,, \\ (r,r) &= \cos\Delta \,, \end{aligned} \qquad (64)$$

and Δ is the difference in azimuth between $\mathbf{n}$ and $\mathbf{n}'$. $\mathbf{P}$ can be derived from $\tilde{\mathbf{P}}$ by a simple matrix transformation.

Although the Stokes parameters are independent of azimuth, it is necessary to integrate numerically with respect to azimuth in Equation (61) if the full angle dependence of the redistribution matrix is taken into account. Details of this general case are given by Dumont *et al.* for $a = 0$ and Faurobert (1987a,

b) for $a \neq 0$. However, motivated by developments in scalar redistribution theory, one can envisage a hierarchy of approximations to $\mathbf{R}_{\rm II}$ which greatly simplify the numerical solution, without compromising the essential physics of the model.

To begin we perform an analytic angle average over the frequency dependent part of Equation (62) to obtain

$$\bar{\mathbf{R}}_{\rm II}(x',\mathbf{n}';x,\mathbf{n}) = \frac{1}{4\pi}\mathbf{P}(\mathbf{n}';\mathbf{n})\frac{1}{4\pi}R_{\rm II}(x';x)\,, \tag{65}$$

a hybrid model which retains the angular correlation in $\mathbf{P}(\mathbf{n}';\mathbf{n})$ necessary for scattering polarization and mimics the frequency correlation via the usual angle averaged scalar redistribution function for *isotropic* scattering,

$$R_{\rm II}(x';x) = \frac{1}{\pi^{\frac{3}{2}}}\int\limits_{\frac{1}{2}|\bar{x}-\underline{x}|}^{\infty} e^{-u^2}\left[\tan^{-1}\left(\frac{\underline{x}+u}{a}\right) - \tan^{-1}\left(\frac{\bar{x}-u}{a}\right)\right]du\,, \tag{66}$$

with $\bar{x} = \max(x,x')$ and $\underline{x} = \min(x,x')$. The advantage of using Equation (65) is that now one can carry out the azimuthal integration in Equation (61) analytically. Naturally some information about the polarization is lost in the averaging procedure (see Faurobert's discussion of the errors incurred). Finally, after some simple algebra, one can rewrite Equation(58) as a pair of coupled transfer equations for I and Q,

$$\mu\frac{d}{d\tau}I(x,\mu) = (\beta+\varphi(x))(I(x,\mu) - S_I(x,\mu))\,, \tag{67}$$

$$\mu\frac{d}{d\tau}Q(x,\mu) = (\beta+\varphi(x))(Q(x,\mu) - S_Q(x,\mu))\,, \tag{68}$$

where the total source functions for I and Q are

$$S_I(x,\mu) = (\varphi(x)S_L(x,\mu) + \beta B_\nu)/(\beta+\varphi(x))\,, \tag{69}$$

$$S_Q(x,\mu) = \varphi(x)(1-\mu^2)P(x)/(\beta+\varphi(x))\,. \tag{70}$$

The line source function for I is

$$S_L(x,\mu) = S_L^u(x) + (1/3-\mu^2)P(x)\,, \tag{71}$$

where

$$(1+\epsilon)\,S_L^u(x) = \frac{1}{2}\int\limits_{-\infty}^{\infty}\frac{R_{\rm II}(x';x)}{\varphi(x)}\,dx'\int\limits_{-1}^{1}I(x',\mu')\,d\mu' + \epsilon\,B_\nu\,, \tag{72}$$

and

$$(1+\epsilon)\,P(x) =$$

$$\frac{3}{16}\int_{-\infty}^{\infty} \frac{R_{\rm II}(x';x)}{\varphi(x)}\,dx' \int_{-1}^{1} \left[(1-3\mu'^2)I(x',\mu') + 3(1-\mu'^2)Q(x',\mu')\right]\,d\mu' . \quad (73)$$

$S_L^u(x)$ is equivalent to the frequency dependent partial redistribution (PRD) line source function when polarization is neglected(the superscript u denotes "unpolarized") and $P(x)$ is a perturbation term which produces the polarization.

An obvious next step in the modelling hierarchy is to adopt Jefferies and White's (1960) approximation to $R_{\rm II}(x';x)$, i.e. to assume complete redistribution for scattering in the Doppler core and coherent scattering in the wings of the line. Rees and Saliba (1982) and Saliba (1985) used Kneer's (1975) version of this approximation,

$$R_{\rm II}^K(x';x) = < a >_x \delta(x'-x)\varphi(x') + (1-a_{x',x})\varphi(x')\varphi(x)\,, \quad (74)$$

with

$$< a >_x = \int_{-\infty}^{\infty} a_{x',x}\,\varphi(x')\,dx'$$

$$a_{x',x} = 1-\exp\left(-(x_{max}-2)^2/4\right)\,; \qquad x_{max} = \max(|x'|,|x|)\,. \quad (75)$$

This PRD model predicts intensity and linear polarization profiles in good qualitative agreement with those observed near the solar limb in resonance lines such as Ca I 4227 Å. For very precise analysis, however, Faurobert has shown that it is essential to use the fully angle dependent redistribution matrix in Equation (62).

3.2 RESONANCE POLARIZATION – PERTURBATION METHOD

Several numerical methods have been used to solve polarization transfer problems of the type posed in Section 3.1:

(*i*) non-iterative solution of the coupled differential equations by Feautrier's method, assuming PRD (Dumont *et al.* and Faurobert);

(*ii*) non-iterative solution of an equivalent pair of coupled integral equations, assuming CRD (Rees 1978) and PRD (McKenna 1984);

(*iii*) Feautrier's method treating Q as a perturbation, assuming PRD (Rees and Saliba 1982, Saliba 1985, and Faurobert);

(*iv*) Rybicki's (1971) core-saturation method treating Q as a perturbation, assuming CRD (Stenflo and Stenholm 1976).

In this Section we describe a new approach, related to Stenflo and Stenholm's, which is an adaptation of Scharmer's approximate lambda operator method. The method has been successfully applied to both CRD and PRD problems (see Rees *et al.* 1987b). For simplicity we consider here only the CRD case. This is equivalent to setting $a_{x',x} \equiv 0$ in Equation (75) and then S_L^u and P in Equations (71) and (72) are *frequency independent*. The discussion closely parallels Scharmer and Nordlund's paper which contains the details about numerical construction of the exact and approximate operators introduced below. Their computer code requires very few modifications to handle resonance polarization.

The atmosphere is assumed to be semi-infinite and the boundary conditions are

$$\begin{aligned} \mathbf{I}(x,\mu) &= (0,0)^{\dagger}\,; \quad \tau = 0\,; \quad \mu < 0\,, \\ \lim_{\tau\to\infty} \mathbf{I}(x,\mu) &= \mathbf{S_c}\,. \end{aligned} \tag{76}$$

Let $\Lambda_{x\mu}$ denote the monochromatic lambda operator for the direction μ (the optical depth dependence will be kept implicit in the following discussion). Then the formal solutions of Equations (67) and (68) are

$$I(x,\mu) = \Lambda_{x\mu}\left[S_I(x,\mu)\right]\,, \tag{77}$$

$$Q(x,\mu) = \Lambda_{x\mu}\left[S_Q(x,\mu)\right]\,. \tag{78}$$

Substituting these in Equations (72) and (73) and defining the double integral operation

$$\int\!\!\int = \frac{1}{2}\int_{-\infty}^{\infty} \varphi(x)\,dx \int_{-1}^{1} d\mu\,, \tag{79}$$

we obtain

$$(1+\epsilon)S_L^u = \int\!\!\int \Lambda_{x\mu} S_I(x,\mu) + \epsilon B_\nu\,, \tag{80}$$

$$(1+\epsilon)P = \frac{3}{8}\int\!\!\int \Lambda_{x\mu}\left[(1-3\mu^2)S_I(x,\mu) + 3(1-\mu^2)S_Q(x,\mu)\right]\,. \tag{81}$$

We now *pre-condition* the problem to account analytically for cancellation effects at large optical depths (Rybicki). Letting

$$\Lambda_{x\mu} = 1 + \delta\Lambda_{x\mu}\,, \tag{82}$$

and using the fact that

$$\int_{-1}^{1} (1/3 - \mu^2)\, d\mu = 0\,, \tag{83}$$

we can write Equation (80) as

$$(\epsilon + \bar{\delta})\, S_L^u - \int\int \delta\Lambda_{x\mu}\, [S_I(x,\mu)] = (\epsilon + \bar{\delta})\, B_\nu\,, \tag{84}$$

where

$$\bar{\delta} = \int\int \left(\frac{\beta}{\beta + \varphi(x)}\right)\,. \tag{85}$$

These equations can be solved iteratively by a linearization method. We shall use the superscript n to denote the n-th iterate of a function. From Equation (84),

$$(\epsilon + \bar{\delta})\, S_L^{u\,(n)} - \int\int \delta\Lambda_{x\mu} \left[S_I^{(n)}(x,\mu)\right] = (\epsilon + \bar{\delta})\, B_\nu + E^{(n)}\,, \tag{86}$$

where $E^{(n)}$ is an error term, and, by Equations (69) and (71),

$$S_I^{(n)}(x,\mu) = \left(\varphi(x)\, S_L^{u\,(n)} + \beta B_\nu + \varphi(x)(1/3 - \mu^2)\, P^{(n-1)}\right) / (\beta + \varphi(x))\,. \tag{87}$$

Note that we use the *lagged* value $P^{(n-1)}$ to estimate $S_I^{(n)}(x,\mu)$. To estimate the correction $\delta S_L^{u\,(n)}$ such that

$$S_L^{u\,(n+1)} = S_L^{u\,(n)} + \delta S_L^{u\,(n)} \tag{88}$$

would satisfy Equation (84) exactly, we linearize Equation (84) treating $P^{(n-1)}$ as *fixed* during the linearization process, i.e. $\delta P^{(n-1)} = 0$. This gives the approximation

$$(\epsilon + \bar{\delta})\, \delta S_L^{u\,(n)} - \int\int \delta\Lambda_{x\,\mu} \left[\frac{\varphi(x)\,\delta S_L^{u\,(n)}}{(\beta + \varphi(x))}\right] = -E^{(n)}\,. \tag{89}$$

Substantial savings in computer time are gained at this stage by solving Equation (89) for $\delta S_L^{u\,(n)}$ for *given* $E^{(n)}$ using an approximate operator $\delta\Lambda^*_{x\,\mu}$.

In summary the algorithm for computing the successive iterates $S_L^{u\,(n)}$ and $P^{(n)}$ is :

(i) Initialize the calculation with

$$S_L^{u\,(0)} = P^{(0)} = 0 \tag{90}$$

so that, from Equation (69),

$$S_I^{(0)}(x,\mu) = \beta B_\nu / (\beta + \varphi(x)) , \tag{91}$$

and estimate $E^{(0)}$ from Equation (86) using $\delta\Lambda^*_{x\,\mu}$ for $\delta\Lambda_{x\,\mu}$, i.e.

$$E^{(0)} = -(\epsilon + \bar{\delta})\, B_\nu - \int\int \delta\Lambda^*_{x\,\mu} \left[S_I^{(0)}(x,\mu) \right] . \tag{92}$$

(ii) For $n \geq 0$ compute $\delta S_L^{u\,(n)}$ using

$$(\epsilon + \bar{\delta})\, \delta S_L^{u\,(n)} - \int\int \delta\Lambda^*_{x\,\mu} \left[\frac{\varphi(x)\delta S_L^{u\,(n)}}{(\beta + \varphi)} \right] = -E^{(n)} . \tag{93}$$

$S_L^{u\,(n+1)}$ follows from Equation (88) and

$$S_I^{(n+1)}(x,\mu) = \left(\varphi(x)\, S_L^{u\,(n+1)} + \beta\, B_\nu + \varphi(x)\, (1/3 - \mu^2)\, P^{(n)} \right) / (\beta + \varphi(x)) \tag{94}$$

$$S_Q^{(n+1)}(x,\mu) = \varphi(x)\, (1 - \mu^2)\, P^{(n)} / (\beta + \varphi(x)) , \tag{95}$$

from Equations (87) and (70) respectively.

Estimate $E^{(n+1)}$ and $P^{(n+1)}$ from Equations (86) and (81) using the *exact* operators $\delta\Lambda_{x\,\mu}$ and $\Lambda_{x\,\mu}$, and many angles and frequencies in the quadrature that replaces the double integral:

$$E^{(n+1)} = (\epsilon + \bar{\delta})\, S_L^{u\,(n+1)} - \int\int \delta\Lambda_{x\,\mu} \left[S_I^{(n+1)}(x,\mu) \right] - (\epsilon + \bar{\delta})\, B_\nu , \tag{96}$$

and

$$(1 + \epsilon)\, P^{(n+1)} = \frac{3}{8} \int\int \Lambda_{x\mu} \left[(1 - 3\mu^2) S_I^{(n+1)}(x,\mu) + 3(1 - \mu^2) S_Q^{(n+1)}(x,\mu) \right] . \tag{97}$$

(*iii*) Loop back to step (*ii*) until convergence is achieved, i.e. $|\delta S_L^{u\,(n)}/S_L^{u\,(n)}|$ and $|\left(P^{(n+1)} - P^{(n)}\right)/P^{(n)}|$ are sufficiently small everywhere.

For CRD typically about 6 iterations are needed to achieve an accuracy of a few percent. We also tested a PRD version of this method and found the convergence to be considerably slower. The reason for this probably lies in the use of an approximate CRD operator to drive the solution in the line wings where an approximate coherent scattering lambda operator would be more suitable (Scharmer 1983). Currently we are working on methods to accelerate the convergence (Rees *et al.* 1987b).

4. CONCLUSION

We have considered two of the simplest problems in non-LTE polarized radiative transfer in spectral lines: (*i*) formal integration of the Stokes transfer equations *given* the opacities and source functions in the presence of a strong magnetic field; (*ii*) self-consistent solution of the coupled equations of transfer and statistical equilibrium (more specifically the line source functions) for resonance scattering polarization in the absence of magnetic fields.

The Feautrier method for problems of type (*i*) has been known for many years (AHH), but it is not in common usage despite its legendary computational advantages (Mihalas 1985). The *formulation* of the second-order finite difference equations is algebraically complex and requires a detailed understanding of the symmetries of the absorption matrix and emission vector, especially when magneto-optical effects are included (Section 2.2). On the other hand the *implementation* of the method is straightforward (RMD). It is arguably the best method for formal integration of the Stokes vector transfer equation.

In Section 2.3 we introduced the lambda operator associated with the diagonal element of the absorption matrix and used it as the basis of a new *one-way* formal integration method. In the absence of a magnetic field this method reduces to the formal solution for the intensity where the source function is represented by linear segments on the line optical depth grid. In speed and accuracy the method competes well with the Feautrier method (RMD) and it is algebraically much simpler.

Each of these methods can be used for rapid computation of the contribution functions for the Stokes parameters. The polarization contribution functions deserve more detailed study given their potential importance in determining line-of-sight gradients in the magnetic field.

In Section 3.2 we applied Scharmer's operator perturbation method to a problem of type (*ii*). We restricted attention to a two-level atom, but in principle the method should work equally well for multi-level problems.

Self-consistent solution of the coupled equations of polarized radiative transfer and statistical equilibrium in the presence of a magnetic field is one of the major challenges in computational astrophysics. AHH applied the complete linearization method to the strong magnetic field case and Van Ballegooijen (Chapter 12) has generalized the equivalent-two-level-atom technique to handle the same problem.

In Section 7 of Chapter 9 we speculated about the possibility of using an operator perturbation method, again for the strong field case. Given the success of the perturbation approach for non-magnetic resonance scattering, it is logical to speculate further that the entire range of line polarization phenomena will succumb to a perturbation solution. Furthermore, the fact that in every case (zero field, weak field (Hanle effect) and strong field (Zeeman effect)) the diagonal elements of the absorption matrix are all the same suggests that the lambda operator technique of Section 2.3 could be exploited more generally.

Indeed, we foresee the imminent possibility of a grand unified approach to the transfer of polarized light in spectral lines employing a perturbation method that involves the diagonal element lambda operator.

ACKNOWLEDGMENTS

We thank Gary Saliba, Andy Skumanich, Bruce Lites and Marianne Faurobert, each of whom has contributed significantly to the development of our ideas about computations in polarized radiative transfer. We are especially grateful to Lawrence Cram and Chris Durrant for constructive criticism of the manuscript. The work was supported by ARGS Grant B8415543 and by funds from the High Altitude Observatory Visiting Scientist and Graduate Student Programs.

REFERENCES

Auer, L.H., Heasley, J.N. & House, L.L. 1977, *Astrophys. J.*, **216**, 531.

Borra, E.F., Landstreet, J.D. & Mestel, L. 1982, *Ann. Rev. Astron. Astrophys.*, **20**, 191.

Carlsson, M. 1986, *Uppsala Astronomical Observatory Report*, **33**.

Chandrasekhar, S. 1950, *Radiative Transfer*, Oxford Uni. Press.

Coyne, G.V. & McLean, I.S. 1982, In *Be Stars*, IAU Symposium **98**, Ed. M. Jaschek & H.-G. Groth, p. 77.

Dumont, S., Omont, A., Pecker, J.-C. & Rees, D.E. 1977, *Astron. Astrophys.*, **54**, 675.

Faurobert, M. 1987a, *Astron. Astrophys.*, in press.

__________ 1987b, *Astron. Astrophys.*, in press.

Feautrier, P. 1964, *C. R. Acad. Sci. Paris*, **258**, 3189.

Gantmacher, F.R. 1959, *The Theory of Matrices*, Vol. II, Chelsea, New York.

Harvey, J.W. 1985, In *Measurements of Solar Vector Magnetic Fields* (NASA-CP2374), Ed. M.J. Hagyard, p. 109.

Hummer, D.G. 1962, *Monthly Notices Roy. Astron. Soc.*, **125**,21.

__________ 1969, *Monthly Notices Roy. Astron. Soc.*, **145**, 95.

Humlicek, J. 1982, *J. Quant. Spectrosc. Rad. Transfer*, **27**, 437.

Jefferies, J.T. & White, O.R. 1960, *Astrophys. J.*, **132**, 767.

Kalkofen, W. 1974, *Astrophys. J.*, **188**, 105.

Kneer, F. 1975, *Astrophys. J.*, **200**, 975.

Landi degl'Innocenti, E. 1976, *Astron. Astrophys. Suppl.*, **25**, 379.

__________ 1984, *Solar Phys.*, **91**, 1.

Lites, B.W., Skumanich, A., Rees, D.E. & Murphy, G.A. 1987a, *Astrophys. J.*, in press

__________ 1987b, in preparation.

McKenna, S.J. 1984, *Astrophys. Space Sci.*, **106**, 283.

Mihalas, D. 1985, *J. Comput. Phys.*, **57**, 1.

Rees, D.E. 1969, *Solar Phys.*, **10**, 268.

__________ 1978, *Publ. Astron. Soc. Japan*, **30**, 405.

Rees, D.E. & Saliba, G.J. 1982, *Astron. Astrophys.*, **115**, 1.

Rees, D.E. , Murphy, G.A. & Durrant, C.J. 1987a, in preparation.

Rees, D.E. , Murphy, G.A. & Faurobert, M. 1987b, in preparation.

Rybicki, G.B. 1971, in *Line Formation in the Presence of Magnetic Fields*,

NCAR, Boulder, CO, p. 146.
Saliba, G.J. 1985, *Solar Phys.*, **98**, 1.
Scharmer, G.B. 1981, *Astrophys. J.*, **249**, 720.
__________ 1983, *Astron. Astrophys.*, **117**, 83.
Scharmer, G.B. & Nordlund, Å 1982, *Stockolm Observatory Report*, **19**.
Staude, J. 1969, *Solar Phys.*, **8**, 264.
__________ 1970, *Solar Phys.*,**15**, 102.
Stenflo, J.O. & Stenholm, L.G. 1976, *Astron. Astrophys.*, **46**, 69.
Streater, A., Cooper, J. & Rees, D.E. 1987, in preparation.
van Ballegooijen, A.A. 1985, In *Measurements of Solar Vector Magnetic Fields* (NASA-CP2374), Ed. M.J. Hagyard, p. 322.

TRANSFER OF POLARIZED RADIATION, USING 4X4 MATRICES

E. Landi Degl'Innocenti
Istituto di Astronomia, Università di Firenze

ABSTRACT: The main characteristics of the radiative transfer equations for polarized radiation, including the symmetry properties and the physical interpretation of the various terms of the absorption matrix, are briefly reviewed. A formal solution of the transfer equations is presented for the case where both the absorption matrix and the source-function vector are given functions of optical depth. A suitable algorithm is presented to obtain from the formal solution a numerical solution, and a brief comparison is presented with different numerical methods that are often used for the solution of the same set of equations. Finally, an iteration scheme is suggested for attacking a large variety of non-LTE problems for polarized radiation according to a perturbative scheme.

1. INTRODUCTION

Polarized radiation is usually described in astrophysical problems by means of the so-called Stokes parameters, I, Q, U, V, whose operational definition is given in various text-books and needs not to be repeated here (see for instance Shurcliff 1962). In strict analogy with the usual transfer equation for unpolarized radiation:

$$\frac{dI}{ds} = - k \ (I - S), \tag{1}$$

the Stokes parameters obey a similar differential equation of the form:

$$\frac{d}{ds} \mathbf{I} = - \mathbf{K} \ (\mathbf{I} - \mathbf{S}), \tag{2}$$

where $\mathbf{I} = (I, Q, U, V)^{\dagger}$ is the Stokes vector, $\mathbf{K}$ is a 4×4 matrix describing the absorption or, more generally, the modification of the Stokes parameters along the ray-path coordinate s, and, finally $\mathbf{S} = (S_I, S_Q, S_U, S_V)^{\dagger}$ is the source function vector in the four Stokes parameters.

Equation (2) can be derived from quantum electrodynamics through a general method that consists in finding the time evolution of the quantum operators $\hat{\mathbf{I}}(\nu,\vec{\Omega})$ whose expectation values, in the quantum-mechanical sense, are the Stokes parameters of a radiation pencil propagating at frequency ν along the direction $\vec{\Omega}$. In the formalism of second quantization these operators are given by:

$$\begin{aligned}
\hat{I}(\nu,\vec{\Omega}) &= b_\nu \left[a^{\dagger}(\nu,\vec{\Omega},\vec{e}_+)\, a(\nu,\vec{\Omega},\vec{e}_+) + a^{\dagger}(\nu,\vec{\Omega},\vec{e}_-)\, a(\nu,\vec{\Omega},\vec{e}_-) \right],\\
\hat{Q}(\nu,\vec{\Omega}) &= b_\nu \left[-a^{\dagger}(\nu,\vec{\Omega},\vec{e}_+)\, a(\nu,\vec{\Omega},\vec{e}_-) - a^{\dagger}(\nu,\vec{\Omega},\vec{e}_-)\, a(\nu,\vec{\Omega},\vec{e}_+) \right],\\
\hat{U}(\nu,\vec{\Omega}) &= i\ b_\nu \left[-a^{\dagger}(\nu,\vec{\Omega},\vec{e}_+)\, a(\nu,\vec{\Omega},\vec{e}_-) + a^{\dagger}(\nu,\vec{\Omega},\vec{e}_-)\, a(\nu,\vec{\Omega},\vec{e}_+) \right],\\
\hat{V}(\nu,\vec{\Omega}) &= b_\nu \left[a^{\dagger}(\nu,\vec{\Omega},\vec{e}_+)\, a(\nu,\vec{\Omega},\vec{e}_+) - a^{\dagger}(\nu,\vec{\Omega},\vec{e}_-)\, a(\nu,\vec{\Omega},\vec{e}_-) \right],
\end{aligned} \tag{3}$$

where $b_\nu = 2h\nu^3 c^{-2}$ is a constant, and where $a^\dagger(\nu,\vec{\Omega},\vec{e}_\gamma)$ and $a(\vec{\nu},\vec{\Omega},\vec{e}_\gamma)$ are the creation and annihilation operators of a photon of frequency ν, direction $\vec{\Omega}$ and polarization described by the unit vector $\vec{e}_\gamma$. The unit vectors $\vec{e}_+$ and $\vec{e}_-$ introduced in the formula above are given by:

$$\begin{aligned} \vec{e}_+ &= (-\vec{e}_a + i\,\vec{e}_b)\ /\ \sqrt{2}, \\ \vec{e}_- &= (\vec{e}_a + i\,\vec{e}_b)\ /\ \sqrt{2}, \end{aligned} \tag{4}$$

where $\vec{e}_a$ and $\vec{e}_b$ are two real unit vectors, both perpendicular to the direction $\vec{\Omega}$, and such that $\vec{e}_a, \vec{e}_b, \vec{\Omega}$, in this order, form a right-handed orthogonal system. $\vec{e}_a$ is the unit vector defining the positive Q direction.

Evaluating the time evolution of the Stokes-parameters through the principles of quantum electrodynamics and introducing a set of simplifying assumptions on the density-matrix operator of the coupled system (consisting of the radiation field interacting with the atomic system), Equation (2) is finally recovered and the elements of the absorption matrix **K** and those of the source-function vector **S** can be expressed through the density-matrix elements of the atomic system. From the quantum-mechanical derivation, whose details can be found in Landi Degl'Innocenti (1983), it follows that the matrix **K** can be expressed in the remarkably symmetrical form:

$$\mathbf{K} = \begin{pmatrix} \eta_I & \eta_Q & \eta_U & \eta_V \\ \eta_Q & \eta_I & \rho_V & -\rho_U \\ \eta_U & -\rho_V & \eta_I & \rho_Q \\ \eta_V & \rho_U & -\rho_Q & \eta_I \end{pmatrix} . \tag{5}$$

This symmetry property of the matrix **K** is a typical characteristic of radiative transfer for polarized radiation and it can be shown, by means of a number of "Gedankenexperimente" that it is intimately connected with the reversible character of the interaction of radiation with matter and with the basic property of the Stokes parameters:

$$I^2 \geq Q^2 + U^2 + V^2 \ . \tag{6}$$

Indeed, if one supposes, for instance, that for a particular medium the two matrix elements K_{12} and K_{21} are different, it is possible to show that a radiation beam having a well-defined polarization signature will end up, propagating through such a medium, with having unphysical Stokes-parameters that do not obey any longer to the basic restriction implicit in Equation (6). For a formal derivation of these statements the reader is referred to Landi Degl'Innocenti and Landi Degl'Innocenti (1981).

The physical interpretation of the various coefficients appearing in Equation (5) is the following: η_I describes the absorption properties of the medium irrespective of the polarization state of the radiation beam and is the generalization of the absorption coefficient appearing in Equation (1). The other three terms, η_Q, η_U, η_V, describe the coupling, due to absorption, of the intensity I of the beam with the Stokes parameters Q, U, and V, respectively. Finally, the three terms ρ_Q, ρ_U, and ρ_V, describe the cyclical coupling of Q, U, and V among each other as a result of anomalous dispersion effects.

In the particular case that is more commonly met in astrophysical applications, namely the case where the polarized radiation beam is interacting with a LTE Zeeman triplet split by a magnetic field, the various coefficients are given, as functions of wavelength, by the expressions:

$$\begin{aligned}
\eta_I &= k_c + \frac{1}{2} k_L \left[\eta_p \sin^2\psi + \frac{(\eta_b + \eta_r)}{2} (1 + \cos^2\psi) \right] \\
\eta_Q &= \frac{1}{2} k_L \left[\eta_p - \frac{\eta_b + \eta_r}{2} \right] \sin^2\psi \cos 2\phi \\
\eta_U &= \frac{1}{2} k_L \left[\eta_p - \frac{\eta_b + \eta_r}{2} \right] \sin^2\psi \sin 2\phi \\
\eta_V &= \frac{1}{2} k_L (\eta_r - \eta_b) \cos\psi \\
\rho_Q &= \frac{1}{2} k_L \left[\rho_p - \frac{\rho_b + \rho_r}{2} \right] \sin^2\psi \cos 2\phi \\
\rho_U &= \frac{1}{2} k_L \left[\rho_p - \frac{\rho_b + \rho_r}{2} \right] \sin^2\psi \sin 2\phi \\
\rho_V &= \frac{1}{2} k_L (\rho_r - \rho_b) \cos\psi
\end{aligned} \tag{7}$$

where k_c and k_L are the continuous and line absorption coefficients, respectively; $\eta_p, \eta_b, \eta_r, \rho_p, \rho_b, \rho_r$ are the absorption and the anomalous dispersion profiles centered at line center (p-components) and line center plus or minus Zeeman splitting (r- and b-components, respectively), and, finally, ψ and ϕ are the polar and azimuthal angles defining the magnetic field direction with respect to the line of sight (see Landi Degl'Innocenti 1976, for a more detailed explanation of the various symbols).

In this physical situation, the term ρ_V describes the so-called phenomenon of Faraday rotation, being responsible, through the coupling of Q with U, of a rotation of the plane of linear polarization, while the terms ρ_Q and ρ_U describe the analogous phenomenon of Faraday pulsation transforming linear into circular polarization and vice versa.

The particular case considered above is also suitable for introducing some remarks on the source function vector **S** appearing in Equation (2). Indeed, when the LTE approximation is valid, the emission vector in the four Stokes parameters can be written, according to the principle of detailed balance, in the form:

$$\begin{aligned}
\epsilon_I &= \eta_I \; B \\
\epsilon_Q &= \eta_Q \; B \\
\epsilon_U &= \eta_U \; B \\
\epsilon_V &= \eta_V \; B
\end{aligned} \tag{8}$$

where B is the local Planck function.

In LTE we then have:

$$\mathbf{S} = (B, 0, 0, 0)^{\dagger} \tag{9}$$

which means that the source function in the Stokes-parameters Q, U, and V is identically zero.

This property of the source function remains valid even in the non-LTE situation, provided that phenomena of atomic polarization are neglected (non-LTE of the 1st kind). In this respect, it might be necessary to remind the reader that an atomic system is said to be polarized when its magnetic sublevels (degenerate with respect to the energy in the non-magnetic situation) are unevenly populated or when non-negligible phase relationships (or coherences) are present among the sublevels themselves.

In the presence of an anisotropic, or a polarized field (or both), various phenomena of atomic polarization become essential in determining the absorption and emission of polarized radiation, unless the presence of collisions with neutral or charged perturbers is able of destroying the atomic polarization itself. In this situation, which is referred to as non-LTE of the 2nd kind, the source-function vector has, in general, all its four components different from zero, and their detailed expressions can be found, as a function of the density-matrix elements of the atomic system, through the theory developed by Landi Degl'Innocenti (1983).

When non-LTE effects are important, the radiative transfer equations for polarized radiation have to be coupled with the statistical equilibrium equations for the density-matrix elements of the atomic system. These equations, where the presence of a magnetic field plays a very important role in destroying coherences between near-lying levels (a phenomenon responsible for the Hanle effect), can be derived from the general principles of quantum electrodynamics through a method strictly analogous to the one previously outlined for deducing the radiative transfer equations for polarized radiation. Up to the present time, the system of coupled equations, consisting of the radiative transfer equations for polarized radiation and the statistical equilibrium equations for the density matrix operator, has never been solved self-consistently, although in some simplified physical situations a perturbative approach has shown to be possible (Landi Degl'Innocenti et al. 1986). For this last application, as well as for many different problems that arise in the interpretation of polarimetric observations from solar active regions, it is particularly important to obtain a solution of the radiative transfer equations for polarized radiation. This solution, which can be considered a direct generalization to the "polarized case" of the familiar solution of Equation (1):

$$\begin{aligned} I(s) = I(s_0)\ \exp\left(-\int_{s_0}^{s} k(s')ds'\right) + \\ + \int_{s_0}^{s} S(s')\exp\left(-\int_{s'}^{s} k(s'')\,ds''\right) k(s')ds' \quad , \end{aligned} \tag{10}$$

is the natural keystone of the development of more sophisticated transfer methods for polarized radiation.

2. FORMULATION

To give a formal solution of Equation (2), we start by considering the homogeneous equation:

$$\frac{d}{ds} \mathbf{I}(s) = -\mathbf{K}(s)\ \mathbf{I}(s). \tag{11}$$

We define the evolution operator $\mathbf{O}(s, s')$ as the linear operator that, acting on the Stokes vector at the point s', gives the Stokes vector at point s:

$$\mathbf{I}(s) = \mathbf{O}(s, s')\ \mathbf{I}(s')\ . \tag{12}$$

Obviously, this operator exists for any given function $\mathbf{K}(s)$ that is well-behaved in the physical sense. Moreover, it obeys the limiting condition:

$$\mathbf{O}(s, s) = 1\ , \tag{13}$$

1 being the 4×4 identity matrix, and the composition law:

$$\mathbf{O}(s, s') = \mathbf{O}(s, s'')\ \mathbf{O}(s'', s')\ . \tag{14}$$

Taking the derivative of Equation (12) with respect to s and comparing the result with Equation (11) we obtain the differential equation for the evolution operator:

$$\frac{d}{ds} \mathbf{O}(s, s') = -\ \mathbf{K}(s)\ \mathbf{O}(s, s')\ ; \tag{15}$$

while differentiating Equation (12) with respect to s' and taking into account Equation (15) we obtain:

$$\frac{d}{ds'} \mathbf{O}(s, s') = \mathbf{O}(s, s')\ \mathbf{K}(s')\ . \tag{16}$$

Returning now to the inhomogeneous equation (2), it is easy to prove, by means of a direct substitution, that this equation is solved by the expression:

$$\mathbf{I}(s) = \int_{s_0}^{s} \mathbf{O}(s, s')\ \mathbf{K}(s')\ \mathbf{S}(s')\ ds' + \mathbf{O}(s, s_0)\ \mathbf{I}(s_0)\ . \tag{17}$$

For a semi-infinite atmosphere, provided that the source function does not increase above reasonable limits when $s \rightarrow -\infty$, or, in mathematical terms, provided that:

$$\lim_{s' \rightarrow -\infty} \mathbf{O}(s, s')\ \mathbf{K}(s')\ \mathbf{S}(s') = 0\ , \tag{18}$$

the formal solution of Equation (2) is simply given by:

$$\mathbf{I}(s) = \int_{-\infty}^{s} \mathbf{O}(s, s')\ \mathbf{K}(s')\ \mathbf{S}(s')\ ds'\ . \tag{19}$$

On the other hand, a closed analytic expression for the operator $\mathbf{O}(s, s')$ can be found by first integrating Equation (15) to obtain:

$$\mathbf{O}(s,s') = 1 - \int_{s'}^{s} \mathbf{K}(s_1)\,\mathbf{O}(s_1,s')\,ds_1 \;, \tag{20}$$

and then substituting, for the operator, $\mathbf{O}(s_1,s')$ in the right-hand side, its expression as given by the left-hand side of the same equation. Iterating this procedure, we obtain:

$$\mathbf{O}(s,s') = 1 + \sum_{n=1}^{\infty} (-1)^n \int_{s'}^{s} ds_1 \int_{s'}^{s_1} ds_2 \ldots \int_{s'}^{s_{n-1}} ds_n \; \mathbf{K}(s_1)\,\mathbf{K}(s_2) \ldots \mathbf{K}(s_n), \tag{21}$$

or, alternatively:

$$\mathbf{O}(s,s') = 1 + \sum_{n=1}^{\infty} \frac{(-1)^n}{n!} \int_{s'}^{s} ds_1 \int_{s'}^{s} ds_2 \cdots \int_{s'}^{s} ds_n \; P\,\{\mathbf{K}(s_1)\,\mathbf{K}(s_2) \cdots \mathbf{K}(s_n)\} \tag{22}$$

where P is the chronological operator that orders the various matrices according to the following law:

$$P\,\{\mathbf{K}(s_1)\;\mathbf{K}(s_2) \ldots \mathbf{K}(s_n)\} = \mathbf{K}(s_{j1})\,\mathbf{K}(s_{j2}) \ldots \mathbf{K}(s_{jn}) \;. \tag{23}$$

when $s_{j1} \geq s_{j2} \geq \ldots \geq s_{jn}$.

The formal operator $\mathbf{O}(s,s')$ is nothing but the generalization to the radiative transfer problem for polarized radiation of the familiar attenuation operator:

$$\exp\left(-\int_{s'}^{s} k(s'')\,ds''\right) \tag{24}$$

that is usually met in the solution of Equation (1). The fact that a chronological ordering operator appears in Equation (22) is not surprising, and is intimately connected to the physical fact that two different slabs, a and b, acting differently on the polarization properties of a radiation beam, do not "commute" in the sense that the emerging polarization is in general different if slab a is located in front of slab b, or vice versa. On the other hand, in the situation where the polarization properties of the radiation beam are neglected, the ordering of the slabs is not improtant.

Unfortunately, it is impossible to reduce to a simpler form the operator $\mathbf{O}(s,s')$ except in some particular cases. One possibility is given by the case where the matrix $\mathbf{K}$ is constant in the interval (s',s), or, more generally, when $\mathbf{K}(s)$ can be written in the form $\mathbf{K}(s) = f(s)\,\mathbf{K}'$, with $\mathbf{K}'$ constant. This last case can be brought back to the previous one by just rescaling the coordinate s. In the following, we will derive a closed form for the evolution operator $\mathbf{O}(s,s')$ in the case $\mathbf{K}$ = constant. This closed form is very useful in practice as it saves the eigenvalue and eigenvector determination for the matrix $\mathbf{K}$ that would otherwise be necessary in more conventional solutions of Equation (3). For $\mathbf{K}$ = constant, we obtain from Equation (22):

$$\mathbf{O}(s,s') = 1 + \sum_{n-1} (-1)^n \frac{(s-s')^n}{n!} \mathbf{K}^n \;, \tag{25}$$

so that we can write:

$$\mathbf{O}(s, s') = \exp[-(s - s')\,\mathbf{K}] \tag{26}$$

where the meaning of the exponential of a matrix is given by its Taylor expansion. This expression can be evaluated in terms of the elements of the matrix **K**. This is done by writing **K** as a linear combination of 4 x 4 matrices whose properties enables the reduction of the infinite products appearing in Equation (25) to a closed form. The two sets of matrices that are introduced in the following have a close similarity in their commutation properties with the well-known Pauli matrices. We introduce the following formalism:

$$\mathbf{K} = \eta_I 1 + \vec{a} \cdot \vec{\mathbf{A}} + \vec{b} \cdot \vec{\mathbf{B}}$$

where $\vec{a}$ and $\vec{b}$ are the formal vectors:

$$\begin{aligned} \vec{a} &= \vec{\eta} + i\vec{\rho} = \frac{1}{2}(\eta_Q + i\rho_Q, \eta_U + i\rho_U, \eta_V + i\rho_V) \\ \vec{b} &= \vec{\eta} - i\vec{\rho} = \frac{1}{2}(\eta_Q - i\rho_Q, \eta_U - i\rho_U, \eta_V - i\rho_V) \ , \end{aligned} \tag{27}$$

and $\vec{\mathbf{A}}$ and $\vec{\mathbf{B}}$ are the matrices defined by:

$$\mathbf{A}_1 = \begin{bmatrix} 0 & 1 & 0 & 0 \\ 1 & 0 & 0 & 0 \\ 0 & 0 & 0 & -i \\ 0 & 0 & i & 0 \end{bmatrix} \qquad \mathbf{A}_2 = \begin{bmatrix} 0 & 0 & 1 & 0 \\ 0 & 0 & 0 & i \\ 1 & 0 & 0 & 0 \\ 0 & -i & 0 & 0 \end{bmatrix}$$
$$\mathbf{A}_3 = \begin{bmatrix} 0 & 0 & 0 & 1 \\ 0 & 0 & -i & 0 \\ 0 & i & 0 & 0 \\ 1 & 0 & 0 & 0 \end{bmatrix} \ , \tag{28}$$

$$\mathbf{B}_i = \mathbf{A}_i^* \ . \tag{29}$$

By means of this formalism, taking into account the following properties of the matrices $\mathbf{A}_i$ and $\mathbf{B}_j$:

$$
\begin{aligned}
[\mathbf{A}_i \ , \mathbf{B}_j] &= \mathbf{A}_i \mathbf{B}_j - \mathbf{B}_j \mathbf{A}_i = 0 \ , \\
\mathbf{A}_i \mathbf{A}_j &= \delta_{ij} 1 + i \sum_k \epsilon_{ijk} \mathbf{A}_k \ , \\
\mathbf{B}_i \mathbf{B}_j &= \delta_{ij} 1 - i \sum_k \epsilon_{ijk} \mathbf{B}_k \ ,
\end{aligned}
\tag{30}
$$

where ϵ_{ijk} is the complete antisymmetrical tensor, and the further property:

$$
\exp(\mathbf{A}+\mathbf{B}) = \exp(\mathbf{A}) \exp(\mathbf{B}) \ \text{ if } \ [\mathbf{A}, \mathbf{B}] = 0 \ , \tag{31}
$$

one can write, wiwith the substitution $s - s' = x$:

$$
\mathbf{O}(x) = \exp[-\eta_I \ x \ 1] \exp[-\vec{a} \cdot \vec{\mathbf{A}} \ x] \exp[-\vec{b} \cdot \vec{\mathbf{B}} \ x] \ . \tag{32}
$$

Developing the exponentials in their Taylor expansions, the following expressions are easily obtained:

$$
\begin{aligned}
\exp[-\eta_I \ x \ 1] &= \exp(-\eta_I \ x) \ 1 \\
\exp[-\vec{a} \cdot \vec{\mathbf{A}} \ x] &= \cosh(ax) \ 1 - \frac{\sinh(ax)}{a} \ \vec{a} \cdot \vec{\mathbf{A}} \\
\exp[-\vec{b} \cdot \vec{\mathbf{B}} \ x] &= \cosh(bx) \ 1 - \frac{\sinh(bx)}{b} \ \vec{b} \cdot \vec{\mathbf{B}}
\end{aligned}
\tag{33}
$$

where

$$
\begin{aligned}
a &= (\vec{a} \cdot \vec{a})^{1/2} \ , \\
b &= (\vec{b} \cdot \vec{b})^{1/2} \ .
\end{aligned}
\tag{34}
$$

Finally, substituting Equations (33) into Equation (32) and taking into account the conjugation properties of the matrices **A** and **B**, the following expression is obtained for the evolution operator (for further details the reader is referred to Landi Degl'Innocenti and Landi Degl'Innocenti 1985):

$$\begin{aligned}\mathbf{O}(s\,,s'\,) &= \exp[-x\,\mathbf{K}] = \\ &= \exp(-\eta_I\,x\,)\left\{\frac{1}{2}\,[\cosh(\Lambda_1 x\,) + \cos(\Lambda_2 x\,)]\;\mathbf{1} - \sin(\Lambda_2 x\,)\,\mathbf{M}_2\right. \\ &\left. - \sinh(\Lambda_1 x\,)\,\mathbf{M}_3 + \frac{1}{2}\,[\cosh(\Lambda_1 x\,) - \cos(\Lambda_2 x\,)]\,\mathbf{M}_4\right\}\end{aligned} \tag{35}$$

where:

$$\mathbf{M}_2 = \frac{1}{\theta}\begin{bmatrix} 0 & \Lambda_2\eta_Q - \sigma\Lambda_1\rho_Q & \Lambda_2\eta_U - \sigma\Lambda_1\rho_U & \Lambda_2\eta_V - \sigma\Lambda_1\rho_V \\ \Lambda_2\eta_Q - \sigma\Lambda_1\rho_Q & 0 & \sigma\Lambda_1\eta_V + \Lambda_2\rho_V & -\sigma\Lambda_1\eta_U - \Lambda_2\rho_U \\ \Lambda_2\eta_U - \sigma\Lambda_1\rho_U & -\sigma\Lambda_1\eta_V - \Lambda_2\rho_V & 0 & \sigma\Lambda_1\eta_Q + \Lambda_2\rho_Q \\ \Lambda_2\eta_V - \sigma\Lambda_1\rho_V & \sigma\Lambda_1\eta_U + \Lambda_2\rho_U & -\sigma\Lambda_1\eta_Q - \Lambda_2\rho_Q & 0 \end{bmatrix} \tag{36}$$

$$\mathbf{M}_3 = \frac{1}{\theta}\begin{bmatrix} 0 & \Lambda_1\eta_Q + \sigma\Lambda_2\rho_Q & \Lambda_1\eta_U + \sigma\Lambda_2\rho_U & \Lambda_1\eta_V + \sigma\Lambda_2\rho_V \\ \Lambda_1\eta_Q + \sigma\Lambda_2\rho_Q & 0 & -\sigma\Lambda_2\eta_V + \Lambda_1\rho_V & \sigma\Lambda_2\eta_U - \Lambda_1\rho_U \\ \Lambda_1\eta_U + \sigma\Lambda_2\rho_U & \sigma\Lambda_2\eta_V - \Lambda_1\rho_V & 0 & -\sigma\Lambda_2\eta_Q + \Lambda_1\rho_Q \\ \Lambda_1\eta_V + \sigma\Lambda_2\rho_V & -\sigma\Lambda_2\eta_U + \Lambda_1\rho_U & \sigma\Lambda_2\eta_Q - \Lambda_1\rho_Q & 0 \end{bmatrix} \tag{37}$$

$$\mathbf{M}_4 = \frac{2}{\theta}\begin{bmatrix} (\eta^2+\rho^2)/2 & \eta_V\,\rho_U - \eta_U\,\rho_V & \eta_Q\,\rho_V - \eta_V\,\rho_Q & \eta_U\,\rho_Q - \eta_Q\,\rho_U \\ \eta_U\,\rho_V - \eta_V\,\rho_U & \eta_Q^2+\rho_Q^2-(\eta^2+\rho^2)/2 & \eta_Q\,\eta_U + \rho_Q\,\rho_U & \eta_V\,n_Q + \rho_V\,\rho_Q \\ \eta_V\,\rho_Q - \eta_Q\,\rho_V & \eta_Q\,\eta_U + \rho_Q\,\rho_U & \eta_U^2+\rho_U^2-(\eta^2+\rho^2)/2 & \eta_U\,\eta_V + \rho_U\,\rho_V \\ \eta_Q\,\rho_U - \eta_U\,\rho_Q & \eta_V\,\eta_Q + \rho_V\,\rho_Q & \eta_U\,\eta_V + \rho_U\,\rho_V & \eta_V^2+\rho_V^2-(\eta^2+\rho^2)/2 \end{bmatrix} \tag{38}$$

and where:

$$\Lambda_1 = \left\{\,[(\eta^2-\rho^2)/4 + (\vec{\eta}\cdot\vec{\rho})^2]^{1/2} + (\eta^2-\rho^2)/2\,\right\}^{1/2}, \tag{39}$$

$$\Lambda_2 = \left\{\,[(\eta^2-\rho^2)/4 + (\vec{\eta}\cdot\vec{\rho})^2]^{1/2} - (\eta^2-\rho^2)/2\,\right\}^{1/2}, \tag{40}$$

$$\sigma = (\vec{\eta}\cdot\vec{\rho})/\;|\;(\vec{\eta}\cdot\vec{\rho})\;|\; = \mathrm{sign}(\vec{\eta}\cdot\vec{\rho})\,, \tag{41}$$

$$\theta = \Lambda_1^2+\Lambda_2^2 = 2[(\eta^2-\rho^2)^2/4+(\vec{\eta}\cdot\vec{\rho})^2]^{1/2}\,. \tag{42}$$

The last expression that we have found for the operator $\mathbf{O}(s, s')$ (Equation 35) can be conveniently applied to find a numerical solution of the radiative transfer equations for polarized radiation by means of the following algorithm: (a) we divide the path of integration ($-\infty \leq s' \leq s$) into n intervals; for each of the them, we suppose the matrix $\mathbf{K}$ to be constant and equal, for instance, to its value at the mid-point of the interval itself. Introducing for the grid-points the notation: $s_0, s_1, s_2 \ldots s_{n-1}, s_n$ with $s_0 = -\infty$, and $s_n = s$, we have, in the i-th interval:

$$\mathbf{K}(t) = \mathbf{K}((s_{i-1} + s_i)/2) = \mathbf{K}_i \qquad s_{i-1} \leq t \leq s_i \;, \tag{43}$$

(b) for each interval, we calculate numerically the operator $\mathbf{O}_i$ defined according to Equation (35):

$$\mathbf{O}_i = \exp\left[-(s_i - s_{i-1})\,\mathbf{K}_i\right] \;, \tag{44}$$

and we calculate, by means of a standard numerical integration (for instance a Laguerre integration), the vector $\mathbf{V}_i$ defined by:

$$\mathbf{V}_i = \int_{s_{i-1}}^{s_i} \exp\left[-(s_i - s')\,\mathbf{K}_i\right] \mathbf{K}_i\, \mathbf{S}(s')\, ds' \;; \tag{45}$$

(c) finally, we recover the Stokes parameters $\mathbf{I}(s)$ by means of the equation:

$$\mathbf{I}(s) = \sum_{i=1}^{n} \mathbf{O}_n \mathbf{O}_{n-1} \cdots \mathbf{O}_{i+1} \mathbf{V}_i \;. \tag{46}$$

An alternative form for $\mathbf{I}(s)$ can also be obtained by noticing that the vector $\mathbf{V}_i$ can be written in the form:

$$\begin{aligned} \mathbf{V}_i &= \int_{s_{i-1}}^{s_i} \left\{ \frac{d}{ds'} \exp\left[-(s_i - s')\,\mathbf{K}_i\right] \right\} \mathbf{S}(s')\, ds' = \\ &= \mathbf{S}_i - \mathbf{O}_i \mathbf{S}_{i-1} + \mathbf{V}_i' \;, \end{aligned} \tag{47}$$

where: $\mathbf{S}_i = \mathbf{S}(s_i)$, and

$$\mathbf{V}_i' = \int_{s_{i-1}}^{s_i} \exp\left[-(s_i - s')\,\mathbf{K}_i\right] \left(\frac{d}{ds'}\,\mathbf{S}(s') \right) ds' \quad . \tag{48}$$

Substituting into Equation (45) and supposing the validity of Equation (18) for the source function at large depths in the atmosphere, we obtain an alternative form for $\mathbf{I}(s)$:

$$\mathbf{I}(s) = \mathbf{S}_n + \sum_{i=1}^{n} \mathbf{O}_n\,\mathbf{O}_{n-1} \cdots \mathbf{O}_{i+1}\,\mathbf{V}_i' \quad . \tag{49}$$

3. COMPARISON WITH ALTERNATIVE METHODS

The integration method that has been presented in the previous section suffers from the main limitation consisting in the fact that the absorption matrix $\mathbf{K}$ is considered constant in each of the n intervals in which the integration path is divided. Obviously, the method gives a better approximation to the true solution for the emerging Stokes parameters the larger is the value that is chosen for n. The main advantage of this method, with respect to alternative methods that are often used for the solution of the radiative transfer equations for polarized radiation, has to be sought in the fact that the intervals defining the integration grid can be selected a-priori taking into account the behavior of the matrix $\mathbf{K}$ with s, as expected in the physical problem investigated. In other words, the grid can be conveniently adjusted to the gradient of $\mathbf{K}$ and can be made finer where this gradient is larger and vice-versa.

In alternative numerical methods, like for instance the fourth-order Runge-Kutta technique, the integration grid has to be primarily controlled by taking into account the local eigenvalues for the matrix $\mathbf{K}$, and it is found that the number of grid-points necessary to have a meaningful integration is of the order of some units times r, where r is the ratio between the maximum and the minimum of the absolute values of the eigenvalues (Landi Degl'Innocenti 1976). As in some cases this ratio can attain extremely large values, the fourth-order Runge-Kutta technique becomes, accordingly, a very inefficient method for the solution of the radiative transfer equations for polarized radiation, and the method proposed in this paper is to be preferred in these cases.

4. APPLICATION TO NON-LTE PROBLEM

In Section 2 we have obtained a *formal* solution of the equations of radiative transfer for polarized radiation and we have shown how it is possible to transform the formal solution into a suitable algorithm for obtaining a *numerical* solution of Equation (2). As we have supposed that both the absorption matrix **K** and the source vector **S** are known functions of the coordinate s , a direct application of the formalism of Section 2 can be found in the usual problem of recovering the emerging Stokes parameters from a LTE atmosphere in the presence of a magnetic field.

When non-LTE phenomena enter into play, the source function **S** contains an integral over the solid angle and frequencies of the Stokes vector **I** and the problem in its generality becomes much more involved. There are, however, several particular cases where the method outlined above can still be applied taking into account of a number of results that suggest the introduction of an iterative scheme.

Considering for instance the problem of non-LTE line formation in the presence of a magnetic field, it has been shown, for a two-level atom under the assumption of Complete Redistribution (CRD) and in the hypothesis of neglecting atomic polarization of the upper level, that the scalar source function in the magnetic situation does not differ by large amounts from the analogous source function that is calculated for the same atmosphere in the absence of a magnetic field (Rees 1969). More quantitatively, a maximum difference of the order of 20% is reached between the two source functions, this maximum value being obtained for a particular combination of the relevant parameters.

These results have been subsequently confirmed by Domke and Staude (1973) and by Landi Degl'Innocenti (1978) and suggest that possibility of attacking the non-LTE problem for polarized radiation by first calculating the source function, S_0 through one of the usual methods that are commonly employed in the transfer problem for unpolarized radiation (see for instance, Mihalas, 1978), and then calculating the zero-order Stokes parameters $\mathbf{I}^{(0)}(P,\nu,\vec{\Omega})$, at any point P, frequency ν, and direction $\vec{\Omega}$, through the methods of Section 2, with a zero-order vector source function, $\mathbf{S}^{(0)}$, given by:

$$\mathbf{S}^{(0)} = (S_0,0,0,0)^{\dagger} \ .$$

If necessary, this procedure can be iterated, by introducing the Stokes parameters previously calculated into the equations of statistical equilibrium, finding a first order source function $\mathbf{S}^{(1)}$ and first-order Stokes parameters $\mathbf{I}^{(1)}(P,\vec{\Omega},\nu)$, and going on along the same scheme. Although a mathematical proof of the convergence of this perturbative procedure has never been given, it has to be expected on general physical grounds that the correct solution might be finally obtained.

This perturbative scheme, which has been outlined here in connection with the problem of non-LTE line formation in a magnetic field, can also be applied with a certain degree of confidence to all those cases where it is a-priori known that polarization phenomena act as a minor perturbation in the overall physical picture of the transfer problem. To be specific on a particular example we can consider the formation of Hydrogen Balmer-lines in optically thick prominences, where the observed linear polarization (due to the anisotropic illumination of the prominence plasma by the photospheric radiation field and to the accompanying action of the magnetic field through the Hanle effect) is of the order of a fraction of a percent. In this case, a perturbative approach appears to be fully justified and a detailed calculation along these lines has indeed been attempted (Landi Degl'Innocenti et al. 1986).

In conclusion, due to the extreme complexity that non-LTE problems of radiative transfer for polarized radiation can easily attain even in the more schematic physical situations, we think that perturbative approaches may become of extreme importance, in the near future, for clarifying the underlying physics and for finding appropriate solutions of various interesting astrophysical problems.

REFERENCES

Domke, H., Staude, J. 1973, *Solar Phys.* **31**,279.

Landi Degl'Innocenti, E. 1976, *Astron. Astrophys. Suppl.* **25**,379.

___________ 1978, *Astron. Astrophys.* **66,** 119.

___________ 1983, *Solar Phys.* **85,** 3.

Landi Degl'Innocenti, E., Landi Degl'Innocenti, M. 1981, *Nuovo Cimento* **62B,** 1.

___________ 1985, *Solar Phys.* **97,** 239.

Landi Degl'Innocenti, E., Bommier, V., Sahal-Brechot, S. 1986, *Astron. Astrophys.*, in press.

Mihalas, D. 1978, *Stellar Atmospheres*, 2nd ed., W. H. Freeman, San Francisco.

Rees, D. E. 1969, *Solar Phys.* **10,** 268.

Shurcliff, W. A. 1962, *Polarized Light*, Harvard Univ. Press, Cambridge, MA.

RADIATIVE TRANSFER IN THE PRESENCE OF STRONG MAGNETIC FIELDS

A. A. van Ballegooijen
Harvard-Smithsonian Center for Astrophysics, Cambridge, USA

ABSTRACT: We describe a method for the solution of non-LTE radiative transfer problems in a plan-parallel, semi-infinite atmosphere in the presence of large Zeeman splitting (i.e., Zeeman splitting larger than the Doppler width of the spectral lines). The transport equation for polarized light is written in terms of complex 2×2 matrices, and a formal solution analogous to the well-known integral expression for the intensity of unpolarized light is obtained. We then use this formal solution together with the statistical equilibrium equations to derive an integral equation for the line source function S_L. Finally, we outline a numerical method for the computation of the integral operator Λ which describes the relation between the line source function S_L and the mean integrated intensity $\bar{J}$. The method applies to line formation in sunspots (and starspots), and in magnetic Ap stars, where magnetic field strengths are several kG.

1. INTRODUCTION

The problem of the transfer of radiation through a plasma or gas in the presence of a magnetic field has received considerable attention in astrophysics. Magnetic fields are found in a great variety of astrophysical objects, and the most direct identification of a magnetic field is via its effect of the spectrum and state of polarization of the light emitted by the object. The first detection of an extraterrestrial magnetic field was the observation of Zeeman splitting in the spectrum of sunspots (Hale, 1908). Since then, the sun has been studied extensively using a variety of techniques based on the Zeeman effect (cf. Harvey 1977; Rees 1982; Stenflo 1985). From these studies we know that the solar magnetic field is highly structured and variable: strong fields are present in sunspots ($B \sim 3000$ G), and in small-scale magnetic structures outside sunspots ($B \sim 1500$ G). Magnetic activity occurs also on other late-type stars (for a recent review, see Baliunas & Vaughan 1985). For some of the more active stars it is possible to detect magnetic fields directly, using the Zeeman effect (Boesgaard 1974; Robinson et al. 1980; Brown & Landstreet 1981; Marcy 1983, 1984; Borra et al. 1984; Gray 1984). However, in addition there are many indirect indicators of magnetic activity, such as the existence of starspots (e.g. Vogt 1982), chromospheric emissions in the CaII H and K lines (e.g. Wilson 1978), and emissions at UV and X-ray wavelengths (e.g. Ayres et al. 1981; Hartmann et al. 1982; Vaiana et al. 1981). As on the sun, the magnetic field is attributed to an as yet poorly understood dynamo mechanism operating in the convective envelopes of these stars.

Magnetic fields are also found in some main-sequence stars of spectral type earlier than the sun, in particular the Ap stars (cf. Borra et al. 1982). In contrast with the magnetically active stars like the sun, these so-called "magnetic" stars have well-ordered magnetic fields which are approximately dipolar. Field strengths range from about 20,000 G down to undetectably small (less than 200 G). Similar dipolar fields, but with much larger field strengths (B = 10^6 - 10^8 G), have been detected on some white dwarf stars (cf. Angel 1978; Angel et al. 1981). These fields are believed to be "fossil", i.e., they are not maintained by some dynamo action presently operating in the star.

In this paper we discuss how Zeeman splitting affects the process of spectral line formation in a stellar atmosphere. Zeeman splitting causes changes in the absorption profiles of spectral lines, and introduces polarization and magneto-optical (or birefringence) effects. The theory of line formation in the presence of a magnetic field was first discussed by Unno (1956), who considered the simplest case of a Zeeman triplet formed in Local Thermodynamic Equilibrium (LTE), neglecting magneto-optical effects. The LTE theory was established on a firm quantum-mechanical basis by Landi Degl'Innocenti & Landi Degl'Innocenti (1972). Various generalizations can be made to account for departures from LTE, in which case the populations of the energy levels are no longer according to Boltzmann's law. In its simplest form, the NLTE theory allows for departures from LTE in the *total* populations of the upper- and lower levels, while retaining the assumption that the individual magnetic sublevels within each level are equally populated (Rees 1969). In this case the departures from LTE are fully described by the departure coefficients b_ℓ and b_u of the lower and upper levels.

Further generalizations of the NLTE theory allow for departures from LTE in the populations of magnetic sublevels (Landi Degl'Innocenti & Landi Degl'Innocenti 1974; Sidlichovsky 1974; House and Steinitz 1975; Auer et al. 1977), and for quantum interferences between magnetic sublevels (cf. Landi Degl'Innocenti 1983, and references therein). If the energy separation of the magnetic sublevels is large compared with their natural width (the so-called "strong field" limit), quantum interferences between sublevels can be neglected, and the sublevels can be treated independently. However, if the Zeeman splitting is smaller or of the order of the natural broadening width, interference effects must be taken into account, and a more complicated density matrix formalism must be used. In the latter case the magnetic field gives rise to the so-called Hanle effect, which affects the linear polarization of resonantly scattered radiation. The Hanle effect has been used, recently, to study weak magnetic fields in solar prominences (Bommier et al. 1981; Landi Degl'Innocenti 1982).

In this paper we discuss magnetic line formation in the simplest form of the NLTE theory, namely, when differences in sublevel populations and interferences are neglected (Rees 1969; Auer et al. 1977), i.e., when the collisional coupling between magnetic sublevels is strong. The transfer equation for this case is the same as in LTE, except that the line source function S_L is no longer given by the local Planck function $B_\nu(T)$. In general, the line source function depends on the departure coefficients b_ℓ and b_u, which are determined by the radiative and collisional excitation rates for the atom in question. The key problem is, therefore, to determine b_ℓ and b_u as function of position in the atmosphere; this must be done by solving the coupled equations of radiative transfer and statistical equilibrium. We will outline a numerical method for solving this problem for arbitrary large Zeeman splitting (splitting larger than the Doppler width). The method is based on the integral equation method for solving radiative transfer problems (Avrett and Hummer 1965; Athay and Skumanich 1967; Mihalas 1978).

There are several limitations to the present NLTE model. In addition to effects of sublevel populations and interferences discussed above, we also assume complete redistribution in the scattering process. This implies that the angle- and frequency distributions of a scattered photon are independent of the direction and frequency of the incoming photon, hence the line source function S_L is angle- and frequency independent. The assumption of complete redistribution is reasonable if the the natural broadening of the energy levels (due to their finite lifetime) is smaller than the collisional broadening. In the zero-field limit the model reduces to the usual NLTE problem with complete redistribution.

The primary applications we have in mind are starspots and magnetic Ap stars, where magnetic field strengths are several kG. Since the magnetic field in these objects varies on a spatial scale which is large compared with the thickness of the line-forming region, we will restrict our discussion to plan-parallel models, and assume that physical quantities such as temperature, density and magnetic field depend only on height in the atmosphere.

The paper is organized as follows. In section 2 we review the transfer equation for polarized radiation in a Zeeman-split spectral line. Then in section 3 we rewrite the transfer equation in terms of complex 2×2 matrices, and we derive a formal solution analogous to the familiar integral expression for unpolarized light. This solution is used in section 4 to formulate an integral equation for the line source function. Finally, in section 5 we describe a numerical method for the evaluation of the integral operator, and a matrix equation for the source function is obtained.

2. THE TRANSFER EQUATION

In the following we formulate the equation of radiative transfer for a Zeeman-sensitive spectral line formed in a magnetic stellar atmosphere. We assume that the atmosphere has a plan-parallel stratification, i.e., physical quantities depend only on height z above some reference level in the atmosphere. The intensity and state of polarization of the light are described by the Stokes parameters I, Q, U and V, and the transfer equation determines how these quantities vary with position. Let $\hat{\mathbf{n}}$ be the unit vector describing the direction of a ray. Then I, Q, U and V are in general functions of direction, $\hat{\mathbf{n}}$, of frequency difference from line center, $\Delta\nu = \nu - \nu_0$, and of height z.

Before writing down the transport equation, let us first define the Stokes parameters more precisely. Let $\hat{\mathbf{z}}$ be the unit vector in the vertical outward direction, and let θ be the angle between $\hat{\mathbf{n}}$ and $\hat{\mathbf{z}}$. As usual we define $\mu \equiv \cos\theta$. Then we define two orthogonal unit vectors $\hat{\mathbf{e}}_1$ and $\hat{\mathbf{e}}_2$ such that:

$$\hat{\mathbf{e}}_1 \equiv \frac{\hat{\mathbf{z}} - \hat{\mathbf{n}}\cos\theta}{\sin\theta}, \tag{2.1}$$

$$\hat{\mathbf{e}}_2 \equiv \hat{\mathbf{n}} \times \hat{\mathbf{e}}_1, \tag{2.2}$$

i.e., $\hat{\mathbf{e}}_1$ and $\hat{\mathbf{e}}_2$ are perpendicular to $\hat{\mathbf{n}}$, and $\hat{\mathbf{e}}_1$ lies in the vertical plane that contains $\hat{\mathbf{n}}$. Note that the reference frame $(\hat{\mathbf{e}}_1, \hat{\mathbf{e}}_2, \hat{\mathbf{n}})$ is *right-handed*. Now let the electric field $\mathbf{E}$ associated with a light wave be given by:

$$\mathbf{E}(t) \propto \mathrm{Re}\left[(E_1\hat{\mathbf{e}}_1 + E_2\hat{\mathbf{e}}_2)\, e^{2\pi i\nu t}\right], \tag{2.3}$$

where t is the time, ν is the nominal frequency of the wave, and E_1 and E_2 are complex quantities describing the amplitude and phase of the two wave components. Then the Stokes parameters are defined as:

$$\begin{aligned} I &= <|E_1|^2> + <|E_2|^2>, \\ Q &= <|E_1|^2> - <|E_2|^2>, \\ U &= \pm 2\,\mathrm{Re}\,[<E_1E_2^*>], \\ V &= 2\,\mathrm{Im}\,[<E_1E_2^*>], \end{aligned} \tag{2.4}$$

where $< \cdots >$ denotes a statistical average over all possible values of E_1 and E_2. The sign in the definition of U refers to outgoing rays ($\mu > 0$) and incoming rays ($\mu < 0$), respectively. This sign convention, which is different from the one usually adopted in the literature, is used here because it leads to a certain symmetry between incoming and outgoing rays (see below).

As discussed in the Introduction we will focus on the Zeeman effect, and neglect phenomena arising from differences in sublevel population (atomic polarization), or interferences between sublevels (Hanle effect). Furthermore, we assume complete redistribution of the scattering process, i.e., the frequency and direction of photons scattered by an atom are assumed to be independent of the frequency and direction of the incoming photons. These assumptions imply that the line emission is characterized by a single, frequency-independent line source function, and the corresponding transport equation is a generalization of the transport equation for the LTE case (e.g. Beckers 1969a):

$$\mu\frac{d}{dz}\begin{pmatrix} I \\ Q \\ U \\ V \end{pmatrix} = -\begin{pmatrix} \kappa_c+\kappa_I & \kappa_Q & \kappa_U & \kappa_V \\ \kappa_Q & \kappa_c+\kappa_I & \rho_V & -\rho_U \\ \kappa_U & -\rho_V & \kappa_c+\kappa_I & \rho_Q \\ \kappa_V & \rho_U & -\rho_Q & \kappa_c+\kappa_I \end{pmatrix}\begin{pmatrix} I \\ Q \\ U \\ V \end{pmatrix} + \begin{pmatrix} \kappa_c S_c + \kappa_I S_L \\ \kappa_Q S_L \\ \kappa_U S_L \\ \kappa_V S_L \end{pmatrix}, \tag{2.5}$$

where S_L and S_c are the line and continuum source functions, κ_c is the continuum opacity, $\kappa_I, \cdots, \kappa_V$ are the line opacities in each polarization mode (corrected for stimulated emission), and $\rho_Q, \cdots, \rho_V$ describe magneto-optical effects. Unlike the opacities encountered in the transfer of unpolarized light, the κ's and ρ's are in general direction dependent. This is a result of the anisotropy imposed on the medium by the magnetic field. For simplicity we assume that the background opacity κ_c is pure absorption (no background scattering). This implies that the background source function does not depend on the radiation field in the line, so that S_c is frequency independent.

Detailed expressions for the κ's and ρ's have been derived by a number of authors (e.g. Beckers 1969a; Landi Degl'Innocenti & Landi Degl'Innocenti 1972; Wittmann 1974). To compensate for the $\pm$ sign in our definition of U, we must include similar $\pm$ signs in our expressions for the off-diagonal matrix elements κ_U, ρ_V and ρ_Q, which couple U with the other Stokes parameters. For electric dipole transitions, the κ's can then be written in the form:

$$\kappa_{I,Q,U,V} = C\phi_{I,Q,U,V}, \tag{2.6}$$

where $\phi_{I,Q,U,V}$ are angle- and frequency dependent functions given by:

$$\phi_I = \frac{1}{2}\left(\phi_0 - \frac{\phi_{+1}+\phi_{-1}}{2}\right)\sin^2\gamma + \frac{1}{2}\left(\phi_{+1}+\phi_{-1}\right), \tag{2.7}$$

$$\phi_Q = \frac{1}{2}\left(\phi_0 - \frac{\phi_{+1}+\phi_{-1}}{2}\right)\sin^2\gamma\ \cos 2\chi, \tag{2.8}$$

$$\phi_U = \pm\frac{1}{2}\left(\phi_0 - \frac{\phi_{+1}+\phi_{-1}}{2}\right)\sin^2\gamma\ \sin 2\chi, \tag{2.9}$$

$$\phi_V = \frac{1}{2}\left(\phi_{+1}-\phi_{-1}\right)\cos\gamma. \tag{2.10}$$

Here γ is the angle between $\hat{\mathbf{n}}$ and the magnetic field vector $\mathbf{B}$, and χ is the angle between $\hat{\mathbf{e}}_1$ and the projection of $\mathbf{B}$ onto the plane $(\hat{\mathbf{e}}_1, \hat{\mathbf{e}}_2)$, with χ measured positive in the direction towards $\hat{\mathbf{e}}_2$. In expression (2.9), the + sign should be used for outgoing rays ($\mu > 0$), and the − sign for incoming rays ($\mu < 0$). The functions ϕ_i ($i = -1, 0, +1$) are the absorption profiles of π and σ components. They are given by:

$$\phi_i(z, \hat{\mathbf{n}}, \Delta\nu) = \frac{1}{\Delta\nu_D\sqrt{\pi}} \sum_M \mathrm{S}_M^{(i)} \, H(a, v - v_M^{(i)} - v_s), \tag{2.11}$$

where the summation is over the M quantum number of the lower level. The $v_M^{(i)}$ describe the frequency position of each subcomponent in units of the Doppler width $\Delta\nu_D$:

$$v_M^{(i)} = \frac{eB}{4\pi mc\Delta\nu_D} \left[G_u(M+i) - G_\ell M\right], \tag{2.12}$$

where B is the magnetic field strength, and G_ℓ and G_u are the Landé factors of the lower and upper levels. The strengths $\mathrm{S}_M^{(i)}$ of the Zeeman components are given, for example, by Beckers (1969b), who also discusses magnetic dipole and electric quadrupole radiation. The strengths are normalized such that:

$$\sum_M \mathrm{S}_M^{(i)} = 1. \tag{2.13}$$

The quantity v_s in equation (2.11) describes Doppler effects due to systematic flows, and is given by:

$$v_s = -\frac{\nu_0}{\Delta\nu_D} \frac{\mathbf{u} \cdot \hat{\mathbf{n}}}{c}, \tag{2.14}$$

where $\mathbf{u}(z)$ is the velocity vector. The function $H(a, v)$ is the Voigt function:

$$H(a, v) = \frac{a}{\pi} \int_{-\infty}^{+\infty} \frac{e^{-y^2}}{(v-y)^2 + a^2} \, dy, \tag{2.15}$$

with $v \equiv \Delta\nu/\Delta\nu_D$, $a \equiv \Gamma/(4\pi\Delta\nu_D)$ and Γ the damping constant. Finally, the quantity C appearing in equation (2.6) is $\Delta\nu_D\sqrt{\pi}$ times the line center opacity without a magnetic field, and with $a = 0$:

$$C(z) = \frac{\pi e^2}{mc} f N_\ell \left(1 - \frac{b_u}{b_\ell} e^{-h\nu_0/kT_e}\right), \tag{2.16}$$

with f the oscillator strength, $N_\ell(z)$ the number density of atoms in the lower level, $T_e(z)$ the electron temperature, and $b_\ell(z)$ and $b_u(z)$ are factors by which the population densities of the lower and upper levels depart from the values in Local Thermodynamic Equilibrium (LTE). The term in brackets in equation (2.16) represents the correction for stimulated emission.

The ρ's appearing in equation (2.5), which describe magneto-optical effects, are written in terms of profile functions $\xi_{Q,U,V}$:

$$\rho_{Q,U,V} = C\,\xi_{Q,U,V}, \tag{2.17}$$

where:

$$\xi_Q = \pm\frac{1}{2}\left(\xi_0 - \frac{\xi_{+1}+\xi_{-1}}{2}\right)\sin^2\gamma\ \cos 2\chi, \tag{2.18}$$

$$\xi_U = \frac{1}{2}\left(\xi_0 - \frac{\xi_{+1}+\xi_{-1}}{2}\right)\sin^2\gamma\ \sin 2\chi, \tag{2.19}$$

$$\xi_V = \pm\frac{1}{2}\left(\xi_{+1}-\xi_{-1}\right)\cos\gamma. \tag{2.20}$$

The two cases for the sign again refer to $\mu > 0$ and $\mu < 0$. The functions ξ_i are given by an expression similar to equation (2.11), except that the Voigt function $H(a,v)$ is replaced by $2F(a,v)$, where F is the plasma or line dispersion function:

$$F(a,v) = \frac{1}{2\pi}\int_{-\infty}^{+\infty}\frac{(v-y)e^{-y^2}}{(v-y)^2+a^2}\,dy. \tag{2.21}$$

A method for the numerical evaluation of $H(a,v)$ and $F(a,v)$ is described by Reichel (1968).

We now consider the symmetry properties of the profile functions under inversion of the direction of propagation, $\hat{\mathbf{n}} \rightarrow -\hat{\mathbf{n}}$. First, note that the Zeeman effect implies a symmetric splitting pattern:

$$v_{-M}^{(-i)} = -v_M^{(i)}, \tag{2.22}$$

$$\mathrm{S}_{-M}^{(-i)} = \mathrm{S}_M^{(i)}, \tag{2.23}$$

and that v_s changes sign under the transformation $\hat{\mathbf{n}} \rightarrow -\hat{\mathbf{n}}$. Second, note that $H(a,v)$ is symmetric in v, whereas $F(a,v)$ is antisymmetric. It follows that the functions ϕ_i and ξ_i satisfy:

$$\phi_{-i}(z,-\hat{\mathbf{n}},-\Delta\nu) = \phi_i(z,\hat{\mathbf{n}},\Delta\nu), \tag{2.24}$$

$$\xi_{-i}(z,-\hat{\mathbf{n}},-\Delta\nu) = -\xi_i(z,\hat{\mathbf{n}},\Delta\nu). \tag{2.25}$$

Reversing the direction of $\hat{\mathbf{n}}$ implies $\hat{\mathbf{e}}_2 \rightarrow -\hat{\mathbf{e}}_2$, whereas $\hat{\mathbf{e}}_1$ remains unchanged; it follows that $\gamma \rightarrow (\pi-\gamma)$ and $\chi \rightarrow -\chi$. Inserting these relations into equations (2.7) - (2.10) and (2.18) - (2.20), we find:

$$\phi_{I,Q,U,V}(z,-\hat{\mathbf{n}},-\Delta\nu) = \phi_{I,Q,U,V}(z,\hat{\mathbf{n}},\Delta\nu), \tag{2.26}$$

$$\xi_{Q,U,V}(z,-\hat{\mathbf{n}},-\Delta\nu) = \xi_{Q,U,V}(z,\hat{\mathbf{n}},\Delta\nu), \tag{2.27}$$

i.e., the functions $\phi_{I,Q,U,V}$ and $\xi_{Q,U,V}$ are *invariant* with respect to the transformation $(\hat{\mathbf{n}}, \Delta\nu) \rightarrow (-\hat{\mathbf{n}}, -\Delta\nu)$.

For later reference we note that the profile functions ϕ_i are normalized such that

$$\int_{-\infty}^{+\infty} \phi_i(z, \hat{\mathbf{n}}, \Delta\nu)\, d(\Delta\nu) = 1, \tag{2.28}$$

which follows from equations (2.11) and (2.13), and the normalization of Voigt functions. Also, the angle average of ϕ_I is given by:

$$\begin{aligned} \frac{1}{4\pi} \oint \phi_I(z, \hat{\mathbf{n}}, \Delta\nu)\, d\Omega &= \frac{1}{2} \int_0^{\pi} \phi_I(z, \gamma, \Delta\nu) \sin\gamma\, d\gamma \\ &= \frac{1}{3} \left(\phi_{-1} + \phi_0 + \phi_{+1}\right). \end{aligned} \tag{2.29}$$

It follows that the integral over both angle and frequency of ϕ_I is given by:

$$\frac{1}{4\pi} \int_{-\infty}^{+\infty} \oint \phi_I(z, \hat{\mathbf{n}}, \Delta\nu)\, d\Omega\, d(\Delta\nu) = 1, \tag{2.30}$$

i.e., the function ϕ_I is normalized.

3. FORMULATION IN TERMS OF COMPLEX 2 × 2 MATRICES

The symmetries and anti-symmetries in the 4×4 matrix of equation (2.5) represent a certain redundancy in the formalism. This redundancy can be removed by describing the state of polarization of the light, and the absorption and birefringence properties of the medium, in terms of complex 2×2 matrices (Jones' formalism, see Shurcliff 1962). We define:

$$A(z, \hat{\mathbf{n}}, \Delta\nu) = \frac{1}{2} \begin{pmatrix} \kappa_c + (\phi_I + \alpha_Q)C & (\alpha_U + i\alpha_V)C \\ (\alpha_U - i\alpha_V)C & \kappa_c + (\phi_I - \alpha_Q)C \end{pmatrix}, \tag{3.1}$$

where the α's are complex quantities:

$$\alpha_{Q,U,V} \equiv \phi_{Q,U,V} - i\xi_{Q,U,V}. \tag{3.2}$$

The polarization state of the light is described, not with the Stokes 4-vector, but with a similar 2×2 matrix D defined by:

$$D(z, \hat{\mathbf{n}}, \Delta\nu) = \frac{1}{2} \begin{pmatrix} I + Q & U + iV \\ U - iV & I - Q \end{pmatrix}. \tag{3.3}$$

The transport equation (2.5) can then be written in the form:

$$\mu \frac{dD}{dz} = -AD - DA^\dagger + H, \tag{3.4}$$

where † denotes Hermitian conjugation (transposition plus complex conjugation), and the source matrix H is given by:

$$H(z, \hat{\mathbf{n}}, \Delta\nu) = \frac{1}{2} \begin{pmatrix} \kappa_c S_c + (\phi_I + \phi_Q) C S_L & (\phi_U + i\phi_V) C S_L \\ (\phi_U - i\phi_V) C S_L & \kappa_c S_c + (\phi_I - \phi_Q) C S_L \end{pmatrix}. \tag{3.5}$$

Equation (3.4) is derived by adding and subtracting the equations (2.5). Note that D and H are Hermitian matrices, whereas A is in general not Hermitian (because the α's are complex).

A formal solution of equation (3.4) can be obtained as follows (cf. van Ballegooijen 1984; Landi Degl'Innocenti & Landi Degl'Innocenti 1985). We introduce a complex 2×2 matrix $T(z, \hat{\mathbf{n}}, \Delta\nu)$, which is defined as the solution of the first-order differential equation

$$\mu \frac{dT}{dz} = -AT, \tag{3.6}$$

subject to the boundary condition that $T(z, \hat{\mathbf{n}}, \Delta\nu)$ approaches the unity matrix for $z \to \infty$. In the absence of polarization effects ($\alpha_{Q,U,V} = 0$), the solution of equation (3.6) is given by:

$$T(z, \hat{\mathbf{n}}, \Delta\nu) = \tilde{I} \exp[\pm\tau(z, \hat{\mathbf{n}}, \Delta\nu)/2], \tag{3.7}$$

where $\tilde{I}$ is the unity 2×2 matrix, $\tau(z, \hat{\mathbf{n}}, \Delta\nu)$ is the optical depth along a ray:

$$\tau(z, \hat{\mathbf{n}}, \Delta\nu) = \frac{1}{|\mu|} \int_z^\infty [\kappa_c(z') + \phi_I(z', \hat{\mathbf{n}}, \Delta\nu) C(z')] \, dz', \tag{3.8}$$

and the $\pm$ sign in (3.7) refers to outgoing rays ($\mu > 0$) and incoming rays ($\mu < 0$), respectively. Hence, the matrix T plays the role of the exponential factor encountered in the transfer of unpolarized light. The Hermitian conjugate of T satisfies the equation:

$$\mu \frac{dT^\dagger}{dz} = -T^\dagger A^\dagger. \tag{3.9}$$

We also introduce a matrix $E(z, \hat{\mathbf{n}}, \Delta\nu)$, related to $D(z, \hat{\mathbf{n}}, \Delta\nu)$ via:

$$D = TET^\dagger. \tag{3.10}$$

Inserting this into equation (3.4), we obtain:

$$\mu \left\{ \frac{dT}{dz} ET^\dagger + T \frac{dE}{dz} T^\dagger + TE \frac{dT^\dagger}{dz} \right\} = -ATET^\dagger - TET^\dagger A^\dagger + H. \tag{3.11}$$

Using (3.6) and (3.9), we see that the terms involving dT/dz on the left-hand-side cancel against the terms with A on the right. It follows that E satisfies the equation:

$$\mu \frac{dE}{dz} = T^{-1} H \left[T^{-1}\right]^{\dagger}. \tag{3.12}$$

Integration over z yields, for an *outgoing* ray ($\mu > 0$):

$$E_{out}(z) = E_{out}(z_0) + \frac{1}{\mu} \int_{z_0}^{z} T_{out}^{-1}(z') H(z') \left[T_{out}^{-1}(z')\right]^{\dagger} dz', \tag{3.13}$$

where z_0 denotes some level at the base of the atmosphere, and where we suppress the $\hat{\mathbf{n}}$ and $\Delta\nu$ dependence of the various matrices for notational convenience. From equation (3.7) we see that the T matrix increases exponentially with optical depth for outgoing rays. Hence the contribution from the lower boundary, $E_{out}(z_0)$, can be neglected, provided z_0 is chosen such that the optical depth between z and z_0 is large compared to unity at all frequencies, angles, and modes of polarization. The D matrix, which contains the four Stokes parameters, then follows from equation (3.10):

$$D_{out}(z) = \frac{1}{\mu} T_{out}(z) \left\{ \int_{z_0}^{z} T_{out}^{-1}(z') H(z') \left[T_{out}^{-1}(z')\right]^{\dagger} dz' \right\} T_{out}^{\dagger}(z). \tag{3.14}$$

Similarly, for an *incoming* ray ($\mu < 0$):

$$D_{in}(z) = \frac{1}{|\mu|} T_{in}(z) \left\{ \int_{z}^{\infty} T_{in}^{-1}(z') H(z') \left[T_{in}^{-1}(z')\right]^{\dagger} dz' \right\} T_{in}^{\dagger}(z). \tag{3.15}$$

Expressions (3.14) and (3.15) are generalizations of the familiar integral expressions for the intensity of unpolarized light, and represent a formal solution of the transfer equation.

4. STATISTICAL EQUILIBRIUM

Our assumption that the populations of the lower and upper levels are fully characterized by departure coefficients b_ℓ and b_u implies that the line source function can be written as:

$$S_L = \frac{2h\nu_0^3}{c^2} \left(\frac{b_\ell}{b_u} e^{h\nu_0/kT_e} - 1 \right)^{-1}. \tag{4.1}$$

The computation of the departure coefficients b_ℓ and b_u requires the solution of the equations of statistical equilibrium for the atom in question. Since the radiative excitation rates necessary for such a calculation depend themselves on the radiation fields in lines and continua, the equations of radiative transfer and statistical equilibrium are coupled. One method for dealing with this coupling is lambda iteration, which is a computational process consisting of the following steps: 1) given estimates of the line source function, solve the transfer equation and determine the radiative fields in all relevant lines and continua; 2) determine the radiative and collisional rates of all transitions; 3) solve the statistical equilibrium equations for the atom in question, and find new values of the departure coefficients; 4) compute new estimates of the source function, and repeat the process. This method works well if the collisional rates dominate, i.e., if the deviations from LTE are small. However, in many practical applications the opposite case is encountered, with radiative rates much larger than collisional rates (strong scattering). In this case the coupling between the radiation field and the gas is essentially nonlocal, and the lambda iteration method converges very slowly (Mihalas 1978).

A better method for dealing with the NLTE problem is to treat the effects of scattering implicitly in the transfer equation. For complex, multi-level atoms this may be achieved by taking the "equivalent two-level atom" approach. In this method the multi-level problem is broken down into a set a two-level problems; using the rate equations, the line source function for each transition is written in the form:

$$S_L = \frac{\bar{J} + \epsilon B^S}{1 + \epsilon}, \tag{4.2}$$

where $\bar{J}$ is the mean integrated intensity (to be defined below). In the two-level case, B^S is the Planck function, and ϵ is the ratio of collisional and spontaneous de-excitation rates from upper to lower level. For atoms with more than two levels, and when bound-free transitions are taken into account, the expressions for ϵ and B^S are more complicated, and depend on the net rates to and from other levels and continua (cf. Mihalas 1978). In the case without Zeeman splitting, $\bar{J}$ is given by:

$$\bar{J} = \int_{-\infty}^{+\infty} \phi_0(\Delta\nu) J(\Delta\nu)\, d(\Delta\nu), \tag{4.3}$$

where $J(\Delta\nu) = (1/4\pi) \oint I(\hat{\mathbf{n}}, \Delta\nu)\, d\Omega$ is the mean intensity, and $\phi_0(\Delta\nu)$ is the (normalized) absorption profile. The generalization of (4.3) to the case with Zeeman splitting is:

$$\bar{J} = \frac{1}{4\pi} \int_{-\infty}^{+\infty} \oint \left[\phi_I I + \phi_Q Q + \phi_U U + \phi_V V\right]\, d\Omega\, d(\Delta\nu), \tag{4.4}$$

which follows, for example, from the first component of equation (2.5). Introducing the matrix Φ,

$$\Phi(z,\hat{\mathbf{n}},\Delta\nu) \equiv \frac{1}{2}\begin{pmatrix} \phi_I+\alpha_Q & \alpha_U+i\alpha_V \\ \alpha_U-i\alpha_V & \phi_I-\alpha_Q \end{pmatrix}, \tag{4.5}$$

with the α's given by equation (3.2), the term in square brackets in equation (4.4) can be written as:

$$\phi_I I+\phi_Q Q+\phi_U U+\phi_V V = \mathrm{Tr}\{\Phi D + D\Phi^\dagger\}. \tag{4.6}$$

Here Tr denotes the trace of a matrix (sum of diagonal elements), and D is the "intensity" matrix defined in equation (3.3).

Using equation (4.5), the source matrix H of equation (3.5) can be written as:

$$H(z,\hat{\mathbf{n}},\Delta\nu) = \frac{1}{2}\kappa_c S_c \tilde{I} + \frac{1}{2} C S_L \left(\Phi+\Phi^\dagger\right), \tag{4.7}$$

where $\tilde{I}$ denotes the unity matrix. Inserting this expression into the formal solution of equations (3.14) and (3.15), and integrating over angle and frequency, we obtain the following expression for the mean integrated intensity $\bar{J}$:

$$\bar{J}(z) = \bar{J}_c(z) + \bar{J}_L(z), \tag{4.8}$$

where:

$$\bar{J}_c(z) \equiv \frac{1}{4\pi}\oint_{\mu>0}\frac{d\Omega}{\mu}\int_{-\infty}^{+\infty} d(\Delta\nu)\int_{z_0}^{\infty} dz'\; f_c(z,z',\hat{\mathbf{n}}_o,\Delta\nu)\kappa_c(z')S_c(z'), \tag{4.9}$$

and

$$\bar{J}_L(z) \equiv \frac{1}{4\pi}\oint_{\mu>0}\frac{d\Omega}{\mu}\int_{-\infty}^{+\infty} d(\Delta\nu)\int_{z_0}^{\infty} dz'\; f_L(z,z',\hat{\mathbf{n}}_o,\Delta\nu)C(z')S_L(z'). \tag{4.10}$$

Note that the angle integrals are taken only over outgoing directions $\hat{\mathbf{n}}_o$. The kernel functions f_c and f_L are defined by:

$$f_c(z,z',\hat{\mathbf{n}}_o,\Delta\nu) = \frac{1}{2}\mathrm{Tr}\left\{\Phi(z)M(z,z')M^\dagger(z,z')+M(z,z')M^\dagger(z,z')\Phi^\dagger(z)\right\}, \tag{4.11}$$

and

$$\begin{aligned} f_L(z,z',\hat{\mathbf{n}}_o,\Delta\nu) = \frac{1}{2}\mathrm{Tr}\Big\{&\Phi(z)M(z,z')\left[\Phi(z')+\Phi^\dagger(z')\right]M^\dagger(z,z') \\ &+ M(z,z')\left[\Phi(z')+\Phi^\dagger(z')\right]M^\dagger(z,z')\Phi^\dagger(z)\Big\}. \end{aligned} \tag{4.12}$$

where we suppressed the dependence on $\hat{\mathbf{n}}_o$ and $\Delta\nu$ on the right hand side of the equation, and where the matrix M is defined by:

$$M(z, z', \hat{\mathbf{n}}_o, \Delta\nu) \equiv \begin{cases} T(z, \hat{\mathbf{n}}_o, \Delta\nu) T^{-1}(z', \hat{\mathbf{n}}_o, \Delta\nu), & \text{if } z' < z; \\ T(z, -\hat{\mathbf{n}}_o, -\Delta\nu) T^{-1}(z', -\hat{\mathbf{n}}_o, -\Delta\nu), & \text{if } z' > z. \end{cases} \tag{4.13}$$

In deriving the above equations we made use of the symmetry relations (2.26) and (2.27). Expression (4.10) defines a so-called lambda operator:

$$\bar{J}_L(z) \equiv \Lambda\{S_L(z)\}. \tag{4.14}$$

This lambda operator involves an integration over height z, as well as integrations over angle and frequency. From equations (4.2), (4.8) and (4.14) we obtain the following integral equation for the line source function:

$$\left[1 - \frac{1}{1+\epsilon}\Lambda\right] S_L = \frac{\bar{J}_c + \epsilon B^S}{1+\epsilon}. \tag{4.15}$$

Hence, the solution of each "two-level atom" problem involves the inversion of the operator on the left hand side of this equation.

The solution of the full, multi-level NLTE problem can now be obtained as follows. Given certain estimates of the level populations as function of height z, we separately consider all transitions in the atom. For each transition we determine the functions $\epsilon(z)$ and $B^S(z)$, and we solve the two-level problem discussed above, which yields the radiation field $\bar{J}(z)$. Then, after all transitions have been considered, the $\bar{J}$'s are used to compute radiative transition rates, and the equations of statistical equilibrium are solved separately for each height. This yields improved values of the level populations throughout the atmosphere, and the computation may be repeated until convergence is achieved. Since the transition rates between levels are often dominated by the direct (radiative and collisional) transitions, this process is expected to converge rapidly, even in atoms with a complicated level structure.

5. NUMERICAL METHOD

In practical applications, physical quantities such as temperature and level populations are defined only at a discrete set of heights z_i, $i = 1, 2, \cdots, N$ (we take z_i to be monotonically increasing). The lambda operator for a certain transition can then be represented by a matrix Λ_{ij}, which describes the radiative coupling between any two grid points i and j. The computation of Λ_{ij} involves the solution of the transport equation (3.6) for the matrix $T(z, \hat{\mathbf{n}}, \Delta\nu)$; this matrix describes the integrated polarization and birefringence properties of the atmospheric layers above height z. In the case *without* Zeeman splitting, the T matrix is proportional to the unity matrix, and the proportionality factor is simply the exponential of the optical depth (see equation [3.7]). However, in the case *with* Zeeman splitting, the matrix character of equation (3.6) prevents such a simple, analytic solution. Hence, our first task is to provide a numerical method for the solution of (3.6); this is done in the next section. In section 5.2 we describe a method for the computation of the matrix Λ_{ij} and the vector $\bar{J}_{c,i}$. Finally, in section 5.3 we derive the appropriate equation for the line source function $S_{L,i}$.

5.1. Solution of the Transfer Equation

To obtain a numerical solution of equation (3.6), we write the absorption matrix A of equation (3.1) as:

$$A(z, \hat{\mathbf{n}}, \Delta\nu) = \frac{1}{2}\left[(\kappa_c + \phi_I C)\tilde{I} + \alpha C X\right], \tag{5.1}$$

where $\tilde{I}$ is the unity matrix, and X is a traceless matrix defined by:

$$X(z, \hat{\mathbf{n}}, \Delta\nu) \equiv \frac{1}{\alpha}\begin{pmatrix} \alpha_Q & \alpha_U + i\alpha_V \\ \alpha_U - i\alpha_V & -\alpha_Q \end{pmatrix}. \tag{5.2}$$

The quantity α is a complex number describing the magnitude of polarization and birefringence effects:

$$\alpha(z, \hat{\mathbf{n}}, \Delta\nu) \equiv \sqrt{\alpha_Q^2 + \alpha_U^2 + \alpha_V^2}. \tag{5.3}$$

Note that the *eigenvalues* of the A matrix are given by $[\kappa_c + (\phi_I \pm \alpha)C]/2$; hence α is related to the difference between the eigenvalues. On the other hand, X contains information about the *eigenvectors* of the A matrix (i.e., the orientation of polarization and birefringence axes). Note that X has the convenient property $X^2 = \tilde{I}$, so that X is equal to its own inverse, $X^{-1} = X$.

The essence of our numerical scheme is to neglect the z-dependence of the X matrix in the neighborhood of each grid point; for each height interval $z_{j-1/2} < z < z_{j+1/2}$ we approximate $X(z) = X_j$, the value of X at grid point j. Here $z_{j+1/2}$ denotes a point in between two grid points. A possible choice for $z_{j+1/2}$ is the mean position, $z_{j+1/2} \equiv (z_{j+1} + z_j)/2$, but this choice is not essential for what follows. The assumption that $X(z)$ is piecewise constant will allow us to perform certain height integrals analytically. This is advantageous when the optical depth between grid points is large, or when Faraday rotation is strong. However, it should be kept in mind that the discontinuities in $X(z)$ may lead to numerical errors if X_j varies rapidly from one grid point to the next.

Let $T_j(\hat{\mathbf{n}}, \Delta\nu)$ denote the T matrix at grid point j. Then the solution of equation (3.6) within the interval $[z_{j-1/2}, z_{j+1/2}]$ is given in terms of T_j as follows:

$$T(z, \hat{\mathbf{n}}, \Delta\nu) = \exp[\pm(\Delta\tau_c + \Delta\tau_L)/2] \cdot \left\{\cosh(\Delta\tau_s/2)\tilde{I} \pm \sinh(\Delta\tau_s/2)X_j\right\} T_j(\hat{\mathbf{n}}, \Delta\nu), \tag{5.4}$$

where the two cases for the sign in equation (5.4) refer to outgoing rays ($\mu > 0$) and incoming rays ($\mu < 0$), respectively. The $\Delta\tau$'s are optical depth differences with respect to z_j:

$$\Delta\tau_c(z, \hat{\mathbf{n}}) = \tau_c(z) - \tau_c(z_j), \tag{5.5}$$

$$\Delta\tau_L(z, \hat{\mathbf{n}}, \Delta\nu) = \tau_L(z) - \tau_L(z_j), \tag{5.6}$$

$$\Delta\tau_s(z, \hat{\mathbf{n}}, \Delta\nu) = \tau_s(z) - \tau_s(z_j), \tag{5.7}$$

and τ_c, τ_L and τ_s are defined by:

$$\tau_c(z, \hat{\mathbf{n}}) = \frac{1}{|\mu|}\int_z^\infty \kappa_c(z')\, dz', \tag{5.8}$$

$$\tau_L(z, \hat{\mathbf{n}}, \Delta\nu) = \frac{1}{|\mu|}\int_z^\infty \phi_I(z', \hat{\mathbf{n}}, \Delta\nu)C(z')\, dz', \tag{5.9}$$

$$\tau_s(z, \hat{\mathbf{n}}, \Delta\nu) = \frac{1}{|\mu|}\int_z^\infty \alpha(z', \hat{\mathbf{n}}, \Delta\nu)C(z')\, dz'. \tag{5.10}$$

Note that τ_s is a complex quantity, whereas τ_c and τ_L are real-valued. The real and imaginary parts of τ_s can be interpreted as the difference in optical depth and phase angle between the two eigenmodes of the absorption matrix (although, strictly speaking, this interpretation is valid only when the eigenvectors of A are constant in z). We recall that the quantities ϕ_I and α are invariant under the transformation $(\hat{\mathbf{n}}, \Delta\nu) \rightarrow (-\hat{\mathbf{n}}, -\Delta\nu)$ (see equations [2.26] and [2.27]). It follows that τ_L and τ_s also exhibit this symmetry.

We now make the additional approximation that the ratios of the various optical depth differences are constant within each interval:

$$\Delta\tau_c(z) = r_j \Delta\tau_L(z), \tag{5.11}$$

$$\Delta\tau_s(z) = s_j \Delta\tau_L(z), \tag{5.12}$$

where r_j and s_j are opacity ratios at grid point j:

$$r_j(\hat{\mathbf{n}}, \Delta\nu) \equiv \frac{\kappa_c(z_j)}{\phi_I(z_j, \hat{\mathbf{n}}, \Delta\nu)C(z_j)}, \tag{5.13}$$

$$s_j(\hat{\mathbf{n}}, \Delta\nu) \equiv \frac{\alpha(z_j, \hat{\mathbf{n}}, \Delta\nu)}{\phi_I(z_j, \hat{\mathbf{n}}, \Delta\nu)}. \tag{5.14}$$

Then equation (5.4) may be written as:

$$T(z,\hat{\mathbf{n}},\Delta\nu) = \exp[\pm(1+r_j)\Delta\tau_L/2] \cdot \left\{\cosh(s_j\Delta\tau_L/2)\tilde{I} \pm \sinh(s_j\Delta\tau_L/2)X_j\right\} T_j(\hat{\mathbf{n}},\Delta\nu). \tag{5.15}$$

Note that s_j is again a complex quantity, whereas r_j is real.

To find the matrix T_j in terms of T_{j+1} we have to apply equation (5.15) twice, first over the interval $[z_{j+1/2}, z_{j+1}]$ and then over the interval $[z_j, z_{j+1/2}]$. In the following we write indices "1" and "2" to denote half-intervals above and below a certain grid point. Furthermore, we introduce the *total* optical depths of the half-intervals $[z_j, z_{j+1/2}]$ and $[z_{j-1/2}, z_j]$ as:

$$\Delta\tau_{1,j} = \tau_L(z_j) - \tau_L(z_{j+1/2}) \qquad \text{for } j \neq N, \tag{5.16}$$

$$\Delta\tau_{2,j} = \tau_L(z_{j-1/2}) - \tau_L(z_j) \qquad \text{for } j \neq 1. \tag{5.17}$$

Thus the quantities $\Delta\tau_{1,j}$ and $\Delta\tau_{2,j}$ are positive by definition. Then, using the fact that the inverse of the matrix $[\cosh(x)\tilde{I} + \sinh(x)X]$ is given by $[\cosh(x)\tilde{I} - \sinh(x)X]$, we obtain from (5.15):

$$\begin{aligned} T_j = &\exp[\pm(1+r_{j+1})\Delta\tau_{2,j+1}/2 \pm (1+r_j)\Delta\tau_{1,j}/2] \\ &\cdot \left\{\cosh(s_j\Delta\tau_{1,j}/2)\tilde{I} \pm \sinh(s_j\Delta\tau_{1,j}/2)X_j\right\} \\ &\cdot \left\{\cosh(s_{j+1}\Delta\tau_{2,j+1}/2)\tilde{I} \pm \sinh(s_{j+1}\Delta\tau_{2,j+1}/2)X_{j+1}\right\} T_{j+1}. \end{aligned} \tag{5.18}$$

Together with a boundary condition at $z = z_N$, equation (5.18) can be used recursively to find T_j at any grid point. Since by definition $T(z \to \infty) = \tilde{I}$ (see section 3), the appropriate boundary condition is:

$$T_N = \exp[\pm(1+r_N)\tau_L(z_N)/2] \cdot \left\{\cosh[s_N\tau_L(z_N)/2]\tilde{I} \pm \sinh[s_N\tau_L(z_N)/2]X_N\right\}. \tag{5.19}$$

Here $\tau_L(z_N)$ is the optical depth of the highest grid point, which must be found from a suitable extrapolation of the atmospheric model.

Note, that the recursion (5.18) should be carried out separately for outgoing and incoming rays, since in spite of the symmetry relations (2.27) and (2.28), the matrices $T(z,\hat{\mathbf{n}},\Delta\nu)$ and $T(z,-\hat{\mathbf{n}},-\Delta\nu)$ are not simply related; in particular, they are not each others inverse, $T(z,\hat{\mathbf{n}}_o,\Delta\nu) \neq T^{-1}(z,-\hat{\mathbf{n}}_o,-\Delta\nu)$. This is a result of the fact that the matrices X_j of different grid points do in general not commute.

5.2. Computation of Λ_{ij} and $\bar{J}_{c,i}$

We now derive a numerical approximation for the lambda operator. Equation (4.10) is written in the form:

$$\bar{J}_{L,i} = \sum_{j=1}^{N} \Lambda_{ij} S_{L,j}, \tag{5.20}$$

where:

$$\Lambda_{ij} = \frac{1}{4\pi} \oint_{\mu>0} d\Omega \int_{-\infty}^{+\infty} d(\Delta\nu) L_{ij}(\hat{\mathbf{n}}_o, \Delta\nu), \tag{5.21}$$

and $L_{ij}(\hat{\mathbf{n}}_o, \Delta\nu)$ denotes the contribution to Λ_{ij} from a given angle and frequency. To compute L_{ij} we use the following approximations for the line source function, $S_L(z')$, in equation (4.10). For $z' < z_{i-1}$ or $z' > z_{i+1}$ we use *linear* interpolation between grid points:

$$S_L(z') = S_{L,j} \left[\frac{z_{j+1} - z'}{z_{j+1} - z_j} \right] + S_{L,j+1} \left[\frac{z' - z_j}{z_{j+1} - z_j} \right], \tag{5.22}$$

where z_j is the grid point immediately below z':

$$z_j \le z' < z_{j+1}. \tag{5.23}$$

For $z_{i-1} \le z' \le z_{i+1}$ we use *quadratic* interpolation:

$$S_L(z') = S_{L,i} + a_i(z' - z_i) + b_i(z' - z_i)^2, \tag{5.24}$$

where the coefficients a_i and b_i are determined from the conditions:

$$S_{L,i+1} \equiv S_{L,i} + a_i(z_{i+1} - z_i) + b_i(z_{i+1} - z_i)^2,$$
$$S_{L,i-1} \equiv S_{L,i} - a_i(z_i - z_{i-1}) + b_i(z_i - z_{i-1})^2.$$

Solving for a_i and b_i yields:

$$\begin{aligned} S_L(z') = & \frac{(z' - z_{i+1})(z' - z_i)}{(z_i - z_{i-1})(z_{i+1} - z_{i-1})} S_{L,i-1} + \frac{(z' - z_{i-1})(z_{i+1} - z')}{(z_i - z_{i-1})(z_{i+1} - z_i)} S_{L,i} \\ & + \frac{(z' - z_{i-1})(z' - z_i)}{(z_{i+1} - z_i)(z_{i+1} - z_{i-1})} S_{L,i+1}. \end{aligned} \tag{5.25}$$

For $j \le i - 2$ or $j \ge i + 2$, L_{ij} is determined entirely by the linearly interpolated part of the source function:

$$L_{ij} = g_{ij} + h_{ij}, \tag{5.26}$$

where g_{ij} and h_{ij} are the contributions from the intervals below and above grid point j:

$$g_{ij}(\hat{\mathbf{n}}_o, \Delta\nu) = \frac{1}{\mu} \int\limits_{z_{j-1}}^{z_j} f_L(z_i, z', \hat{\mathbf{n}}_o, \Delta\nu) C(z') \left[\frac{z' - z_{j-1}}{z_j - z_{j-1}} \right] dz', \tag{5.27}$$

$$h_{ij}(\hat{\mathbf{n}}_o, \Delta\nu) = \frac{1}{\mu} \int\limits_{z_j}^{z_{j+1}} f_L(z_i, z', \hat{\mathbf{n}}_o, \Delta\nu) C(z') \left[\frac{z_{j+1} - z'}{z_{j+1} - z_j} \right] dz'. \tag{5.28}$$

For $i - 1 \leq j \leq i + 1$, L_{ij} has contributions from the region of quadratic interpolation:

$$L_{ii-1} = g_{ii-1} + q_{ii-1}, \tag{5.29}$$

$$L_{ii} = q_{ii}, \tag{5.30}$$

$$L_{ii+1} = q_{ii+1} + h_{ii+1}, \tag{5.31}$$

where:

$$q_{ii-1} = \frac{1}{\mu} \int\limits_{z_{i-1}}^{z_{i+1}} f_L(z_i, z', \hat{\mathbf{n}}_o, \Delta\nu) C(z') \left[\frac{(z' - z_{i-1})(z_{i+1} - z')}{(z_i - z_{i-1})(z_{i+1} - z_i)} \right] dz', \tag{5.32}$$

$$q_{ii} = \frac{1}{\mu} \int\limits_{z_{i-1}}^{z_{i+1}} f_L(z_i, z', \hat{\mathbf{n}}_o, \Delta\nu) C(z') \left[\frac{(z' - z_{i-1})(z' - z_i)}{(z_{i+1} - z_i)(z_{i+1} - z_{i-1})} \right] dz', \tag{5.33}$$

$$q_{ii+1} = \frac{1}{\mu} \int\limits_{z_{i-1}}^{z_{i+1}} f_L(z_i, z', \hat{\mathbf{n}}_o, \Delta\nu) C(z') \left[\frac{(z' - z_{i+1})(z' - z_i)}{(z_i - z_{i-1})(z_{i+1} - z_{i-1})} \right] dz', \tag{5.34}$$

and g_{ii-1} and h_{ii+1} are given by (5.27) and (5.28). Quadratic interpolation is used only for the interior grid points, $1 < i < N$; for $i = 1$ and $i = N$ we use linear interpolation. The contributions of the layers below the first grid point ($z' < z_1$) are accounted for by taking $\Delta\tau_{2,1} \to \infty$.

To evaluate these expressions we approximate the interpolations in z by interpolations in *optical* depth. For example, using equations (4.12), (4.13) and (5.15), expression (5.27) can be written as:

$$\begin{aligned} g_{ij}(\hat{\mathbf{n}}_o, \Delta\nu) = &\frac{1}{2} \mathrm{Tr}\left\{ \Phi_i M_{ij-1} \mathcal{G}_{1,j} M^{\dagger}_{ij-1} + M_{ij-1} \mathcal{G}_{1,j} M^{\dagger}_{ij-1} \Phi^{\dagger}_i \right\} \\ &+ \frac{1}{2} \mathrm{Tr}\left\{ \Phi_i M_{ij} \mathcal{G}_{2,j} M^{\dagger}_{ij} + M_{ij} \mathcal{G}_{2,j} M^{\dagger}_{ij} \Phi^{\dagger}_i \right\}. \end{aligned} \tag{5.35}$$

The two terms in this equation correspond to the contributions from the two half-intervals $[z_{j-1}, z_{j-1/2}]$ and $[z_{j-1/2}, z_j]$, respectively. The matrix M_{ij} is defined by:

$$M_{ij} \equiv \begin{cases} T_i(\hat{\mathbf{n}}_o, \Delta\nu) T_j^{-1}(\hat{\mathbf{n}}_o, \Delta\nu) & \text{for } j < i, \\ T_i(-\hat{\mathbf{n}}_o, -\Delta\nu) T_j^{-1}(-\hat{\mathbf{n}}_o, -\Delta\nu) & \text{for } j > i, \end{cases} \tag{5.36}$$

and the matrices $\mathcal{G}_{1,j}$ and $\mathcal{G}_{2,j}$ are given by integrals over optical depth:

$$\mathcal{G}_{1,j} \equiv \int_0^{\Delta\tau_{1,j-1}} \mathcal{N}_{j-1}(\Delta\tau_L, +\sigma) w_{1,j-1}\left(\frac{\Delta\tau_L}{\Delta\tau_{1,j-1}}\right) d(\Delta\tau_L), \tag{5.37}$$

$$\mathcal{G}_{2,j} \equiv \int_0^{\Delta\tau_{2,j}} \mathcal{N}_{j}(\Delta\tau_L, -\sigma) \left[1 - w_{2,j}\left(\frac{\Delta\tau_L}{\Delta\tau_{2,j}}\right)\right] d(\Delta\tau_L). \tag{5.38}$$

Here σ is defined by:

$$\sigma \equiv \begin{cases} +1 & \text{for } j < i, \\ -1 & \text{for } j > i, \end{cases} \tag{5.39}$$

and the matrix $\mathcal{N}_j(\Delta\tau_L, \sigma)$ is defined by:

$$\begin{aligned} \mathcal{N}_j(\Delta\tau_L, \sigma) = & \exp[+\sigma(1 + r_j)\Delta\tau_L] \left\{\cosh(s_j \Delta\tau_L/2)\tilde{I} + \sigma \sinh(s_j \Delta\tau_L/2) X_j\right\} \\ & \cdot \left\{\tilde{I} + \frac{1}{2} s_j X_j + \frac{1}{2} s_j^* X_j^\dagger\right\} \left\{\cosh(s_j^* \Delta\tau_L/2)\tilde{I} + \sigma \sinh(s_j^* \Delta\tau_L/2) X_j^\dagger\right\}. \end{aligned} \tag{5.40}$$

Furthermore, we have introduced relative weights of the half-intervals:

$$w_{1,j} = \frac{\Delta\tau_{1,j}}{\Delta\tau_{1,j} + \Delta\tau_{2,j+1}}, \tag{5.41}$$

$$w_{2,j} = \frac{\Delta\tau_{2,j}}{\Delta\tau_{2,j} + \Delta\tau_{1,j-1}}. \tag{5.42}$$

In the above derivation we have used the fact that the matrix Φ of equation (4.5) can be written as:

$$\Phi(z') = \frac{1}{2}\left\{\phi_I(z')\tilde{I} + \alpha(z') X(z')\right\} = \frac{1}{2}\phi_I(z')\left\{\tilde{I} + s_j X_j\right\}. \tag{5.43}$$

Expressions similar to (5.35) can be derived for the quantities h_{ij} and q_{ij}.

The advantage of the present method is that integrals such as (5.37) and (5.38) can be evaluated analytically. The evaluation makes use of the fact that the matrices X_j have the property $X_j^2 = \tilde{I}$, and that $\sinh(x)$ and $\cosh(x)$ can be written in terms of exponential functions, $e^{\pm x}$. Thus equation (5.40) yields:

$$\mathcal{N}_j(\Delta\tau_L, \sigma) = \frac{1}{4}\sum_{k=1}^{4}(1 + t_{k,j}) \exp[\sigma(1 + r_j + t_{k,j})\Delta\tau_L]\, Y_{k,j}, \tag{5.44}$$

where the scalar quantities $t_{k,j}$ $(k = 1, \cdots, 4)$ are defined by:

$$\begin{aligned} t_{1,j} &= +\mathrm{Re}[s_j], \\ t_{2,j} &= -\mathrm{Re}[s_j], \\ t_{3,j} &= +i\mathrm{Im}[s_j], \\ t_{4,j} &= -i\mathrm{Im}[s_j], \end{aligned} \tag{5.45}$$

and the matrices $Y_{k,j}$ are defined by:

$$\begin{aligned} Y_{1,j} &= \tilde{I} + X_j + X_j^\dagger + X_j X_j^\dagger, \\ Y_{2,j} &= \tilde{I} - X_j - X_j^\dagger + X_j X_j^\dagger, \\ Y_{3,j} &= \tilde{I} + X_j - X_j^\dagger - X_j X_j^\dagger, \\ Y_{4,j} &= \tilde{I} - X_j + X_j^\dagger - X_j X_j^\dagger. \end{aligned} \tag{5.46}$$

The integrals over the exponentials in (5.44) can be expressed in terms of functions

$$m_n(x,\sigma) \equiv \int_0^x e^{\sigma x'} \left(\frac{x'}{x}\right)^n dx', \tag{5.47}$$

which are written out in Appendix A. Furthermore, the "trace" operators such as in equation (5.35) all reduce to trace operations on the $Y_{k,j}$ matrices:

$$y_{k,ij} \equiv \frac{1}{2}\mathrm{Tr}\left\{\Phi_i M_{ij} Y_{k,j} M_{ij}^\dagger + M_{ij} Y_{k,j} M_{ij}^\dagger \Phi_i^\dagger\right\}, \tag{5.48}$$

where M_{ij} for $i \neq j$ is given by equation (5.36), and $M_{ii} \equiv \tilde{I}$. The final result can then be written as a sum over half-intervals:

$$g_{ij} = g_{1,ij} + g_{2,ij}, \tag{5.49}$$

$$h_{ij} = h_{1,ij} + h_{2,ij}, \tag{5.50}$$

$$q_{ij} = q_{1,ij} + q_{2,ij} + q_{3,ij} + q_{4,ij}. \tag{5.51}$$

The detailed expressions for these quantities are given in the Appendix. Together with quadrature formulae to represent the angle and frequency integrations in (5.21), the above equations provide a method for the computation of Λ_{ij}.

The computation of $\bar{J}_{c,i}$, the contribution of the background emission to the mean integrated intensity at z_i, proceeds in a similar way. From equation (4.9) we obtain:

$$\bar{J}_{c,i} = \frac{1}{4\pi} \oint_{\mu>0} d\Omega \int_{-\infty}^{+\infty} d(\Delta\nu) \sum_{j=1}^{N} \tilde{L}_{ij} S_{c,j}. \tag{5.52}$$

The expressions for $\tilde{L}_{ij}$ are similar to those of L_{ij}, except that the matrix $\{\tilde{I} + \frac{1}{2}s_j X_j + \frac{1}{2}s_j^* X_j^\dagger\}$ appearing in equation (5.40) must be replaced by $r_j \tilde{I}$. Consequently, the factors $(1 + t_{k,j})$ in equations (5.44) and (A.1) - (A.16) must be replaced by r_j.

5.3. Matrix Equation for the Line Source Function

If the line opacity is dominated by scattering ($\epsilon \ll 1$), a small amount of background absorption can strongly affect the thermalization of photons. Thus the behaviour of the line source function at large optical depth is sensitive to the numerical treatment of the background absorption (see Mihalas 1978). To obtain the correct behaviour of the source function it is necessary to work with the *difference* of Λ_{ij} from the unity matrix, rather than with Λ_{ij} itself. In the following we discuss the asymptotic behaviour of the diagonal elements of the lambda operator in the limit of large optical depth between the grid points. This leads to the proper definition of the difference operator, and to a numerically accurate formulation of the equation for $S_{L,j}$.

When the optical depth between grid points is very large, the radiative coupling between different grid points is weak, and Λ_{ij} is nearly diagonal. Furthermore, the main contributions to the diagonal elements Λ_{ii} come from the two half-intervals closest to grid point z_i, i.e., the terms $q_{2,ii}$ and $q_{3,ii}$ in equation (5.51). Detailed expressions for $q_{2,ii}$ and $q_{3,ii}$ are given in the Appendix (equation [A.10] and [A.11]). The functions $m_n(x,-1)$ ($n = 0,1,2$) appearing in these expressions approach the values:

$$m_0(x,-1) \approx 1, \quad m_1(x,-1) \approx \frac{1}{x}, \quad m_2(x,-1) \approx \frac{2}{x^2}, \quad \text{for } |x| \gg 1. \tag{5.53}$$

Furthermore, for $i = j$ equation (5.48) yields:

$$\begin{aligned} y_{k,ii} &= \frac{1}{2}\mathrm{Tr}\left\{\Phi_i Y_{k,i} + Y_{k,i}\Phi_i^\dagger\right\} \\ &= (1+\eta_{k,i})(1+t_{k,i})\phi_{I,i}, \end{aligned} \tag{5.54}$$

where $\eta_{k,i}$ is defined by:

$$\eta_{1,i} = \eta_{2,i} = -\eta_{3,i} = -\eta_{4,i} = \frac{1}{2}\mathrm{Tr}\{X_i X_i^\dagger\}, \tag{5.55}$$

and where equations (5.43), (5.45) and (5.46) have been used. Inserting (5.53) and (5.54) into (A.10) and (A.11), we obtain:

$$\begin{aligned} q_{2,ii} + q_{3,ii} \approx 2\phi_{I,i}\ &\frac{1}{4}\sum_{k=1}^{4} \frac{(1+\eta_{k,i})(1+t_{k,i})^2}{1+r_i+t_{k,i}} \\ &\cdot\left\{1 - \frac{2}{(1+r_i+t_{k,i})^2(\Delta\tau_{1,i}+\Delta\tau_{2,i+1})(\Delta\tau_{2,i}+\Delta\tau_{1,i-1})}\right\}. \end{aligned} \tag{5.56}$$

The second term in curly brackets is of order $\Delta\tau^{-2}$. Hence in the limit $\Delta\tau \to \infty$ the diagonal element of the lambda operator becomes:

$$\lim_{\Delta\tau\to\infty} \Lambda_{ii} = \frac{1}{4\pi}\oint_{\mu>0} d\Omega \int_{-\infty}^{+\infty} d(\Delta\nu)\ 2\phi_{I,i}\ \frac{1}{4}\sum_{k=1}^{4} \frac{(1+\eta_{k,i})(1+t_{k,i})^2}{1+r_i+t_{k,i}}. \tag{5.57}$$

We now define the difference operator $\mathcal{L}_{ij}$ as follows:

$$\mathcal{L}_{ij} \equiv \Lambda_{ij} - (1 - \Delta_i)\delta_{ij}, \tag{5.58}$$

where δ_{ij} is the Kronecker-δ, and Δ_i is defined by:

$$\Delta_i \equiv 1 - \lim_{\Delta\tau \to \infty} \Lambda_{ii}. \tag{5.59}$$

Using the normalization of $\phi_{I,i}$ (see equation [2.30]) and the relation

$$\frac{1}{4} \sum_{k=1}^{4} (1 + \eta_{k,i})(1 + t_{k,i}) = 1,$$

which follows from (5.45) and (5.55), equation (5.57) yields:

$$\Delta_i = \frac{1}{4\pi} \oint_{\mu>0} d\Omega \int_{-\infty}^{+\infty} d(\Delta\nu)\; 2r_i\phi_{I,i}\, \frac{1}{4} \sum_{k=1}^{4} \frac{(1+\eta_{k,i})(1+t_{k,i})}{1 + r_i + t_{k,i}}. \tag{5.60}$$

Inserting (5.58) into equation (4.15), we finally obtain the desired matrix equation for the line source function:

$$S_{L,i} - \frac{1}{\epsilon_i + \Delta_i} \sum_{j=1}^{N} \mathcal{L}_{ij} S_{L,j} = \frac{\epsilon_i B_i^S + \bar{J}_{c,i}}{\epsilon_i + \Delta_i}. \tag{5.61}$$

Since the thermal source B_i^S and the continuum source $S_{c,i}$ are known from a previous iteration (see the discussion in section 4), the right hand side of (5.61) can be readily computed. Unless the number of grid point is very large, the solution of (5.61) can be obtained using standard matrix methods.

6. SUMMARY

We formulated a method to compute the line source function S_L for the case in which the Zeeman splitting of the spectral line is larger than the Doppler width, and for arbitrary small values of the collision parameter ϵ. The method is a generalization of the integral equation method for solving NLTE problems (see Avrett and Hummer 1965, Mihalas 1978). The advantage of this method over the differential equation method for solving the equations of polarized radiation transfer (Rees 1969; Auer et al. 1977) is that it allows one to determine the effect of Zeeman splitting on the Λ_{ij} matrix, as well as the effect on the line source function. This is useful for understanding how different atmospheric layers are coupled by radiative transport, and to assess the role of magneto-optical effects in this coupling. Rees (1969) and Auer et al. (1977) found that the effect of Zeeman splitting on the level populations is apparently rather small. With the integral method we hope to clearify the reasons for this result.

Once the source function $S_L(z)$ is known, the emergent intensity and polarization from the atmosphere can be readily computed for any angle and frequency. Hence, the method described here provides a basis for the interpretation of Stokes polarimeter data of sunspots, or other objects with strong magnetic fields.

ACKNOWLEDGEMENT

The author would like to thank E. H. Avrett, G.B. Rybicki and W. Kalkofen for useful comments and suggestions. This work was supported by NASA grant NSG-7054 and NASA contract NAS5-27792.

REFERENCES

Angel, J.R.P. (1978). *Ann. Rev. Astron. & Astrophys.*, **16**, 487.

Angel, J.R.P., Borra, E.F. & Landstreet, J.D. (1981). *Astrophys. J. Suppl.*, **45**, 457.

Athay, R.G. & Skumanich, A. (1967). *Ann. d'Astrophys.*, **30**, 669.

Auer, L.H., Heasley, J.N. & House L.L. (1977). *Astrophys. J.*, **216**, 531.

Avrett, E.H. & Hummer, D.G. (1965). *Monthly Notices Roy. Astron. Soc.* **130**, 295.

Ayres, T.R., Marstad, N.C. & Linsky, J.L. (1981). *Astrophys. J.*, **247**, 545

Baliunas, S.L. & Vaughan, A.H. (1985). *Ann. Rev. Astron. & Astrophys.*, **23**, 379.

Beckers, J.M. (1969a). *Solar Phys.*, **9**, 372.

——— (1969a). *Solar Phys.*, **10**, 262.

Boesgaard, A.M. (1974). *Astrophys. J.*, **188**, 567.

Bommier, V., Leroy, J.L. & Sahal-Brechot, S. (1981). *Astron. Astrophys.*, **100**, 231.

Borra, E.F., Landstreet, J.D. & Mestel, L. (1982). *Ann. Rev. Astron. & Astrophys.*, **20**, 191.

Borra, E.F., Edwards, G. & Mayer, M. (1984). *Astrophys. J.*, **284**, 211.

Brown, D.N. & Landstreet, J.D. (1981). *Astrophys. J.*, **246**, 899

Gray, D.F. (1984). *Astrophys. J.*, **277**, 640.

Hale, G.E. (1908). *Astrophys. J.*, **28**, 315.

Hartmann, L.W., Dupree, A.K. & Raymond, J.C. (1982). *Astrophys. J.*, **252**, 214.

Harvey, J.W. (1977). In *Highlights of Astronomy*, **4**, Part 2, ed. E. Mueller, p. 223. Dordrecht: Reidel.

House, L.L. & Steinitz, R. (1975). *Astrophys. J.*, **195**, 235.

Landi Degl'Innocenti, E. (1982). *Solar Phys.*, **79**, 291.

__________ (1983). *Solar Phys.*, **85**, 3.

Landi Degl'Innocenti, E. & and Landi Degl'Innocenti, M. (1972). *Solar Phys.*, **27**, 319.

__________ (1974). *Il Nuovo Cimento*, **27B**, 134.

__________ (1985). *Solar Phys.*, **97**, 239.

Marcy, G.W. (1983). In *Solar and Stellar Magnetic Fields, IAU Symp. No.*, **102**, ed. J.O. Stenflo, p. 3. Boston: Reidel.

__________ (1984). *Astrophys. J.*, **276**, 786.

Mihalas, D. (1978). *Stellar Atmospheres.* San Francisco: W.H. Freeman and Co..

Rees, D.E. (1969). *Solar Phys.*, **10**, 268.

Rees, D.E. (1982). *Proc. Astron. Soc. Aust.*, **4**, No.4, 335.

Reichel, A. (1968). *J. Quant. Spectrosc. Rad. Transfer*, **8**, 1601.

Robinson, R.D., Worden, S.P. & Harvey, J.W. (1980). *Astrophys. J. Lett.*, **236**, L155.

Shurcliff, W.A. (1962). *Polarized Light.* Cambridge, Mass.: Harvard University Press.

Sidlichovsky, M. (1974). *Bull. Astron. Inst. Czech.*, **25**, 198.

Stenflo, J.O. (1985). *Solar Phys.*, **100**, 189.

Unno, W. (1956). *Publ. Astron. Soc. Japan*, **8**, 108.

Vaiana, G.S., Cassinelli, J.P., Fabbiano, G., Giacconi, R., Golub, L. et al. (1981). *Astrophys. J.*, **245**, 163.

van Ballegooijen, A.A. (1984). In *Measurements of Solar Vector Magnetic Fields*, ed. M. Hagyard (NASA CP-2374), p. 322.

Vogt, S.S. (1982). In *Activity in Red Dwarf Stars*, ed. P.B. Byrne, M. Rodono, p. 137. Boston: Reidel.

Wilson, O.C. (1978). *Astrophys. J.*, **226**, 379.

Wittmann, A. (1974). *Solar Phys.*, **35**, 11.

APPENDIX A

The quantities $g_{n,ij}$ and $h_{n,ij}$ of equations (5.49) and (5.50) are defined as follows ($i \neq j$):

$$g_{1,ij} = \frac{1}{4}\sum_{k=1}^{4}(1+t_{k,j-1})\theta_{k,j-1}^{-1}w_{1,j-1}m_1(\theta_{k,j-1}\Delta\tau_{1,j-1}, +\sigma)y_{k,ij-1}, \quad (A.1)$$

$$g_{2,ij} = \frac{1}{4}\sum_{k=1}^{4}(1+t_{k,j})\theta_{k,j}^{-1}\Big[m_0(\theta_{k,j}\Delta\tau_{2,j}, -\sigma)$$

$$- w_{2,j} m_1(\theta_{k,j}\Delta\tau_{2,j}, -\sigma)\Big] y_{k,ij}, \tag{A.2}$$

$$h_{1,ij} = \frac{1}{4}\sum_{k=1}^{4}(1+t_{k,j})\theta_{k,j}^{-1}\Big[m_0(\theta_{k,j}\Delta\tau_{1,j}, +\sigma) - w_{1,j} m_1(\theta_{k,j}\Delta\tau_{1,j}, +\sigma)\Big] y_{k,ij}, \tag{A.3}$$

$$h_{2,ij} = \frac{1}{4}\sum_{k=1}^{4}(1+t_{k,j+1})\theta_{k,j+1}^{-1} w_{2,j+1} m_1(\theta_{k,j+1}\Delta\tau_{2,j+1}, -\sigma) y_{k,ij+1}, \tag{A.4}$$

The $q_{n,ij}$ $(j = i-1, i, i+1)$ of equation (5.51) are given by:

$$q_{1,ii-1} = \frac{1}{4}\sum_{k=1}^{4}(1+t_{k,i-1})\theta_{k,i-1}^{-1}\Big[m_0(\theta_{k,i-1}\Delta\tau_{1,i-1}, +1) - \frac{2+\xi_i}{1+\xi_i} w_{1,i-1} m_1(\theta_{k,i-1}\Delta\tau_{1,i-1}, +1) + \frac{1}{1+\xi_i} w_{1,i-1}^2 m_2(\theta_{k,i-1}\Delta\tau_{1,i-1}, +1)\Big] y_{k,ii-1}, \tag{A.5}$$

$$q_{2,ii-1} = \frac{1}{4}\sum_{k=1}^{4}(1+t_{k,i})\theta_{k,i}^{-1}\Big[\frac{\xi_i}{1+\xi_i} w_{2,i} m_1(\theta_{k,i}\Delta\tau_{2,i}, -1) + \frac{1}{1+\xi_i} w_{2,i}^2 m_2(\theta_{k,i}\Delta\tau_{2,i}, -1)\Big] y_{k,ii}, \tag{A.6}$$

$$q_{3,ii-1} = \frac{1}{4}\sum_{k=1}^{4}(1+t_{k,i})\theta_{k,i}^{-1}\Big[-\frac{\xi_i^2}{1+\xi_i} w_{1,i} m_1(\theta_{k,i}\Delta\tau_{1,i}, -1) + \frac{\xi_i^2}{1+\xi_i} w_{1,i}^2 m_2(\theta_{k,i}\Delta\tau_{1,i}, -1)\Big] y_{k,ii}, \tag{A.7}$$

$$q_{4,ii-1} = \frac{1}{4}\sum_{k=1}^{4}(1+t_{k,i+1})\theta_{k,i+1}^{-1}\Big[-\frac{\xi_i^2}{1+\xi_i} w_{2,i+1} m_1(\theta_{k,i+1}\Delta\tau_{2,i+1}, +1) + \frac{\xi_i^2}{1+\xi_i} w_{2,i+1}^2 m_2(\theta_{k,i+1}\Delta\tau_{2,i+1}, +1)\Big] y_{k,ii+1}, \tag{A.8}$$

$$q_{1,ii} = \frac{1}{4}\sum_{k=1}^{4}(1+t_{k,i-1})\theta_{k,i-1}^{-1}\Big[\frac{1+\xi_i}{\xi_i} w_{1,i-1} m_1(\theta_{k,i-1}\Delta\tau_{1,i-1}, +1) - \frac{1}{\xi_i} w_{1,i-1}^2 m_2(\theta_{k,i-1}\Delta\tau_{1,i-1}, +1)\Big] y_{k,ii-1}, \tag{A.9}$$

$$q_{2,ii} = \frac{1}{4}\sum_{k=1}^{4}(1+t_{k,i})\theta_{k,i}^{-1}\Big[m_0(\theta_{k,i}\Delta\tau_{2,i}, -1) - \frac{\xi_i - 1}{\xi_i} w_{2,i} m_1(\theta_{k,i}\Delta\tau_{2,i}, -1) - \frac{1}{\xi_i} w_{2,i}^2 m_2(\theta_{k,i}\Delta\tau_{2,i}, -1)\Big] y_{k,ii}, \tag{A.10}$$

$$
\begin{aligned}
q_{3,ii} = \frac{1}{4}\sum_{k=1}^{4}(1+t_{k,i})\theta_{k,i}^{-1}\Big[& m_0(\theta_{k,i}\Delta\tau_{1,i},-1) \\
&+(\xi_i-1)w_{1,i}m_1(\theta_{k,i}\Delta\tau_{1,i},-1) \\
&-\xi_i w_{1,i}^2 m_2(\theta_{k,i}\Delta\tau_{1,i},-1)\Big]y_{k,ii},
\end{aligned} \tag{A.11}
$$

$$
\begin{aligned}
q_{4,ii} = \frac{1}{4}\sum_{k=1}^{4}(1+t_{k,i+1})\theta_{k,i+1}^{-1}\Big[&(\xi_i+1)w_{2,i+1}m_1(\theta_{k,i-1}\Delta\tau_{2,i+1},+1) \\
&-\xi_i w_{2,i+1}^2 m_2(\theta_{k,i-1}\Delta\tau_{2,i+1},+1)\Big]y_{k,ii+1},
\end{aligned} \tag{A.12}
$$

$$
\begin{aligned}
q_{1,ii+1} = \frac{1}{4}\sum_{k=1}^{4}(1+t_{k,i-1})\theta_{k,i-1}^{-1}\Big[&-\frac{1}{\xi_i(1+\xi_i)}w_{1,i-1}m_1(\theta_{k,i-1}\Delta\tau_{1,i-1},+1) \\
&+\frac{1}{\xi_i(1+\xi_i)}w_{1,i-1}^2 m_2(\theta_{k,i-1}\Delta\tau_{1,i-1},+1)\Big]y_{k,ii-1},
\end{aligned} \tag{A.13}
$$

$$
\begin{aligned}
q_{2,ii+1} = \frac{1}{4}\sum_{k=1}^{4}(1+t_{k,i})\theta_{k,i}^{-1}\Big[&-\frac{1}{\xi_i(1+\xi_i)}w_{2,i}m_1(\theta_{k,i}\Delta\tau_{2,i},-1) \\
&+\frac{1}{\xi_i(1+\xi_i)}w_{2,i}^2 m_2(\theta_{k,i}\Delta\tau_{2,i},-1)\Big]y_{k,ii},
\end{aligned} \tag{A.14}
$$

$$
\begin{aligned}
q_{3,ii+1} = \frac{1}{4}\sum_{k=1}^{4}(1+t_{k,i})\theta_{k,i}^{-1}\Big[&\frac{1}{1+\xi_i}w_{1,i}m_1(\theta_{k,i}\Delta\tau_{1,i},-1) \\
&+\frac{\xi_i}{1+\xi_i}w_{1,i}^2 m_2(\theta_{k,i}\Delta\tau_{1,i},-1)\Big]y_{k,ii},
\end{aligned} \tag{A.15}
$$

$$
\begin{aligned}
q_{4,ii+1} = \frac{1}{4}\sum_{k=1}^{4}(1+t_{k,i+1})\theta_{k,i+1}^{-1}\Big[& m_0(\theta_{k,i+1}\Delta\tau_{2,i+1},+1) \\
&-\frac{1+2\xi_i}{1+\xi_i}w_{2,i+1}m_1(\theta_{k,i+1}\Delta\tau_{2,i+1},+1) \\
&+\frac{\xi}{1+\xi_i}w_{2,i+1}^2 m_2(\theta_{k,i+1}\Delta\tau_{2,i+1},+1)\Big]y_{k,ii+1}.
\end{aligned} \tag{A.16}
$$

Here ξ_i and $\theta_{k,j}$ are defined by:

$$
\xi_i \equiv \frac{\Delta\tau_{1,i}+\Delta\tau_{2,i+1}}{\Delta\tau_{2,i}+\Delta\tau_{1,i-1}}, \tag{A.17}
$$

$$
\theta_{k,j} \equiv 1+r_j+t_{k,j}, \tag{A.18}
$$

and the functions $m_n(x,\sigma)$ of equation (5.47) are given by ($\sigma=\pm1$):

$$
m_0(x,\sigma) = \sigma\left(e^{\sigma x}-1\right), \tag{A.19}
$$

$$
m_1(x,\sigma) = \sigma e^{\sigma x}-\frac{1}{x}\left(e^{\sigma x}-1\right), \tag{A.20}
$$

$$
m_2(x,\sigma) = \sigma\left[e^{\sigma x}+\frac{2}{x^2}\left(e^{\sigma x}-1\right)\right]-\frac{2}{x}e^{\sigma x}. \tag{A.21}
$$

AN INTEGRAL OPERATOR TECHNIQUE OF RADIATIVE TRANSFER IN SPHERICAL SYMMETRY

A. Peraiah
Indian Institute of Astrophysics, Bangalore 560034, India.

ABSTRACT

The integral operator technique, by which the solution of radiative transfer equation is obtained, is reviewed. Three applications of this method have been presented: (1) solution of transfer equation with coherent, isotropic scattering (2) solution of line transfer in the comoving frame and (3) solution of transfer equation with aberration and advection terms included.

INTRODUCTION

Integration of the transfer equation is required in many physical situations. A solution that is derived for a given physical situation may not always be useful or appropriate in a different physical context. Therefore it becomes necessary to develop a numerical solution which will be useful in many situations. In the 'CELL' method (see Peraiah 1984) we performed the integration over the angle-radius grid given by

$$[\mu_{j-\frac{1}{2}}, \mu_{j+\frac{1}{2}}] \, [r_n, r_{n+1}]$$

where

$$\mu_{n+\frac{1}{2}} = \sum_{j=1}^{k} c_j, \quad k = 1,2,\ldots,J$$

where C_j's are quadrature weights and J is the total number of angles. This approach was quite useful because the term containing the angle derivative transforms into a single band matrix, the elements of which are derived in such a way that they always satisfy flux conservation in a medium which is in radiative equilibrium. This method had been applied to many problems successfully. However, the positive and negative elements of the matrix obtained by discretization of the curvature term, are distributed unevenly and this becomes a source for generating unphysical negative intensities in the final solution sought in vacuum. Therefore, we felt that there is a need to develop a better numerical scheme which can avoid the difficulty mentioned above while retaining the agreeable features of the 'CELL' method, such as flux conservation, accuracy and stability of the solution.

We selected the angle-radius grid for monochromatic case or the angle-radius-frequency grid in the case of line transfer in the comoving frame in such a way that they are as general as possible (see Figure 1a, 1b). In general we have the radial coordinates (r_i, r_{i-1}), the angle coordinates (μ_j, μ_{j-1}) and frequency coordinates (X_k, X_{k-1}) and these can be chosen from the zeroes of any quadrature formula depending upon the numerical necessity of the problem (Lathrop & Carlson 1967, 1971; Periah & Varghese 1985).

Figure 1a. Schematic diagram of angle-radius grid.

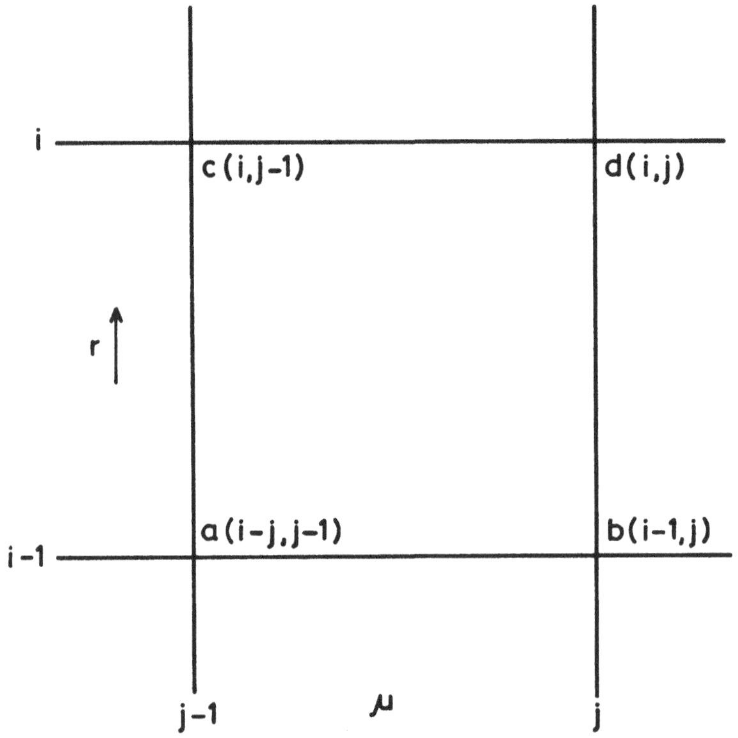

We express the intensity at any point on the grid, in terms of certain interpolation coefficients and weighted means of the grid points. These interpolation coefficients are estimated in terms of the nodal values of the specific intensities. The equation of transfer is integrated on the grid by applying the integral operators. Finally, from the resulting equation, we calculate the reflection and transmission matrices by using the interaction principle - a relationship that connects the input and output intensities through the transmission and reflection properties of the medium. This has been discussed in detail along with diffuse radiation field in Peraiah (1984). In this article we shall derive the transmission and reflection matrices wherever they are required and the reader may refer to the above reference for obtaining scheme for the diffuse internal radiation field or see Appendix V. This method has been applied to three cases and accordingly we divide the article into three parts:

Figure 1b. Schematic diagram of angle-radius-frequency grid.

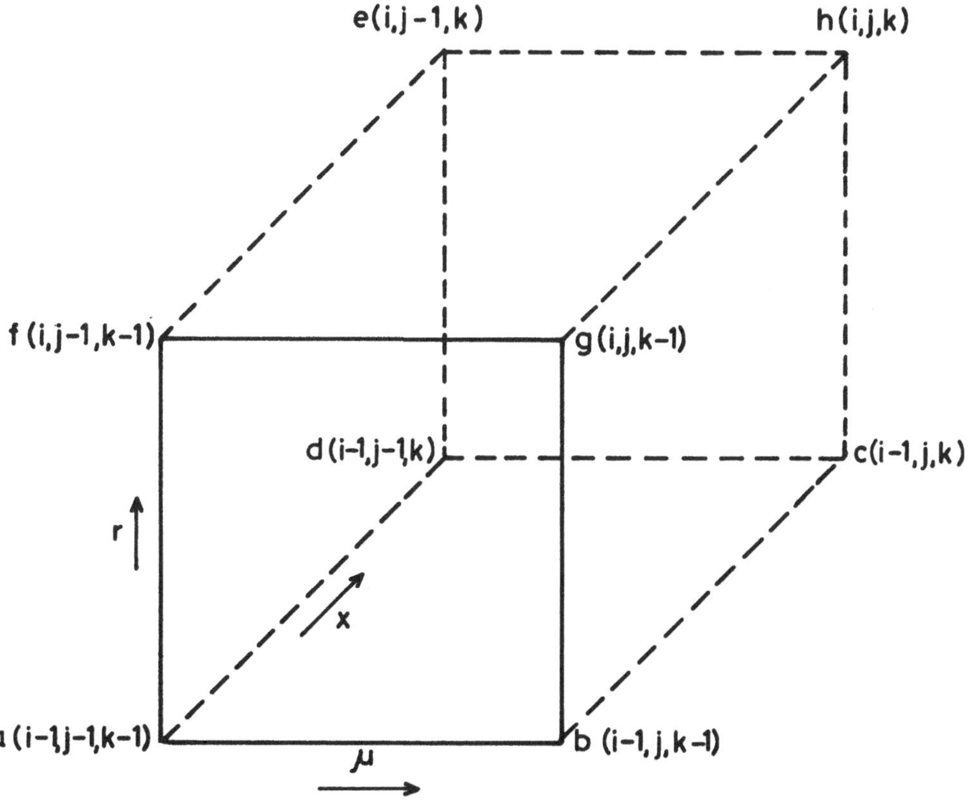

(I) Solution in a Monochromatic Radiation Field

(II) Solution of Line Transfer in the Comoving Frame.

(III) Effects of Aberration and Advection.

I. Solution in a Monochromatic Radiation Field with Isotropic Scattering

As mentioned in the introduction we express the specific intensity in terms of certain interpolation coefficients. Thus, we have,

$$u(r,\mu) \approx u_{oo} + u_{o1}\xi + (u_{1o} + u_{11}\xi)\eta \qquad (1)$$

Here $U(r,\mu)$ is given by

$$u(r,\mu) = 4\pi r^2 I(r,\mu) \qquad (2)$$

$I(r,\mu)$ being the specific intensity of the ray making an angle $\cos^{-1}\mu$ with the radius vector. U_{oo} and U_{o1}, U_{1o} and U_{11} are the interpolation coefficients and the quantities ξ and η are defined as

$$\xi = \frac{r-\bar{r}}{\Delta r/2} \quad , \quad \eta = \frac{\mu-\bar{\mu}}{\Delta\mu/2} \qquad (3)$$

where

$$\bar{r} = \tfrac{1}{2}\,(r_i + r_{i-1}) \ , \ \bar{\mu} = \tfrac{1}{2}(\mu_j + \mu_{j-1}) \qquad (4)$$

and

$$\Delta r = (r_i - r_{i-1}) \ , \quad \Delta\mu = (\mu_j - \mu_{j-1}) \quad . \qquad (5)$$

The quantities U_{oo} etc. are estimated in terms of the nodal values of intensities U_a, U_b, U_c and U_d (see Figure 1a). There are four main steps to obtain the transmission and reflection matrices in a given layer or a spherical shell: (see Peraiah and Varghese 1985).

(1) Substitute equation (1) in the equation of transfer given by

$$\mu \frac{\partial u(r,\mu)}{\partial r} + \frac{1}{r}\frac{\partial}{\partial \mu} \{(1-\mu^2)\ u(r,\mu)\} = K(r)$$

$$\{S(r,\mu) - u(r,\mu)\} \tag{6a}$$

and

$$-\mu \frac{\partial u(r,-\mu)}{\partial r} - \frac{1}{r}\ \frac{\partial}{\partial \mu} \{(1-\mu^2)u(r,-\mu)\} = K(r)$$

$$\{S(r,-\mu) - u(r,-\mu)\} \tag{6b}$$

where $K(r)$ is the absorption coefficient and $S(r,\mu)$ is the source function and $0 < \mu \le 1$.

(2) The resulting equation is operated by the quantities X and Y given by

$$X = \frac{1}{\Delta\mu} \int_{\Delta\mu} \ldots\ d\mu \tag{7}$$

and

$$Y = \frac{1}{V} \int_{\Delta r} \ldots\ 4\pi r^2 dr \tag{8}$$

where

$$V = \frac{4}{3}\pi(r_i^3 - r_{i-1}^3) \quad . \tag{9}$$

(3) The interpolation coefficients U_{oo}, U_{o1}, U_{1o} and U_{11} in the integrand resulted in step (2) are replaced by the nodal values U_a, U_b, U_c and U_d (Figure 1a).

(4) The resulting equation in step (3) is rearranged in the canonical form of the interaction principle (Peraiah 1984) and the transmission and reflection matrices are obtained. These matrices are given in Appendix I. Several tests have been applied to this method to check accuracy and stability. These tests and applications in a monochromatic, isotropic scattering case are described in Peraiah and Varghese (1985). We discuss

the following aspects of the solution:

(a) The invariance of the specific intensity in a non-absorbing and non-emitting medium.

(b) Continuity of the solution both in the angle and radial coordinates.

(c) Uniqueness of the solution.

(d) The condition of zero net flux in the case of a specular reflecting boundary at the inner surface in a purely scattering spherical medium and global conservation of flux in such a medium. These tests are described below:

(a) Invariance of the Specific intensity in vacuum

In the absence of absorbing and emitting sources, the specific intensity should remain constant. This is a crucial test that checks the numerical accuracy of the solution. The transmission and reflection matrices are functions not only of the physics of the medium but also of the geometry of the medium. For example, as we mentioned in the introduction, the 'CELL' method gives negative intensities. This is because even in the absence of any absorbing and emitting medium, ($\tau = 0$, vacuum) the transmission matrices are given by (see Appendix I)

$$\underset{\sim}{t}(i,i-1) = (\underset{\sim}{M}_1 - \bar{\underset{\sim}{\rho}}_+)^{-1} (\underset{\sim}{M}_3 + \bar{\underset{\sim}{\rho}}_.)$$

and

$$\underset{\sim}{t}(i-1,i) = (\underset{\sim}{M}_4 - \bar{\underset{\sim}{\rho}}_-)^{-1} (\underset{\sim}{M}_2 - \bar{\underset{\sim}{\rho}}_+) \quad .$$

In the plane parallel stratification the quantities $t(i,i-1)$ and $t(i-1,i)$ reduce to I, the unit matrix. However in spherically symmetric approximation they do not reduce to I. This is because of the presence of the spherical term given by,

$$\frac{1}{r}\frac{\partial}{\partial\mu}\{(1-\mu^2)u(r,\mu)\}$$

Therefore, the matrices $t(i,i-1)$, and $t(i-1,i)$ contain the equivalent quantities $\bar{\underset{\sim}{\rho}}_{\pm}$ (which represent curvature factors) even when $\tau = 0$. Therefore we have to check whether the presence of these terms in the r & t matrices cause any numerical error. For this purpose we set $\tau = 0$ and $B/A = 1, 2, 5, 10$ and 100 where B and A are the outer and inner radii of the spherical shell, and computed the solution. The results are plotted in Figure 2.

The spherical surface at A is irradiated with unit intensity and no radiation is incident at the surface B.

$$U_A^- = 1$$

and for all μ_js.

$$U_B^+ = 0$$

Figure 2. Invariance of specific intensity. The outward intensities (U^-) are always unity because the incident intensities $U_{N+1}^-(\mu_j) = 1$, and the inward intensities U^+ are always zero. (Peraiah and Varghese 1985).

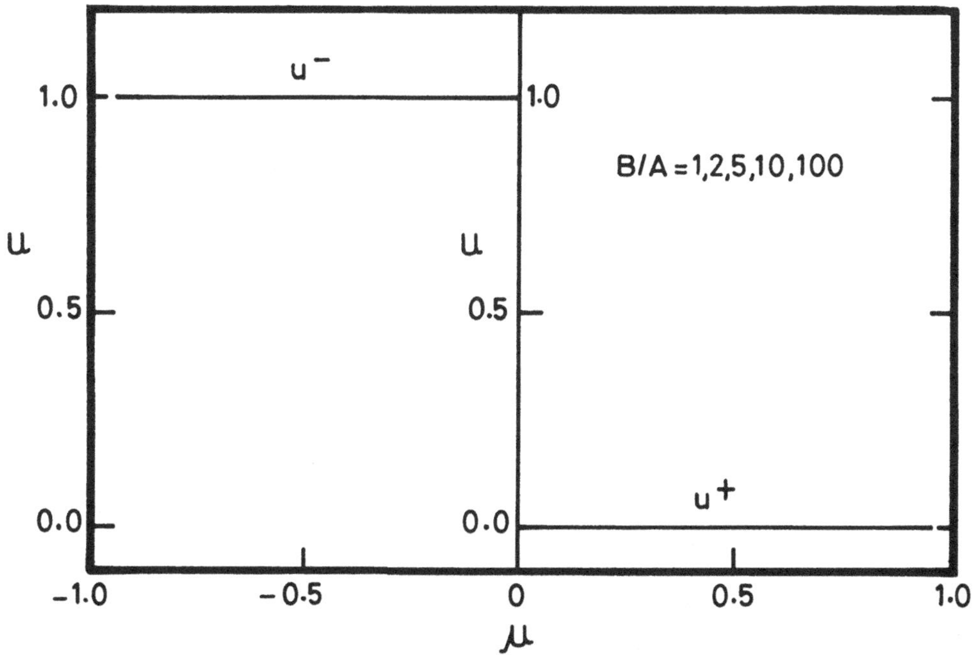

We have calculated the solution for different sizes of the spherical shells (i.e.) $B/A = 1$ (plane parallel case) 2, 5, 10 and 100. As B/A increases the curvature factor $\Delta r/\bar{r}$ increases and introduces more numerical errors. From Figure 2, we see that the outward intensities at all angles are constant $(U^- = 1)$ and equal to the incident intensities. There is no radiation directed inwards $(U^+ = 0)$. This holds true at all radial points for any given B/A. The intensities behave exactly the same way irrespective of the value of B/A, whether it equals 1 or 2 or 5 or 100. In the 'CELL' method we have to reduce the size (or critical stepsize) of the 'CELL' so small (to obtain non-negative t & r matrices) that the number of subshells increases enormously. In this method even when we choose a large $\Delta r/\bar{r}$ $(=\rho)$ in the fundamental shell we always obtain positive intensities. However, one should remember that there are other factors such as the physical nature of the medium that could affect the size of the fundamental shell. Thus, we have verified the invariance of the specific intensity in vacuum.

(b) Continuity of the solution

We must ensure that the solution developed should not show any discontinuities either in the radial coordinate or in the angle coordinate. For this purpose we must have non-negative r & t matrices. This can be achieved if we manage to obtain $\underset{\sim}{\Delta}^+$ and $\underset{\sim}{\Delta}^-$ (see Appendix I) which contain non-negative elements. This requires that the diagonal elements of $[\Delta^+]^{-1}$ and $[\Delta^-]^{-1}$ satisfy the condition that (see Appendix I)

$$\{[\underset{\sim}{\Delta}^+]^{-1}_{jj} \;,\; [\underset{\sim}{\Delta}^-]^{-1}_{jj}\} \geq 0$$

while the off-diagonal elements satisfy the condition that

$$\{[\Delta^+]^{-1}_{jk} \;,\; [\Delta^-]^{-1}_{jk}\} \leq 0$$

If we expand the above inequalities we obtain a condition for the minimum size of the shell (τ_{crit}) for which we can obtain a stable and continuous solution. This restriction contains the constraints on the curvature factor also. This condition of stepsize forces us to divide the medium into several fundamental shells whose thickness is less than or equal to τ_{crit} and we call this as τ_f. However, we are free to divide the medium into larger shells whose thickness is greater than τ_{crit} or τ_f and apply star product algorithm (see Peraiah 1984 or Appendix V).

In an isotropically and coherently scattering medium with $p(r,\mu_j,\mu_k) = 1$ and for Gaussian points with four angles in each quadrant one obtains a maximum optical depth up to 0.5 with curvature factor $\Delta r/\bar{r} < 10^{-3}$. The curvature factor of shells with constant geometrical thickness, increases towards the centre of the sphere and decreases outwards. If the curvature factor increases then also we employ the star algorithm to obtain the non-negative r & t matrices. In Figure 3, we plotted the angular distribution of intensities. We can see that the

Figure 3. Angular distribution of the intensities for B/A = 2. Dotted lines, T = 50; solid lines, T = 20 (Peraiah and Varghese 1985).

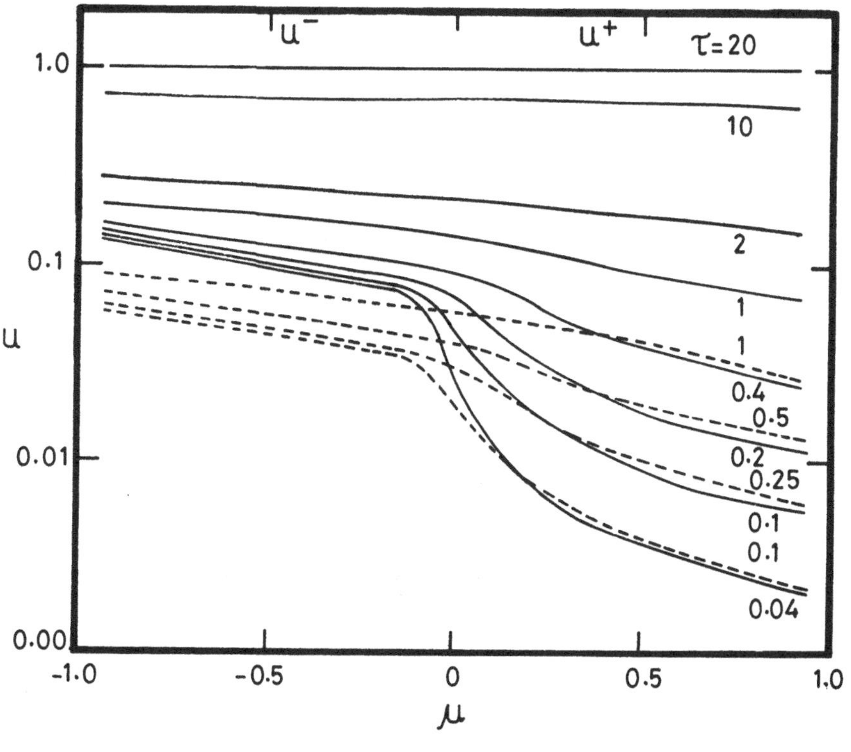

solution is continuous in the domain of $-1 \le \mu \le 1$. The solution is given at various points in the medium. The quantities U^- and U^+ are intensities directed outwards $(\tau \to 0)$ and inwards $(\tau \to \tau_{max} = T)$ respectively. The dotted lines correspond to $T = 50$ and the solid lines correspond to $T = 20$ where T is the total optical depth. In Figure 4 we plotted U^+ with respect to τ for the four angles μ_1, μ_2, μ_3 and μ_4, where

Figure 4. Radial distribution of the intensities directed inwards (U^+) for the rays corresponding to μ_1, μ_2, μ_3 and μ_4 (Peraiah and Varghese 1985).

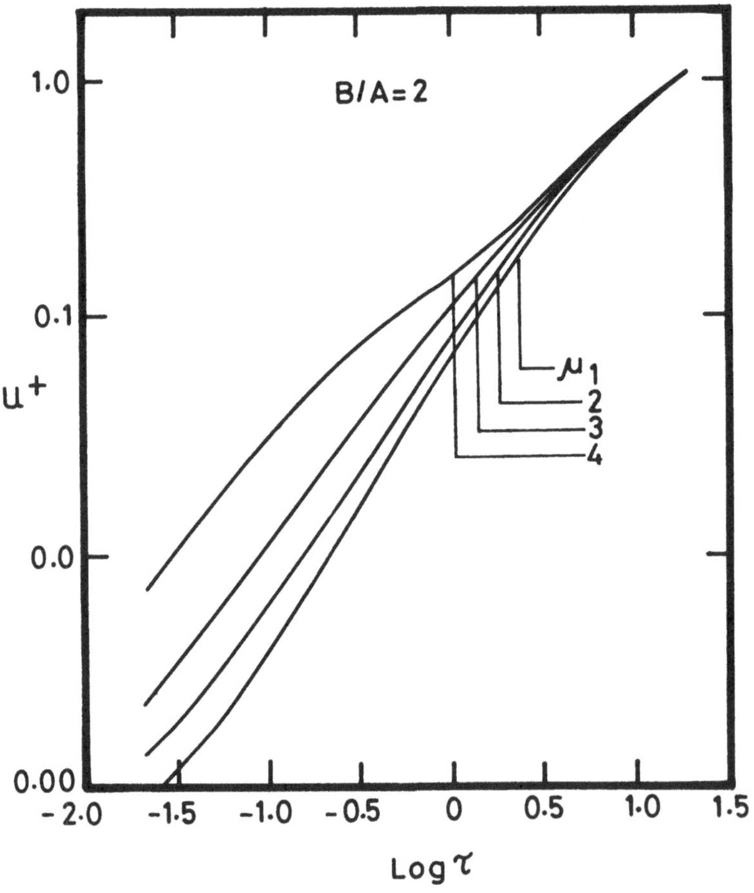

$\mu_1 = 0.06943$
$\mu_2 = 0.3301$
$\mu_3 = 0.66999$
$\mu_4 = 0.93057$

are the Gauss-Legendre points on $\mu \in [0,1]$. We can conventiently use Radau's points also which give equally accurate results. We can see that the intensities or the solution shows continuity in the radial coordinate.

(c) Uniqueness of the solution

The numerical solution must be tested for uniqueness. The solution is unique if we obtain the non-negative r & t operators. This can be achieved if we calculate the r & t operators in a shell whose thickness is equal to or less than τ_{crit} or τ_f. The uniqueness of the numerical solution can be checked as follows. Divide the fundamental shell into two shells, each of equal thickness. Then calculate the r & t matrices in the two shells (1/2 τ_f each) and then estimate the r & t matrices of the compound shell by using the star algorithm. Such operators should be exactly the same as those obtained by treating the shell as a single one with τ_f. In Table 1 we describe the uniqueness. The emergent intensities for μ_1, μ_2, μ_3 and μ_4 are given. We see that there is excellent agreement between the solutions obtained in the two cases.

They differ only in the fourth place because the curvature factor was kept constant in both the shells. As we explained earlier this changes from shell to shell and therefore we can expect the changes in the fourth decimal place as shown in Table 1.

Table 1. Check on uniqueness of solution. (Taken from Peraiah and Varghese 1985).

τ	μ_1	μ_2	μ_3	μ_4
τ_f	.6890	.8806	.9326	.9491
$\frac{1}{2}\tau_f + \frac{1}{2}\tau_f$	.6892	.8805	.9325	.9490

(d) Zero-net flux

If the surface at $\tau = T$ (point A) is a perfectly reflecting surface, then any radiation incident at $\tau = 0$ (at point B) will get completely reflected at A, if the medium between A and B scatters conservatively. Therefore the net flux at any intermediate point will be exactly zero. The net flux $|\Delta F_n|$ is calculated by the relation

$$|\Delta F_n| = \left| \sum_{j=1}^{J} \mu_j c_j (u_n^+ - u_n^-) \right| = 0$$

where subscript n represents any point inside the spherical medium. $|\Delta F_n|$ is given in Table 2.

Table 2. Net flux $|\Delta F_n|$ in specular reflection (Peraiah and Varghese 1985).

τ	T = 20, B/A = 2	τ	T = 50, B/A = 2
20.00	3.90×10^{-7}	50.0	1.40×10^{-7}
18.00	3.60×10^{-6}	45.00	1.90×10^{-6}
16.00	7.30×10^{-6}	40.00	4.22×10^{-6}
14.00	1.13×10^{-6}	35.00	7.30×10^{-6}
12.00	1.53×10^{-6}	30.00	1.17×10^{-6}
10.00	1.60×10^{-5}	25.00	1.58×10^{-5}
8.00	2.42×10^{-5}	20.00	2.12×10^{-4}
6.00	2.81×10^{-4}	15.00	2.73×10^{-4}
4.00	3.40×10^{-5}	10.00	3.43×10^{-5}
2.00	3.93×10^{-6}	5.00	2.23×10^{-5}
1.00	4.20×10^{-6}	2.50	4.66×10^{-5}
0.40	4.36×10^{-6}	1.00	4.92×10^{-5}
0.20	4.40×10^{-5}	0.50	5.01×10^{-4}
0.10	4.21×10^{-5}	0.25	5.03×10^{-5}
0.08	4.40×10^{-5}	0.20	5.04×10^{-4}
0.06	4.30×10^{-4}	0.15	5.03×10^{-5}
0.04	4.30×10^{-5}	0.10	5.02×10^{-5}
0.02	4.39×10^{-5}	0.05	5.01×10^{-5}

We can see from Table 2 that the flux is conserved at all points to the machine accuracy (single precision).

The transmission and reflection matrices computed in each shell can be employed to estimate the diffuse radiation field by using the algorithm given in Appendix V.

II. Solution of Line Transfer in the Comoving Frame

When line radiation is to be treated in the transfer equation we have to include the term containing the frequency derivative. In the comoving frame the following term is added to the transfer equation:

$$\left\{(1 - \mu^2)\frac{V'(r)}{r} + \mu^2 \frac{dV'(r)}{dr}\right\} \frac{\partial u(r,\mu,x)}{\partial x} \qquad (10)$$

where $V'(r)$ is the velocity of expansion and x is the normalized frequency coordinate given by

$$x = \frac{\nu - \nu_o}{\Delta_D} \qquad (11)$$

ν_o, ν being the frequencies at line centre and at any point in the line respectively, Δ_D is some standard frequency interval such as Doppler width. The intensity in this case is expressed as

$$\begin{aligned} U(r,\mu,x) &= U_o + u_r\xi + U_\mu\eta + U_x\chi \\ &+ U_{r\mu}\xi\eta + U_{\mu x}\eta\chi + U_{xr}\chi\xi \\ &+ U_{r\mu x}\xi\eta\chi \quad . \end{aligned} \qquad (12)$$

Here we have included the terms corresponding to the frequency discretization. The quantities U_o, U_r, U_μ etc., are the interpolation coefficients. The corresponding grid is shown in Figure (1b). Here χ is defined as

$$\chi = \frac{x - \bar{x}}{\Delta x/2} \qquad (13)$$

where

$$\bar{x} = \tfrac{1}{2}(x_{k-1} + x_k) \quad , \quad \Delta x = (x_k - x_{k-1}) \tag{14}$$

and ξ and η are defined in equation (3). The equation of line transfer is written as

$$\mu \frac{\partial u(r,\mu,x)}{\partial r} + \frac{1}{r}\frac{\partial}{\partial \mu}\{(1-\mu^2)U(r,\mu,x)\} = K_L[\beta+\phi(r,\mu,x)]$$

$$[S(r,\mu,x) - U(r,\mu,x)] + \{(1-\mu^2)\frac{V'(r)}{r} + \mu^2 \frac{dV'(r)}{dr}]\}$$

$$\frac{\partial u(r,\mu,x)}{\partial x} \tag{15}$$

where K_L is the absorption coefficient at line centre, β is the ratio of the opacity in the continuum to that at line centre and $\phi(r,\mu,x)$ is the Doppler profile function. The source function $S(r,\mu,x)$ is given by

$$S(r,\mu,x) = \frac{\phi(r,\mu,x)}{\beta+\phi(r,\mu,x)} S_L(r) + \frac{\beta}{\beta+\phi(r,\mu,x)} S_C(r,\mu,x) \tag{16}$$

where $S_L(r)$ and $S_C(r,\mu,x)$ are the line and continuum source functions respectively. They are given by

$$S_L(r) = (1-\varepsilon)\int_{-\infty}^{+\infty} J_x \phi_x dx + \varepsilon B \tag{17}$$

and

$$S_C(r,\mu,x) = \rho(r)B \tag{18}$$

where B is the Planck function, ε is the probability per scatter that a photon is thermalized after collisional de-excitation, and J_x is the mean intensity. The quantity $\rho(r)$ is chosen arbitrarily.

Same procedure of integration that is described in part I may be applied on equation (15) and its counterpart for $-\mu$ and obtain the transmission and reflection matrices (see Appendices II and III). We shall briefly describe how this is done. We define the vectors $\underset{\sim}{E}$ and $\underset{\sim}{F}$ corresponding to the interpolation coefficients and nodal values of the intensity respectively, as follows:

$$\underset{\sim}{E} = [U_o, U_r, U_\mu, U_x, U_{r\mu}, U_{\mu x}, U_{xr}, U_{r\mu x}]^T \tag{19}$$

and

$$\underset{\sim}{F} = [U_a, U_b, U_c, U_d, U_e, U_f, U_g, U_h]^T \tag{20}$$

where T represents the transpose of the vector. The two vectors are related by the relation

$$\underset{\sim}{E} = \frac{1}{8} \underset{\sim}{R} \underset{\sim}{F} \quad . \tag{21}$$

The matrix $\underset{\sim}{R}$ is given in Appendix II. From the relation (21) we can obtain $\underset{\sim}{E}$ or $\underset{\sim}{F}$ if we know the other. After substituting equation (12) in (15), the integration is performed on the resulting equation. We apply the operators X and Y successively. The resultant equation is integrated over the frequency domain by applying the operator Z, where Z is given by

$$Z = \frac{1}{\Delta x} \int_{\Delta x} \ldots dx \tag{22}$$

We obtain equations containing the element of the vector $\underset{\sim}{E}$ and by using equation (21) we transform these equations into those which contain the elements of the vector $\underset{\sim}{F}$ which are the nodal values of the specific intensity. This gives us

$$\begin{aligned} & A_a U_a^+ + A_b U_b^+ + A_c U_c^+ + A_d U_d^+ + A_f U_f^+ + A_e U_e^+ + A_g U_g^+ \\ & + A_h U_h^+ = \tau^- (S_a^+ + S_b^+ + S_c^+ + S_d^+) \\ & + \tau^+ (S_e^+ + S_f^+ + S_g^+ + S_h^+) \end{aligned} \tag{23}$$

and

$$A'_a U^-_a + A'_b U^-_b + A'_c U^-_c + A'_d U^-_d + A'_e U^-_e + A'_f U^-_f + A'_g U^-_g$$

$$+ A'_h U^-_h = \bar{\tau}(S^-_a + S^-_b + S^-_c + S^-_d) + \tau^+(S^-_e + S^-_f + S^-_g + S^-_h) \quad (24)$$

The quantities A_a, A'_a etc are given in Appendix II. τ^+ and τ^- are given by

$$\tau^{\pm} = \tau(1 \pm \frac{1}{6}\frac{\Delta A}{\bar{A}}) \quad (25)$$

$$\tau = K_L \cdot \Delta r \quad . \quad (26)$$

The intensities $U^{\pm}_a$, $U^{\pm}_b$ etc (see Figure 1b) are given by

$$U^+_a = U(r_{i-1}, +\mu_{j-1}, x_{k-1}) = U^{i-1,+}_{j-1,k-1} \quad (27)$$

etc.

By comparing equations (23) and (24) with the interaction principle (Peraiah 1984) we obtain the transmission and reflection matrices and these are given in Appendix III.

Boundary Conditions

(a) Radial boundary condition

We set the intensities at $\tau = 0$ and $\tau = T$, where T is maximum of τ, as follows

$$\left.\begin{aligned} U^+(\tau = 0, \mu_j) &= 0 \\ U^-(\tau = T, \mu_j) &= 1 \end{aligned}\right\} \quad \text{for} \quad \varepsilon, \beta = 0 \quad (28)$$

and

$$\left.\begin{aligned} U^+(\tau = 0, \mu_j) &= 0 \\ U^-(\tau = T, \mu_j) &= 0 \end{aligned}\right\} \quad \text{for} \quad \varepsilon, \beta > 0 \quad (29)$$

(b) Frequency boundary condition

We assume that there is no variation of frequencies well away from the centre of the line. This requires that we must have

$$\frac{\partial U^{\pm}}{\partial x} \text{ (at } x \geq |X_{max}|) = 0 \quad . \tag{30}$$

(c) Velocities of expansion

For a medium moving with velocity gradients we set V (at $\tau = T$) $= 0$ and $V(\tau = 0) = V_1$, say.

For a medium moving with constant velocities, we set $V(\tau = T) = V(\tau = 0) = V_2$, say.

III. Effects of Aberration and Advection

In any astrophysical contexts the velocities of expansion or contraction are very large and are a good fraction of the velocity of light. The effects of aberration and advection in such cases are considerably large and we should expect substantial changes in the radiation field. However, Mihalas et al (1976) considered the effects of aberration and advection due to velocities of the order $V/C \approx .01$ (where V and C are velocities of the gas and light) on the line formation and found that these produce negligible changes and that the changes produced by the Doppler shifts in the frequency are of more serious nature. The nature of changes that can be introduced due to aberration and advection into the radiation field will have to be investigated in a variety of physical situations such as coherent and isotropic scattering, plane parallel and spherically symmetric medium etc. Therefore, we extend the method of integral operators to this problem. We shall consider a plane parallel coherent and isotropically scattering medium. The monochromatic, plane parallel, steady state radiative transfer equation in a fluid frame (Castor 1972; Mihalas 1978; Munier and Weaver (1986) with aberration and advection terms is written as

$$(\mu + \beta) \frac{\partial I(z,\mu)}{\partial z} + \frac{\mu(\mu^2-1)}{c} \frac{\partial V}{\partial z} \frac{\partial I(z,\mu)}{\partial \mu}$$

$$+ \frac{3\mu^2}{c} \frac{\partial V}{\partial z} I(z,\mu) = K[S - I(z,\mu)] \tag{31}$$

where

$$\mu = \frac{\mu'-\beta}{1-\mu'\beta}, \quad 0 < \mu' \leq 1 \tag{32}$$

and

$$\beta = V/C \quad . \tag{33}$$

Here μ' is the cosine of the angle made by the ray with the normal along Z-axis. The aberration terms are those which change the direction of μ'. Those terms which arise essentially from gradients or from a "sweeping up" of radiation by the transformation to the moving frame are called advection terms (see Mihalas et al. 1976). K is the absorption coefficient. $I(z,\mu)$ is the specific intensity. S is the source function and is given by

$$S = \tfrac{1}{2} \int_{-1}^{+1} P(z,\mu_1',\mu_2') I(z,\mu_2') d\mu_2' \tag{34}$$

where $P(z,\mu_1',\mu_2')$ is the isotropic phase function. We integrate equation (31) following the procedure given in part I. We express I as (Peraiah 1986)

$$I = I_o + I_z\xi + I_\mu\eta + I_{z\mu}\xi\eta \tag{35}$$

where $I_o, I_z, I_\mu, I_{z\mu}$ are the interpolation coefficients, and

$$\xi = \frac{2(z-\bar{z})}{\Delta z}, \quad \eta = \frac{2(\mu-\bar{\mu})}{\Delta\mu} \tag{36}$$

$$\bar{z} = \tfrac{1}{2}(z_i + z_{i-1}) \quad , \quad \bar{\mu} = \tfrac{1}{2}(\mu_j + \mu_{j-1}) \tag{37}$$

$$\Delta z = (z_i - z_{i-1}), \quad \Delta\mu = (\mu_j - \mu_{j-1}) \tag{38}$$

Z_i, Z_{i-1} and μ_j, μ_{j-1} are the discrete points on the X-μ grid. Equation (35) is substituted into equation (31) and the resulting equation is integrated over the z-μ grid. The interpolation coefficients I_o, I_z etc are replaced by the nodal values (Peraiah 1986).

This gives us

$$A_a I_{j-1}^{i-1,+} + A_b I_j^{i-1,+} + A_c I_{j-1}^{i,+} + A_d I_j^{i,+}$$

$$= \tau(S_{j-1}^{i-1,+} + S_j^{i-1,+} + S_{j-1}^{i,+} + S_j^{i,+}) \tag{39}$$

and

$$A'_a I_{j-1}^{i-1,-} + A'_b I_j^{i-1,-} + A'_c I_{j-1}^{i,-} + A'_d I_j^{i,-}$$

$$= \tau(S_{j-1}^{i-1,-} + S_j^{i-1,-} + S_{j-1}^{i,-} + S_j^{i,-}) \tag{40}$$

where

$$I_{j-1}^{i-1,+} = I(z_{i-1}, + \mu_{j-1}) \tag{41}$$

$$S_{j-1}^{i-1,+} = \tfrac{1}{2}\sum(P^{++}CI^{i-1,+} + P^{+-}CI^{i-1,-})_{j-1} \tag{42}$$

where

$$P^{++} = P(+\mu'_1, + \mu'_2) \tag{43}$$

$$P^{+-} = P(+\mu'_1, - \mu'_2) \tag{44}$$

and C's are the weights of the angle quadrature. τ is the optical depth given by

$$\tau = K.\Delta z \tag{45}$$

we can write the quantities $I^{i,i-1,\pm}_{j,j-1}$ and $S^{i,i-1,\pm}_{j,j-1}$ similarly.

Equations (39) and (40) can be rewritten for J number of angles as

$$[\underset{\sim}{A}^{cd} - \tau\underset{\approx}{Q}\gamma^{++}]I^{+}_{i} + [\underset{\sim}{A}^{ab} - \tau\underset{\approx}{Q}\gamma^{++}]I^{+}_{i-1}$$

$$= \tau\underset{\approx}{Q}\gamma^{+-}\,\underset{\sim}{I}^{-}_{i} + \tau\underset{\approx}{Q}\gamma^{+-}\,\underset{\sim}{I}^{-}_{i-1} \tag{46}$$

and

$$[\underset{\sim}{A}'^{cd} - \tau\underset{\approx}{Q}\gamma^{--}]I^{-}_{i} + [A'^{ab} - \tau\underset{\approx}{Q}\gamma^{--}]I^{-}_{i-1}$$

$$= \tau\underset{\approx}{Q}\gamma^{-+}\,\underset{\sim}{I}^{+}_{i} + \tau\underset{\approx}{Q}\gamma^{-+}\,\underset{\sim}{I}^{+}_{i-1} \tag{47}$$

where

$$I^{\pm}_{i} = [I_{i}(\pm\mu_{i}),\ I_{i}(\pm\mu_{2}),\ldots,I_{i}(\pm\mu_{J})]^{T} \tag{48}$$

$\{Q_{jj},\ Q_{j,j+1}\} = 1$ and all other elements are equal to zero. (49)

and

$$\underset{\sim}{A}^{ab} = \begin{bmatrix} A^{j-1}_{a} & A^{j}_{b} & & & \\ & A^{j}_{a} & A^{j+1}_{b} & & \\ & & \ddots & & \\ & & & A^{J-1}_{a} & A^{J}_{b} \\ & & & & A^{J}_{a} \end{bmatrix} \tag{50}$$

other matrices A^{cd}, A'^{ab}, A'^{cd} are similarly written, and

$$\gamma^{++} = \tfrac{1}{2}\omega P^{++} C \quad \text{etc.} \tag{51}$$

By comparing equations (46) and (47) with the interaction principle we obtain the transmission and reflections matrices. These are given in Appendix IV.

We present sample calculations of the effects produced by the aberration and advection terms in a plane parallel, coherent and isotropically scattering medium.

The boundary conditions are

$$I^{-}(\tau = \tau_{max}, \mu_j) = 1 \tag{52}$$

$$I^{+}(\tau = 0, \mu_j) = 0 \tag{53}$$

The physical meaning of the above conditions is that we irradiate the medium at $\tau = \tau_{max}$ with unit intensity, and that no radiation is incident at $\tau = 0$.

The boundary conditions for velocities are

$$V(\tau = \tau_{max}) = 0 \tag{54}$$

and

$$V(\tau = 0) = V, \text{ say.} \tag{55}$$

where

$$\begin{aligned} V &= 0 \text{ km/sec } (\beta = 0) \\ &= 1000 \text{ km/sec } (\beta = 0.0033) \\ &= 2000 \text{ km/sec } (\beta = 0.0067) \\ &= 3000 \text{ km/sec } (\beta = 0.01) \\ &= 4000 \text{ km/sec } (\beta = 0.013) \\ &= 5000 \text{ km/sec } (\beta = 0.0167) \end{aligned}$$

This means that we are introducing velocity gradients.

The total optical depths τ_{max}'s are taken to be

$$\tau_{max} = 10 \text{ and } 30$$

We estimate from the diffuse radiation field, the mean intensity J at any point, given by

$$J = \frac{1}{2}\int_{-1}^{+1} I(\mu)\, d\mu \qquad (56)$$

In Figures (5) and (6) we plotted $\bar{J}$ versus the run of the optical depth, where $\bar{J}$ is given by

Figure 5. Effects of aberration and advection for $\tau = 10$ shown in terms of $\bar{J}$.

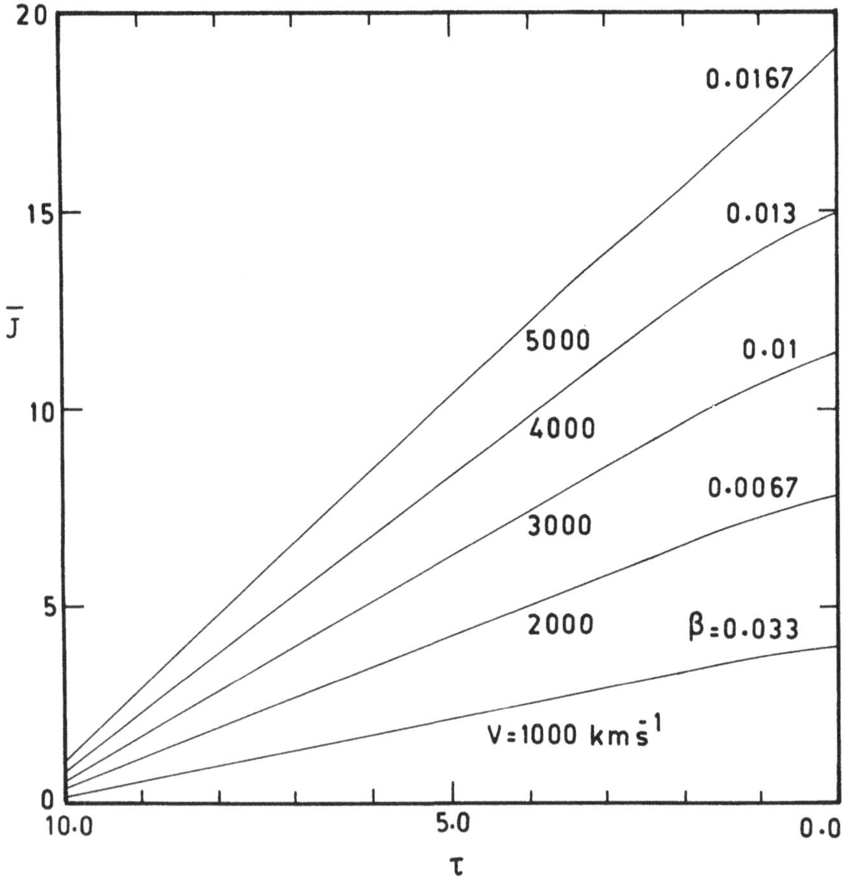

$$\bar{J} = \frac{\Delta J}{J_o} \times 100 \quad , \quad \Delta J = J_o - J_V \tag{57}$$

where

$$J_o = J(V=0) \quad , \quad J_V = J(V>0) \tag{58}$$

Figure 6. Same as those given in Figure 5 but for $\tau = 30$.

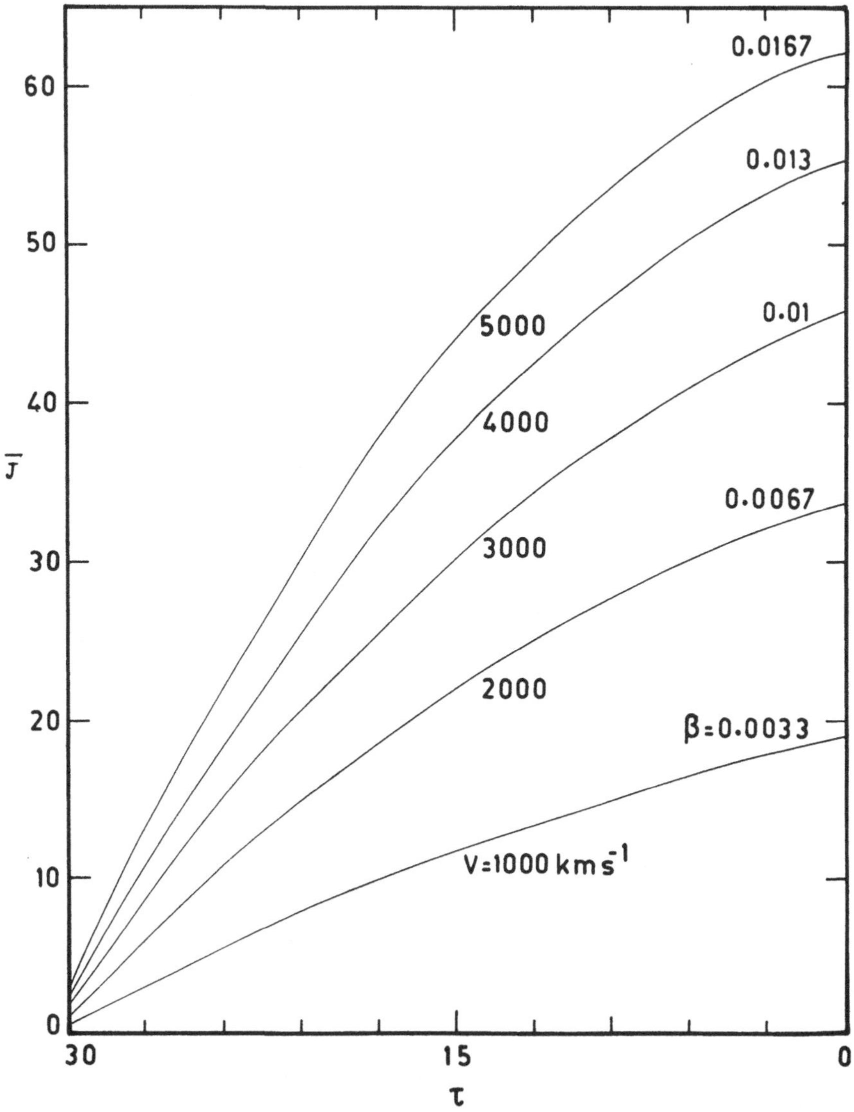

We can immediately see that as the velocity increases the percentage difference i.e. $\bar{J}$ is increasing. When $\tau = 10$, $\bar{J}$ is nearly equal to 20% for V = 5000 km/sec and this increases to 63% when τ is increased to 30. This clearly demonstrates that aberration and advection effects shall have to be taken into account in fast flowing fluids. The results shown in Figure (5) and (6) correspond to a medium in plane parallel symmetry. However we can obtain the solution of the equation of transfer in spherically symmetric media in a comoving frame given by

$$(\mu_o+\beta)\frac{\partial U(r,\mu_o,x)}{\partial r} + \frac{1-\mu_o^2}{r}\left[1 - \mu_o\beta\left(1 - \frac{r}{V}\frac{dV}{dr}\right)\right]\frac{\partial U(r,\mu_o,x)}{\partial \mu_o} = K[S(r,x) - U(r,\mu_o,x)]$$

$$+ \frac{2(\mu_o+\beta)U(r,\mu_o,x)}{r} + \left[\frac{V'(r)}{r}(1-\mu_o^2) + \mu_o\frac{dV'(r)}{dr}\right]\frac{\partial U(r,\mu_o,x)}{\partial x}$$

$$-3\left\{\frac{\beta}{r}(1-\mu_o^2) + \mu_o^2\frac{d\beta}{dr}\right\}U(r,\mu_o,x) \qquad (59)$$

The solution of this equation in the comoving frame is of great importance in investigating the effect of aberration and advection in the spectral lines formed in rapidly expanding spherical media. This is under investigation.

CONCLUSIONS

We presented a method of obtaining a numerical solution of radiative transfer equation by expanding the specific intensity in terms of the local values of the grid points. Solution of radiative transfer equation has been presented in three physical situations: (1) medium with isotropic and coherent scattering of radiation in pherical symmetry with numerical analysis of the solution, (2) medium with isotropic and incoherent scattering of radiation and expanding radially in spherical symmetry and (3) medium with isotropic and coherent scattering of radiation in plane parallel approximation and moving with velocities of the first order in v/c where v and c are the velocities of the gas and light respectively. The effects of aberration and advection are taken into account in such medium. We have considered velocities up to 5000 km/sec in an optically thick medium. The aberration and advection effects generate substantial differences in the mean intensities - as much as 60% to 90% depending upon the size of velocities and optical thicknesses employed. It appears that we shall have to consider radiative transfer by taking aberration and advection into account when calculating radiation from objects such as supergiant stars, novae, supernovae, active galactic nuclei etc.

REFERENCES

Castor, J.I. 1972. Ap. J., 178, 779.

Grant, I.P. & Hunt, G.E. 1969. Proc. Roy. Soc. Lond. A. 313, 183.

Lathrop, K.D. & Carlson, B.G. 1967. J. Comput. Phys. 2, 173.

Lathrop, K.D. & Carlson, B.G. 1971. J. Quant. Spectrosc. Rad. Transf., 11, 921.

Mihalas, D. 1978. Stellar Atmospherics (Freeman: San Francisco).

Mihalas, D., Kunasz, P.B. & Hummer, D.G. 1976. Ap. J., 206, 515.

Munier, A. & Weaver, R. 1986. Preprint in Computer Phys. Reports.

Peraiah, A. 1984. Methods in Radiative Transfer, ed. W. Kalkofen, Cambridge.

Peraiah, A. 1987. Ap. J. in print.

Peraiah, A. & Varghese, B.A. 1985. Ap. J., 290, 411.

APPENDIX I

The two pairs of transmission and reflection operators are given by

$$\underset{\sim}{t}(i,i-1) = \underset{\sim}{R}^{+-} [\underset{\sim}{\Delta}^{+} \underset{\sim}{A} + \underset{\sim}{r}^{+-} \underset{\sim}{\Delta}^{-} \underset{\sim}{C}]$$

$$\underset{\sim}{t}(i-1,i) = \underset{\sim}{R}^{-+} [\underset{\sim}{\Delta}^{-} \underset{\sim}{D} + \underset{\sim}{r}^{-+} \underset{\sim}{\Delta}^{+} \underset{\sim}{B}]$$

$$\underset{\sim}{r}(i,i-1) = \underset{\sim}{R}^{-+} [\underset{\sim}{\Delta}^{-} \underset{\sim}{C} + \underset{\sim}{r}^{-+} \underset{\sim}{\Delta}^{+} \underset{\sim}{A}]$$

$$\underset{\sim}{r}(i-1,i) = \underset{\sim}{R}^{+-} [\underset{\sim}{\Delta}^{+} \underset{\sim}{B} + \underset{\sim}{r}^{+-} \underset{\sim}{\Delta}^{-} \underset{\sim}{D}]$$

where

$$\underset{\sim}{R}^{+-} = [\underset{\sim}{I} - \underset{\sim}{r}^{+-} \underset{\sim}{r}^{-+}]^{-1}$$

$$\underset{\sim}{\Sigma}^{+}_{i-1/2} = \tau_{i-1/2}(1-\omega)\ \underset{\sim}{R}^{+-}[\underset{\sim}{\Delta}^{+} \underset{\sim}{B}^{+} + \underset{\sim}{r}^{+-} \underset{\sim}{\Delta}^{-}\underset{\sim}{B}^{-}] ,$$

$$\underset{\sim}{\Sigma}^{-}_{i-1/2} = \tau(1-\omega)\ \underset{\sim}{R}^{-+} [\underset{\sim}{\Delta}^{-} \underset{\sim}{B}^{-} + \underset{\sim}{r}^{-+} \underset{\sim}{\Delta}^{+} \underset{\sim}{B}^{+}]$$

Similarly $\underset{\sim}{R}^{-+}$ and $\underset{\sim}{\Sigma}^{-}_{i-1/2}$ may be obtained by interchanging the plus and minus signs. Further

$$\underset{\sim}{\Delta}^{+} = [\underset{\sim}{M}_1 - \underset{\sim}{\bar{\rho}}_{+} + \tfrac{1}{2}\tau^{+}(\underset{\sim}{I}-\underset{\sim}{\gamma}^{++})]^{-1} ,$$

$$\underset{\sim}{\Delta}^{-} = [\underset{\sim}{M}_4 + \underset{\sim}{\bar{\rho}}_{-} + \tfrac{1}{2}\tau^{-}(\underset{\sim}{I}-\underset{\sim}{\gamma}^{--})]^{-1} ,$$

$$\underset{\sim}{A} = [\underset{\sim}{M}_3 + \underset{\sim}{\bar{\rho}}_{-} - \tfrac{1}{2}\tau^{-}(\underset{\sim}{I} - \underset{\sim}{\gamma}^{++})],\quad \underset{\sim}{B} = \tfrac{1}{2}\tau^{+} \underset{\sim}{\gamma}^{+-},\quad \underset{\sim}{C} = \tfrac{1}{2}\tau^{-} \underset{\sim}{\gamma}^{-+}$$

$$\underset{\sim}{Q} = \begin{pmatrix} 1 & 1 & & & \\ & 1 & 1 & & \\ & & 1 & 1 & \\ & & & 1 & \end{pmatrix}$$

$$\tau^{+}_{i-1/2} = \tau_{i-1/2}\left(1 + \frac{1}{6}\frac{\Delta A}{\bar{A}}\right) = \tau_{i-1/2}\frac{3r_i^2 + 2r_i r_{i-1} + r_{i-1}^2}{2(r_i^2 + r_i r_{i-1} + r_{i-1}^2)}$$

$$\tau^{-}_{i-1/2} = \tau_{i-1/2}\left(1 - \frac{1}{6}\frac{\Delta A}{\bar{A}}\right) = \tau_{i-1/2}\frac{r_i^2 + 2r_i r_{i-1} + 3r_{i-1}^2}{2(r_i^2 + r_i r_{i-1} + r_{i-1}^2)}$$

$$\rho^{i-1/2,\pm}_{j-1/2} = \rho^{i-1/2}_{j-1/2}\left(1 \pm \frac{1}{6}\frac{\Delta r}{\bar{r}}\right)$$

$$\rho^{i-1/2}_{j-1/2} = \frac{1}{2}\frac{\overline{1-\mu^2}}{\Delta\mu}\frac{\Delta A}{\bar{A}} = \left[\frac{1-1/3(\mu_j^2 + \mu_j\mu_{j-1} + \mu_{j-1}^2)}{2(\mu_j - \mu_{j-1})}\right]\left[\frac{3(r_i^2 - r_{i-1}^2)}{(r_i^2 + r_1 r_{i-1} + r_{i-1}^2)}\right]$$

$$p_i^{++} = p(r_i, +\mu_j, +\mu_{j'})$$

$$p_i^{+-} = p(r_i, +\mu_j, -\mu_{j'})$$

$$\mu^{+}_{j-1/2} = \bar{\mu}\left(1 + \frac{1}{6}\frac{\Delta\mu}{\bar{\mu}}\right) = \frac{1}{3}(2\mu_j + \mu_{j-1})$$

$$\mu^{-}_{j-1/2} = \bar{\mu}\left(1 - \frac{1}{6}\frac{\Delta\mu}{\bar{\mu}}\right) = \frac{1}{3}(\mu_j + 2\mu_{j-1})$$

$$p = \frac{\bar{r}}{\Delta r}\frac{\Delta A}{\bar{A}} - \frac{1}{2}\frac{\Delta A}{\bar{A}} - 2, \quad q = 2 - \frac{1}{2}\frac{\Delta A}{\bar{A}} - \frac{\bar{r}}{\Delta r}\frac{\Delta A}{\bar{A}}$$

$$\underset{\sim}{D} = [\underset{\sim}{M}_2 - \bar{\underset{\sim}{\rho}}_+ - \tfrac{1}{2}\tau^+(\underset{\sim}{I}-\underset{\sim}{\gamma}^{--})], \underset{\sim}{r}^{+-} = \tfrac{1}{2}\tau^-\underset{\sim}{\Delta}^+\underset{\sim}{\gamma}^{+-}, \quad \underset{\sim}{r}^{-+} = \tfrac{1}{2}\tau^+\underset{\sim}{\Delta}^-\underset{\sim}{\gamma}^{-+}$$

$$\underset{\sim}{\gamma}^{++} = \tfrac{1}{2}\omega\underset{\sim}{P}^{++}\underset{\sim}{C}, \quad \underset{\sim}{\gamma}^{--} = \tfrac{1}{2}\omega\underset{\sim}{P}^{--}\underset{\sim}{C}, \quad \underset{\sim}{\gamma}^{+-} = \tfrac{1}{2}\omega\underset{\sim}{P}^{+-}\underset{\sim}{C},$$

$$\underset{\sim}{\gamma}^{-+} = \tfrac{1}{2}\omega\underset{\sim}{P}^{-+}\underset{\sim}{C}, \quad \bar{\underset{\sim}{\rho}}_{\pm} = \underset{\sim}{Q}^{-1}\underset{\sim}{\rho}^{\pm}, \quad \underset{\sim}{M}_1 = \underset{\sim}{Q}^{-1}\underset{\sim}{M}_P^+,$$

$$\underset{\sim}{M}_2 = Q^{-1}\underset{\sim}{M}_P^-, \quad \underset{\sim}{M}_3 = \underset{\sim}{Q}^{-1}\underset{\sim}{M}_q^+, \quad \underset{\sim}{M}_4 = \underset{\sim}{Q}^{-1}\underset{\sim}{M}_q^-,$$

$$\underset{\sim}{M}_P^+ = (1+p)\underset{\sim}{M}, \quad \underset{\sim}{M}_P^- = (1-p)\underset{\sim}{M}, \quad \underset{\sim}{M}_q^+ = (1+q)\underset{\sim}{M}, \quad \underset{\sim}{M}_q^- = (1-q)\underset{\sim}{M},$$

$$\underset{\sim}{M} = \begin{bmatrix} \mu_{1/2}^- & \mu_{1/2}^+ & & & \\ & \mu_{3/2}^- & \mu_{3/2}^+ & & \\ & & \mu_{j-1/2}^- & \mu_{j-1/2}^+ & \\ & & & \mu_j & \end{bmatrix}$$

$$\underset{\sim}{\rho}^{\pm} = \begin{bmatrix} \underset{\sim}{\rho}_{1/2}^{\pm} & -\rho_{1/2}^{\pm} & & \\ & \rho_{3/2}^{\pm} & -\rho_{3/2}^{\pm} & \\ & & \rho_{j-1/2}^{\pm} & -\rho_{j-1/2}^{\pm} \\ & & & 0 \end{bmatrix}$$

APPENDIX II

$$\underset{\sim}{R} = \begin{matrix} 1 & 1 & 1 & 1 & 1 & 1 & 1 & 1 \\ -1 & -1 & -1 & -1 & 1 & 1 & 1 & 1 \\ -1 & 1 & 1 & -1 & -1 & -1 & 1 & 1 \\ -1 & +1 & 1 & 1 & 1 & -1 & -1 & 1 \\ 1 & -1 & -1 & 1 & -1 & -1 & 1 & 1 \\ 1 & -1 & 1 & -1 & -1 & 1 & 1 & 1 \\ 1 & 1 & -1 & -1 & 1 & -1 & -1 & 1 \\ -1 & 1 & -1 & 1 & -1 & 1 & -1 & 1 \end{matrix}$$

$$\underset{\sim}{P} = \begin{matrix} -1 & -1 & 1 & 1 & 1 & -1 & -1 & 1 \\ -1 & 1 & -1 & 1 & 1 & -1 & 1 & -1 \\ -1 & 1 & -1 & 1 & -1 & 1 & -1 & 1 \\ -1 & -1 & 1 & 1 & -1 & 1 & 1 & -1 \\ 1 & -1 & -1 & 1 & -1 & -1 & 1 & 1 \\ 1 & -1 & -1 & 1 & 1 & 1 & -1 & -1 \\ 1 & 1 & 1 & 1 & 1 & 1 & 1 & 1 \\ 1 & 1 & 1 & 1 & -1 & -1 & -1 & -1 \end{matrix}$$

$$\underset{\sim}{A} = \underset{\sim}{P}\,\underset{\sim}{G} \quad , \quad \underset{\sim}{A}' = \underset{\sim}{P}'\underset{\sim}{G}'$$

$$\underset{\sim}{A} = [\underset{\sim}{A}_a,\ \underset{\sim}{A}_b,\ \underset{\sim}{A}_c,\ \underset{\sim}{A}_d,\ \underset{\sim}{A}_e,\ \underset{\sim}{A}_f,\ \underset{\sim}{A}_g,\ \underset{\sim}{A}_h]^T$$

$$\underset{\sim}{A}' = [\underset{\sim}{A}'_a,\ \underset{\sim}{A}'_b,\ \underset{\sim}{A}'_c,\ \underset{\sim}{A}'_d,\ \underset{\sim}{A}'_e,\ \underset{\sim}{A}'_f,\ \underset{\sim}{A}'_g,\ \underset{\sim}{A}'_h]^T$$

$$\underset{\sim}{G} = [\alpha,\beta,\gamma,\delta,\rho,\sigma,\varepsilon,\lambda]^T$$

$$\underset{\sim}{G}' = [\alpha',\beta',\gamma',\delta',\rho,\sigma,\varepsilon,\lambda]^T$$

$$\alpha = \frac{1}{6}\tau_x s - 2\bar{\mu}p,\ \beta = s(m_1 - \frac{1}{6}m),\quad \gamma = 2m_1(1+p) - \frac{1}{3}mp, \delta = \tau_x - \bar{\mu}s$$

$$\alpha' = \frac{1}{6}\tau_x s + 2\bar{\mu}p,\quad \beta' = -\beta,\quad \gamma' = -\gamma,\quad \delta' = \tau_x + \bar{\mu}s,\ \tau_x = \tau\phi(x)$$

$$\rho = X'[\frac{1}{2}\,V'mm_1 s + \overline{\mu^2}\ .\ \Delta V'] \quad , \quad \sigma = X'[V'mm_1(1+p) + \frac{1}{6}\Delta V'.\overline{\mu^2}s]$$

$$\varepsilon = X'[\frac{1}{3}m\bar{\mu}(\Delta V' - \frac{1}{2}V's)],\ \lambda = X'[\frac{1}{3}m\bar{\mu}\{\frac{1}{6}\ \Delta V'.s - V'(1+p)\}]$$

$$X' = \frac{2}{\Delta x},\quad p = 1 - \frac{r}{\Delta r}\ .\ \frac{\Delta A}{A}\quad ,\quad m = \Delta\mu,\qquad = \frac{\Delta A}{\bar{A}},\quad m_1 = \frac{\overline{1-\mu^2}}{\Delta\mu}$$

APPENDIX III

The transmission and reflection matrices are given as follows:

$$\mathbf{t}(i,i-1) = \mathbf{R}^{+-}\, \mathbf{X}_1^{-1}\, [\mathbf{X}_2 \mathbf{X}_4^{-1}\, \mathbf{Y}_3 - \mathbf{Y}_1]$$

$$\mathbf{t}(i-1,i) = \mathbf{R}^{-+}\, \mathbf{X}_4^{-1} [\mathbf{X}_3 \mathbf{X}_1^{-1}\, \mathbf{Y}_2 - \mathbf{Y}_4]$$

$$\mathbf{r}(i,i-1) = \mathbf{R}^{-+}\, \mathbf{X}_4^{-1}\, [\mathbf{Y}_3 - \mathbf{X}_3 \mathbf{X}_1^{-1} \mathbf{Y}_1]$$

$$\mathbf{r}(i-1,i) = \mathbf{R}^{+-} \mathbf{X}_1^{-1}\, [\mathbf{Y}_2 - \mathbf{X}_2 \mathbf{X}_4^{-1} \mathbf{Y}_4]$$

$$\mathbf{R}^{+-} = [\mathbf{I} - \mathbf{X}_1^{-1}\, \mathbf{X}_2 \mathbf{X}_4^{-1} \mathbf{X}_3]^{-1}$$

$$\mathbf{R}^{-+} = [\mathbf{I} - \mathbf{X}_4^{-1} \mathbf{X}_3 \mathbf{X}_1^{-1} \mathbf{X}_2]^{-1}$$

$$\mathbf{\Sigma}^{+} = 2\sigma\tau \mathbf{R}^{+-} \mathbf{X}_1^{-1} (\mathbf{I} + \mathbf{X}_2 \mathbf{X}_4^{-1})\ \mathbf{\Phi} \mathbf{B}$$

$\mathbf{\Phi} = (\rho\beta + \varepsilon\phi)\,\delta_{kk'}$, k,k' being the running indices for frequency points.

B = Planck function, $\sigma = (1-\varepsilon)/2$

$$\mathbf{X}_1 = \mathbf{A}^{fg}_{eh} - \sigma\tau^{+} \mathbf{F}_i,\quad \mathbf{X}_2 = \sigma\tau^{-} \mathbf{F}_i,\quad \mathbf{X}_3 = \sigma\tau^{+} \mathbf{F}_i,$$

$$\mathbf{X}_4 = \mathbf{D}^{ab}_{dc} - \tau\sigma^{-}\, \mathbf{F}_i,\ \mathbf{Y}_1 = \mathbf{A}^{ab}_{dc} - \sigma\tau^{-} \mathbf{F}_{i-1},\ \mathbf{Y}_2 = \sigma\tau^{+} \mathbf{F}_i,$$

$$\mathbf{Y}_3 = \sigma\tau^{-} \mathbf{E}_i,\quad \mathbf{Y}_4 = \mathbf{D}^{fg}_{eh} - \sigma\tau^{+} \mathbf{F}$$

The quantities $\underset{\sim}{A}^{ab}_{dc}$, $\underset{\sim}{A}^{ef}_{eh}$, $\underset{\sim}{D}^{ab}_{dc}$, $\underset{\sim}{D}^{fg}_{eh}$, are written in a similar pattern. For example $\underset{\sim}{A}^{ab}_{dc}$ is obtained as follows:

$$A^{ab}(j,j) = A_a(j), \; A^{ab}(j,j+1) = A_b(j+1) \quad \text{(see Appendix II)}.$$

Similarly $\underset{\sim}{A}^{dc}$ is written

$$A^{ab}_{dc}(k,k) = A^{ab}_{q}(k,k)$$

$$A^{ab}_{dc}(k,k+1) = A^{dc}_{q}(k,k+1)$$

$$\underset{\sim}{A}^{ab}_{dc} = \underset{\sim}{Q}^{-1}_{F} \; \underset{\sim}{A}^{ab}_{dc}$$

$$\underset{\sim}{Q}_F(k,k) = I = \underset{\sim}{Q}_F(k,k+1)$$

$$\underset{\sim}{F}_i = [\underset{\sim}{\phi}_i \; \underset{\sim}{\phi}^T_i \; \underset{\sim}{W}], \; [\underset{\sim}{W} = \underset{\sim}{W}_m \; \underset{\sim}{\delta}_{mn}]$$

$$W_m = a_k c_j, \quad m = j + (k-1)j, \quad 1 \le m \le M = KJ$$

$$\Phi_i = [\rho\beta + \varepsilon\phi]_i \; \delta_{kk} \, ,$$

$$a_k = \frac{A_k \; \phi_k}{\sum\limits_{k=-K}^{k} A_k \; \phi_k}$$

A_k and C_j are the quadrature weights of the frequency and angle point.

APPENDIX IV

$$A_a = \tau + p_1 + q_1 - r_1$$

$$A_b = \tau - p_1 + q_2 - r_1$$

$$A_c = \tau - p_2 + q_1 + r_1$$

$$A_d = \tau + p_2 + q_2 + r_1$$

$$A'_a = \tau - p_2 + q_1 - r_2$$

$$A'_b = \tau + p_2 + q_2 - r_2$$

$$A'_c = \tau + p_1 + q_1 + r_2$$

$$A'_d = \tau - p_2 + q_2 + r_2$$

where

$$p_1 = \Delta\mu\left(\frac{1}{3} - \bar{\mu}\Delta\beta\right), \quad q_1 = \Delta\beta(3\overline{\mu^2} - g), \quad r_1 = 2(\bar{\mu} + \beta)$$

$$p_2 = \Delta\mu\left(\frac{1}{3} + \bar{\mu}\Delta\beta\right), \quad q_2 = \Delta\beta(3\,\overline{\mu^2} + g), \quad r_2 = 2(\beta - \bar{\mu})$$

$$\Delta\beta = (\beta_{n+1} - \beta_n), \quad g = \frac{2\bar{\mu}}{\Delta\mu}\left\{\frac{1}{2}\left(\mu_j^2 + \mu_{j-1}^2\right) - 1\right\}$$

The two pairs of transmission and reflection operators are given by:

$$\underset{\sim}{t}(i,i-1) = \underset{\sim}{R}^{+-}[\underset{\sim}{\Delta}^{+}\underset{\sim}{A} + \underset{\sim}{r}^{+-}\underset{\sim}{\Delta}^{-}\underset{\sim}{C}]$$

$$\underset{\sim}{t}(i-1,i) = \underset{\sim}{R}^{-+}[\underset{\sim}{\Delta}^{-}\underset{\sim}{D} + \underset{\sim}{r}^{-+}\underset{\sim}{\Delta}^{+}\underset{\sim}{B}]$$

$$\underset{\sim}{r}(i,i-1) = \underset{\sim}{R}^{-+}[\underset{\sim}{\Delta}^{-}\underset{\sim}{C} + \underset{\sim}{r}^{-+}\underset{\sim}{\Delta}^{+}\underset{\sim}{A}]$$

$$\underset{\sim}{r}(i-1,i) = \underset{\sim}{R}^{+-}[\underset{\sim}{\Delta}^{+}\underset{\sim}{B} + \underset{\sim}{r}^{+-}\Delta^{-}\underset{\sim}{D}]$$

where

$$\underset{\sim}{\Delta}^{+} = (\underset{\sim}{A}_{cd} - \tau\underset{\sim}{\gamma}^{++})^{-1}, \quad \Delta^{-} = (\underset{\sim}{A}_{ab} - \tau\underset{\sim}{\gamma}^{--})^{-1}$$

$$\underset{\sim}{A}_{ab} = \underset{\sim}{Q}^{-1}\underset{\sim}{A}^{ab}, \quad A = \tau\underset{\sim}{\gamma}^{++} - \underset{\sim}{A}_{ab}', \quad \underset{\sim}{B} = \tau\underset{\sim}{\gamma}^{+-}$$

$$\underset{\sim}{A}_{cd} = \underset{\sim}{Q}^{-1}\underset{\sim}{A}^{cd}, \quad \underset{\sim}{A}'_{ab} = \underset{\sim}{Q}^{-1}\underset{\sim}{A}'_{ab}$$

$$\underset{\sim}{C} = \tau\underset{\sim}{\gamma}^{-+}, \quad \underset{\sim}{D} = \tau\underset{\sim}{\gamma}^{++} - \underset{\sim}{A}'_{cd}$$

$$\underset{\sim}{r}^{+-} = \tau\underset{\sim}{\Delta}^{+}\underset{\sim}{\gamma}^{+-}, \quad \underset{\sim}{R}^{+-} = [\underset{\sim}{I} - \underset{\sim}{r}^{+-}\underset{\sim}{r}^{-+}]^{-1}$$

where $\underset{\sim}{I}$ is the identity matrix and the quantities $\underset{\sim}{r}^{-+}$, $\underset{\sim}{R}^{-+}$ are obtained by interchanging the signs + and − on the right hand side of $\underset{\sim}{r}^{+-}$ and $\underset{\sim}{R}^{-+}$.

APPENDIX V

After obtaining the transmission and reflection matrices together with the source terms in each of the fundamental shells (see Appendices I, III and IV), we can obtain the internal diffuse radiation field by using the following scheme. The intensities at any internal point (this includes diffuse radiation also) are given by (see Grant & Hunt 1969),

$$\underset{\sim}{U}^{+}_{p+1} = \underset{\sim}{r}\,(1,p+1)\,\underset{\sim}{U}^{-}_{p+1} + \underset{\sim}{V}^{+}_{p+\frac{1}{2}}$$

$$\underset{\sim}{U}^{-}_{p} = \underset{\sim}{\bar{t}}\,(p,p+1)\,\underset{\sim}{U}^{-}_{p+1} + \underset{\sim}{V}^{-}_{p+\frac{1}{2}}$$

with the boundary condition

$$\underset{\sim}{U}^{-}_{p+1} = \underset{\sim}{U}^{-}(A)$$

where P is the total number of shells.

The diffuse reflection and transmission matrices are given by

$$\underset{\sim}{r}(1,p+1) = \underset{\sim}{r}(p,p+1) + \underset{\sim}{t}(p+1,p)\ \underset{\sim}{r}(1,p)\ \underset{\sim}{T}_{p}\ \underset{\sim}{t}(p,p+1)$$

$$\underset{\sim}{\bar{t}}(p,p+1) = \underset{\sim}{T}_{p}\ \underset{\sim}{t}\,(p,p+1)$$

And the source vectors $\underset{\sim}{V}^{+}_{p+\frac{1}{2}}$ and $\underset{\sim}{V}^{-}_{p+\frac{1}{2}}$ are given by,

$$\underset{\sim}{V}^{+}_{p+\frac{1}{2}} = \underset{\sim}{\bar{t}}(p+1,p)\ \underset{\sim}{V}^{+}_{p-\frac{1}{2}} + \underset{\sim}{\Sigma}^{+}\,(p+1,p) + \underset{\sim}{\bar{R}}\ \underset{\sim}{\Sigma}^{-}\,(p,p+1)$$

$$\underset{\sim}{V}^{-}_{p+\frac{1}{2}} = \underset{\sim}{\bar{r}}(p+1,p)\ \underset{\sim}{V}^{+}_{p-\frac{1}{2}} + \underset{\sim}{T}_{p}\underset{\sim}{\Sigma}^{-}\,(p,p+1)$$

where

$$\bar{t}(p+1,p) = t(p+1,p)\,\bar{T}_p$$

$$\bar{r}(p+1,p) = r(p+1,p)\,\bar{T}_p$$

$$\bar{R} = \bar{t}(p+1,p)\,r(1,p)$$

$$T_p = I - r(p+1,p)\,r(1,p)^{-1}$$

$$\bar{T}_p = I - r(1,p)\,r(p+1,p)^{-1}$$

with the boundary conditions

$$r(1,1) = 0$$

$$V^+_{\frac{1}{2}} = U^+ (A)$$

If we have a shell bounded by X and Y and if another shell with boundaries Y and Z is to be compounded with the former, we can obtain the r & t operators of the composite shell with boundaries X and Z in terms of those of shell XY and shell YZ. This is called star product algorithm.

These are given by,

$$\begin{aligned}
\underset{\sim}{t}(Z,X) &= \underset{\sim}{t}(Z,Y)\,\underset{\sim}{R}_1\underset{\sim}{t}(Y,Z)\\
\underset{\sim}{t}(X,Z) &= \underset{\sim}{t}(X,Y)\,\underset{\sim}{R}_2\underset{\sim}{t}(Y,Z)\\
\underset{\sim}{r}(Z,X) &= \underset{\sim}{r}(Y,X) + \underset{\sim}{t}(X,Y)\underset{\sim}{r}(Z,Y)\underset{\sim}{R}_1\underset{\sim}{t}(Y,X)\\
\underset{\sim}{r}(X,Z) &= \underset{\sim}{r}(Y,Z) + \underset{\sim}{t}(Z,Y)\underset{\sim}{r}(X,Y)\underset{\sim}{R}_2\underset{\sim}{t}(Y,Z)\\
\underset{\sim}{R}_1 &= [\underset{\sim}{I} - \underset{\sim}{r}(X,Y)\,\underset{\sim}{r}(Z,Y)]^{-1}\\
\underset{\sim}{R}_2 &= [\underset{\sim}{I} - \underset{\sim}{r}(Z,Y)\,\underset{\sim}{r}(X,Y)]^{-1}
\end{aligned}$$

$\underset{\sim}{I}$ denotes the identity operator.

DISCRETE ORDINATE MATRIX METHODS

M. Schmidt
R. Wehrse
Institut für Theoretische Astrophysik,
Im Neuenheimer Feld 561, D 6900 Heidelberg
Federal Republic of Germany

Abstract: Methods for the solution of the angle and frequency discretized one-dimensional transfer equation by means of transition matrices are described. It is shown that expressions for the transition and reflection operators as well as the source terms can be derived which do not contain increasing exponentials and which can be evaluated by numerical standard techniques. They are given by closed formulae if the coefficients for the interaction between the radiation and the matter meet certain reqirements; in the general case by matrix differential equations with initial conditions. Several examples demonstrate the applicability of these methods to complex analytical and numerical problems.

1. INTRODUCTION

Discrete ordinate methods have been used successfully to derive fundamental results in the theory of radiative transfer (see e.g. Chandrasekhar, 1950, Kourganoff, 1963). These largely analytical calculations use the facts that (i) the transfer equation is - at least explicitely - linear in the specific intensities and (ii) the replacement of the integral over angle and/or frequency transforms it from an integro-differential equation into a system of differential equations. Therefore the well known theorems of linear algebra and analysis could be used to study the properties of the solutions.

These calculations showed in particular that the general solution of the transfer equation contains - in addition to terms which decrease with increasing optical depths τ - terms that increase exponentially with τ. In semi-infinite problems the growing terms are not of concern since the corresponding coefficients are required to be zero (cf. Chandrasekhar, 1950). However, in problems of finite optical thickness these coefficients have to be evaluated explicitly from the particular boundary conditions at both ends of the medium by the solution of a linear system of algebraic equations, which (due to the increasing exponentials) is extremely ill-conditioned for moderate and large optical depths. Therefore, such a system cannot be solved numerically on a computer with a finite word length. This disadvantage lead to the belief (see e.g. Mihalas, 1978) that the discrete ordinate method is not suited for numerical evaluations and a number of different algorithms were used instead (for an overview see Cannon, 1984).

But it turns out that these methods, too, encountered a number of difficulties, e.g. for second order difference equations (Feautrier type equations) the handling of the boundary conditions is difficult, accurate error estimates are possible only posteriori, and anisotropic media cannot be treated (in such a case the second order system has to be supplemented by a first order one); first order difference equations of Grant-Hunt-type often face stability and accuracy problems; and the treatment of partial redistribution by means of integral equations is very complicated.

These problems led to the reconsideration of the classical discrete ordinate method. In this paper we show by means of a matrix formulation that the increasing exponentials can be eliminated analytically and the resulting equations are numerically well behaved. In this form the method therefore can be used with advantage for both analytical and numerical problems.

In Chapter 2 the matrix nomenclature is introduced and the general solution of the transfer equation in terms of the transition matrix $\boldsymbol{\Phi}$ is given. If the transfer equations meet some conditions, explicit matrix exponential expressions for $\boldsymbol{\Phi}$ can be derived. They are subsequently transformed into "transition matrices $\mathbf{t}$", "reflection matrices $\mathbf{r}$", and "source vectors $\boldsymbol{\beta}$", which describe the problem completely and are suitable for numerical calculations (Chapter 3). They correspond to formulae given by Wehrse (1985) and (for a fully discrete system) by Wehrse and Kalkofen (1985). The range of applicability is demonstrated by two examples in Section 4; in particular we show that the formalism developped here can be used for a perturbation approach that is very similar to one in quantum-electro-dynamics as introduced by Dyson (1949a, b).

In Chapter 5 we return to the general problem and derive differential equations for the **t** and **r** matrices and the **β** vectors, which are subject to initial values only. We show that again forms can be found that allow on a computer numerical solutions for large optical depths.

In the following Section (6) we give two illustrating examples. The general performance of this method is discussed in the final Chapter 7.

2. BASIC EQUATIONS

We start with the transfer equation for unpolarized radiation of the form

$$DI = \chi\ (- I + \alpha \int RI\ d\nu'\ d\mu' + \beta B) \tag{2.1}$$

where D is a differential operator, I is the specific intensity depending on the frequency ν and the direction specified by μ = cosine of angle between the ray and the normal direction, χ the extinction coefficient, α a factor describing the strength of scattering (in the case of a two level atom it is given by $(1-\epsilon)$ with ϵ = deexcitation parameter); R is the redistribution, β a parameter for the efficiency of photon generation from thermal energy (for the two level atom it is given by ϵ) and finally B is the Planck function. D is given by

$$D = \mu \frac{d}{dz} \tag{2.2}$$

(z = vertical space coordinate) for a plane-parallel medium and by

$$D = \mu \frac{\partial}{\partial r} + \frac{(1 - \mu^2)}{r} \frac{\partial}{\partial \mu} \tag{2.3}$$

(r = radial geometrical coordinate) for spherical configurations.

We discretize the frequency and angle space and move the second term of Eq. (2.3) (if it is considered) to the r.h.s. By this process the transfer equation (2.1) is transformed into a system the ordinary linear differential equations

$$\mu_k \frac{dI_i}{dz} = \chi \left(- I_i + \alpha \sum_j R_{ij} I_j a_j + \sum_j b_j I_j + \beta B \right) \tag{2.4}$$

where a_j are the weights for the scattering integral, b_j the coefficients for the μ differentiationand the indices k and i indicate the angle and angle-frequency of the ray under consideration.

Introducing the intensity vector referring to a standard optical depth τ

$$\mathbf{I}_\tau = (\mathbf{I}_\tau^+, \mathbf{I}_\tau^-)^t = (I_1(\tau)\ldots I_N(\tau))^t \tag{2.5}$$

where the superscript t indicates transposition we can write Eq. (2.4)in compact matrix notation

$$\frac{d\mathbf{I}_\tau}{d\tau} = \mathbf{A}_\tau \mathbf{I}_\tau + \mathbf{B}_\tau \tag{2.6}$$

Note that Eq. (2.6) is more general than Eq. (2.1), since e.g. the transfer of polarized radiation can also be described by an equation of this type.

Eq. (2.6) has to be solved subject to the boundary condition for the incident intensities at the optical depths 0 and τ

$$\mathbf{I}_\tau^+ = \mathbf{I}_{bc}^+ \qquad \mathbf{I}_0^- = \mathbf{I}_{bc}^- \tag{2.7}$$

In this paper we be concerned mainly with the calculation of the emergent intensities $\mathbf{I}_0^+$ and $\mathbf{I}_\tau^-$. This is no significant limitation since, if we are interested in the radiation field at some intermediate optical depth s, we can consider the medium to consist of two shells of optical depths s and τ - s, solve for them the transfer equation separately, and then calculate I_s from the condition of continuity (cf. Peraiah, 1984).

The general solution of Eq. (2.6) can be written (cf. Bronson, 1970)

$$I_\tau = \Phi_{\tau,0}\, I_0 + \int_0^\tau \Phi_{\tau,s}\, B_s \, ds \tag{2.8}$$

where the transition matrix $\Phi_{\tau,s}$ is given by the differential (matrix) equation

$$\frac{d}{d\tau} \Phi_{\tau,s} = A_\tau\, \Phi_{\tau,s} \tag{2.9}$$

with the initial condition

$$\Phi_{\tau,\tau} = E = \text{identity matrix.} \tag{2.10}$$

3. THE MATRIX EXPONENTIAL METHOD

Eq. (2.9) can be integrated analytically if the matrix $\mathbf{A}$ is independent of the optical depth. In this special case Φ is given by the exponential function, which is defined - as in the scalar case - by the power series

$$\exp \Phi_{\tau,s} = \sum_{n=0}^{\infty} \frac{1}{n!} \Phi_{\tau,s}^n \,. \tag{3.1}$$

It is evident from this expansion that for matrices the relation

$$\frac{d}{d\tau} \exp(\Phi_{\tau,s}) = (\frac{d}{d\tau} \Phi_{\tau,s})\, \exp(\Phi_{\tau,s}) \tag{3.2}$$

is only valid, if

$$(\frac{d}{d\tau} \Phi_{\tau,s}) \Phi_{\tau,s} = \Phi_{\tau,s} (\frac{d}{d\tau} \Phi_{\tau,s}) \ . \tag{3.3}$$

This condition is in particular fulfilled, if all elements of Φ depend on τ in the same way, i.e.

$$\Phi_{\tau,s} = f(\tau) \Phi^{*}_{s} \tag{3.4}$$

where $f(\tau)$ is some (differentiable) scalar function of τ and Φ^{*}_{s} is a matrix independent of τ).

With Eqs. (2.8) and (2.9) it therefore follows that for a matrix **A** not depending on the optical depth the solution of Eq. (1.6) can be written in a form that is completely analogous to the scalar case

$$I_{\tau} = \exp(\mathbf{A}\ \tau)\ (I_0 + \int_0^{\tau} \exp(-\mathbf{A}s)\ B_s\ ds) \ . \tag{3.5}$$

However, this equation is not suitable for actual numerical evaluations, since

(i) both vectors I_{τ} and I_0 contain parts of the boundary condition .

(ii) the determination of the exponential functions by means of eq. (3.1) is extremely time consuming even for moderately large optical depths τ (as in the scalar case).

(iii) it contains terms that increase exponentially.

We avoid these three problems simultaneously in the following way:From matrix theory it is known that the application of an exponential operator onto a matrix is equivalent to the application of the scalar exponential function onto its eigenvalues (see e.g. Zurmühl, 1964). Therefore it is easy and fast to evaluate the exponential function, if the matrix **A** is diagonalized with the help of its eigenvalues and eigenvectors. (We will assume in this chapter that **A** is equivalent to a diagonal matrix, i.e. all eigenvalues of **A** are pairwise distinct). By means of the modal matrix **T** and the diagonal matrix **L** containing the eigenvalues l_i the matrix **A** can be written

$$\mathbf{A} = \mathbf{T}\,\mathbf{L}\,\mathbf{T}^{-1} \tag{3.6}$$

so that

$$\exp\left\{\mathbf{A}\,\tau\right\} = \mathbf{T}\,\boldsymbol{\Lambda}\,\mathbf{T}^{-1} \tag{3.7}$$

with

$$\boldsymbol{\Lambda} = (\exp(l_i\tau)\;\delta_{ij}) =: (\lambda_\tau^i\;\delta_{ij}) \tag{3.8}$$

In our case of the radiative transfer equation the elements of the vectors should be arranged according to incoming and outgoing rays. Correspondingly, the eigenvalues can also be ordered and their exponentials be collected in the submatrices λ_τ^+ (positive eigenvalues) and λ_τ^- (negative eigenvalues). After a corresponding splitting of the modal matrix

$$\mathbf{T} = \begin{bmatrix} \mathbf{A} & \mathbf{B} \\ \mathbf{C} & \mathbf{D} \end{bmatrix} = \begin{bmatrix} \mathbf{a} & \mathbf{b} \\ \mathbf{c} & \mathbf{d} \end{bmatrix}^{-1} \tag{3.9}$$

we obtain from Eq. (3.6)

$$\exp\left\{\mathbf{A}\,\tau\right\} = \begin{bmatrix} \mathbf{A} & \mathbf{B} \\ \mathbf{C} & \mathbf{D} \end{bmatrix} \begin{bmatrix} \lambda_\tau^+ & 0 \\ 0 & \lambda_\tau^- \end{bmatrix} \begin{bmatrix} \mathbf{a} & \mathbf{b} \\ \mathbf{c} & \mathbf{d} \end{bmatrix} =: \begin{bmatrix} \mathbf{u}_\tau & \mathbf{v}_\tau \\ \mathbf{w}_\tau & \mathbf{x}_\tau \end{bmatrix} \tag{3.10}$$

In terms of these quantities, Eq. (3.5) can be written

$$\begin{bmatrix} \mathbf{I}_{\tau}^{+} \\ \mathbf{I}_{\tau}^{-} \end{bmatrix} = \begin{bmatrix} \mathbf{u}_{\tau} & \mathbf{v}_{\tau} \\ \mathbf{w}_{\tau} & \mathbf{x}_{\tau} \end{bmatrix} \left\{ \begin{bmatrix} \mathbf{I}_{0}^{+} \\ \mathbf{I}_{0}^{-} \end{bmatrix} + \int_{0}^{\tau} \begin{bmatrix} \mathbf{u}_{-s} & \mathbf{v}_{-s} \\ \mathbf{w}_{-s} & \mathbf{x}_{-s} \end{bmatrix} \begin{bmatrix} \mathbf{B}_{s}^{+} \\ \mathbf{B}_{s}^{-} \end{bmatrix} ds \right\} \tag{3.11}$$

or in abbreviated form

$$\begin{bmatrix} \mathbf{I}_{\tau}^{+} \\ \mathbf{I}_{\tau}^{-} \end{bmatrix} = \begin{bmatrix} \mathbf{u}_{\tau} & \mathbf{v}_{\tau} \\ \mathbf{w}_{\tau} & \mathbf{x}_{\tau} \end{bmatrix} \begin{bmatrix} \mathbf{I}_{0}^{+} \\ \mathbf{I}_{0}^{-} \end{bmatrix} + \begin{bmatrix} \mathbf{L}_{\tau}^{+} \\ \mathbf{L}_{\tau}^{-} \end{bmatrix} \tag{3.12}$$

Since $\mathbf{I}_{\tau}^{+}$ and $\mathbf{I}_{0}^{-}$ are given as boundary values and we want to calculate the outgoing intensities $\mathbf{I}_{0}^{+}$ and $\mathbf{I}_{\tau}^{-}$ we have to transform Eq. (3.11) to the canonical form of the interaction principle (see e.g. Peraiah, 1984)

$$\begin{bmatrix} \mathbf{I}_{0}^{+} \\ \mathbf{I}_{\tau}^{-} \end{bmatrix} = \begin{bmatrix} \mathbf{t}_{\tau}^{+} & \mathbf{r}_{\tau}^{+} \\ \mathbf{r}_{\tau}^{-} & \mathbf{t}_{\tau}^{-} \end{bmatrix} \begin{bmatrix} \mathbf{I}_{\tau}^{+} \\ \mathbf{I}_{0}^{-} \end{bmatrix} + \begin{bmatrix} \boldsymbol{\beta}_{\tau}^{+} \\ \boldsymbol{\beta}_{\tau}^{-} \end{bmatrix} \tag{3.13}$$

where the **r** and **t** matrices refer to the reflection and transmission of the radiation and the vectors $\boldsymbol{\beta}^{+}$ and $\boldsymbol{\beta}^{-}$ represent the internal photon sources. Expressing Eq. (3.12) in the form of Eq. (3.13) we obtain

$$\begin{aligned} &\mathbf{t}_{\tau}^{+} = \mathbf{u}_{\tau}^{-1} && \mathbf{r}_{\tau}^{+} = -\mathbf{u}_{\tau}^{-1}\,\mathbf{v}_{\tau} && \boldsymbol{\beta}_{\tau}^{+} = -\mathbf{u}_{\tau}^{-1}\,\mathbf{L}_{\tau}^{+} \\ &\mathbf{t}_{\tau}^{-} = \mathbf{x}_{\tau} - \mathbf{w}_{\tau}\,\mathbf{u}_{\tau}^{-1}\,\mathbf{v}_{\tau} && \mathbf{r}_{\tau}^{-} = \mathbf{w}_{\tau}\,\mathbf{u}_{\tau}^{-1} && \boldsymbol{\beta}_{\tau}^{-} = \mathbf{L}_{\tau}^{-} - \mathbf{w}_{\tau}\,\mathbf{u}_{\tau}^{-1}\,\mathbf{L}_{\tau}^{+} \end{aligned} \tag{3.14}$$

In the numerical evaluation of the transmission and reflection matrices the following difficulty can occur: The arguments of the exponentials in $\boldsymbol{\lambda}^{+}$ are positive and therefore numerical overflows result for large optical depths (compare Wehrse and Kalkofen, 1985). Note that $\boldsymbol{\lambda}^{-}$ contains only decreasing exponentials so that

$$\lim_{\tau \to \infty} \boldsymbol{\lambda}^-_\tau = \mathbf{O} \ . \tag{3.15}$$

As is shown in the appendix, this problem can be avoided if the $\boldsymbol{\lambda}^+$ matrix is replaced by its inverse $\boldsymbol{\Gamma}$, which tends to $\mathbf{O}$ for large τ's, and the resulting expressions are written in such a way that terms containing $\boldsymbol{\lambda}^-$ or $\boldsymbol{\Gamma}$ either appear in the "numerator" only or are accompanied by terms that reach finite values for $\tau \to \infty$.

The analogous problems for the source terms can be solved if the depth dependence of the Planck function is specified. Let us assume that $\mathbf{B}_\tau$ is given by

$$\mathbf{B}_\tau = \begin{bmatrix} \mathbf{B}^+_\tau \\ \mathbf{B}^-_\tau \end{bmatrix} = \begin{bmatrix} \mathbf{p} \\ \mathbf{q} \end{bmatrix} P_m(\tau) \tag{3.16}$$

where p and q are constants referring essentially to the angle discretisation and $P(\tau)$ is a polynomial of order m.

Inserting into Eq. (3.11) we obtain for the source vector

$$\begin{bmatrix} \mathbf{L}^+_\tau \\ \mathbf{L}^-_\tau \end{bmatrix} = \begin{bmatrix} \mathbf{A} & \mathbf{B} \\ \mathbf{C} & \mathbf{D} \end{bmatrix} \begin{bmatrix} \boldsymbol{\lambda}^+_\tau & \mathbf{O} \\ \mathbf{O} & \boldsymbol{\lambda}^-_\tau \end{bmatrix} \begin{bmatrix} \mathbf{a} & \mathbf{b} \\ \mathbf{c} & \mathbf{d} \end{bmatrix} \begin{bmatrix} \mathbf{A} & \mathbf{B} \\ \mathbf{C} & \mathbf{D} \end{bmatrix} \int_0^\tau \begin{bmatrix} \boldsymbol{\lambda}^+_{-s} & \mathbf{O} \\ \mathbf{O} & \boldsymbol{\lambda}^-_{-s} \end{bmatrix} P(s)\, ds \begin{bmatrix} \mathbf{a} & \mathbf{b} \\ \mathbf{c} & \mathbf{d} \end{bmatrix} \begin{bmatrix} \mathbf{P} \\ \mathbf{q} \end{bmatrix}$$

$$= \begin{bmatrix} \mathbf{A} & \mathbf{B} \\ \mathbf{C} & \mathbf{D} \end{bmatrix} \begin{bmatrix} \boldsymbol{\lambda}^+_\tau & \mathbf{O} \\ \mathbf{O} & \boldsymbol{\lambda}^-_\tau \end{bmatrix} \int_0^\tau \begin{bmatrix} \boldsymbol{\lambda}^+_{-s} & \mathbf{O} \\ \mathbf{O} & \boldsymbol{\lambda}^-_{-s} \end{bmatrix} P(s)\, ds \begin{bmatrix} \mathbf{a} & \mathbf{b} \\ \mathbf{c} & \mathbf{d} \end{bmatrix} \begin{bmatrix} \mathbf{P} \\ \mathbf{q} \end{bmatrix} \tag{3.17}$$

If we denote the kth derivative of P by $P^{(k)}$ and define

$$Q^{\pm}_s = \sum \frac{P^{(k)}(s)}{(\pm\, 1_i)^{k+1}} \tag{3.18}$$

(the sum extends from k=0 to k=m) we find

$$\int_0^{\tau} \exp(\pm l_i s)\, P_m(s)\, ds = \exp(\pm l_i s) \sum_{k=0}^{m} \frac{P^{(k)}(s)}{(\pm l_i)^{k+1}} \Bigg|_0^{\tau}$$

$$=: \quad \exp(\pm l_i s)\, Q_s^{\pm} \Big|_0^{\tau} \tag{3.19}$$

(see Gradshteyn and Ryzhik, 1980,p. 92). The last expression defines Q_s. Therefore, in matix form we can write

$$\int_0^{\tau} \begin{bmatrix} \lambda_{-s}^{+} & 0 \\ 0 & \lambda_{-s}^{-} \end{bmatrix} P(s)\, ds = \begin{bmatrix} \lambda_{-\tau}^{+} & 0 \\ 0 & \lambda_{-\tau}^{-} \end{bmatrix} \begin{bmatrix} Q_{\tau}^{-} & 0 \\ 0 & Q_{\tau}^{+} \end{bmatrix} - \begin{bmatrix} Q_0^{-} & 0 \\ 0 & Q_0^{+} \end{bmatrix} \tag{3.19}$$

The substitution of λ^{+} by Γ and subsequent transformations according to the same lines as above then lead to numerically benign expressions for $L^{\pm}$ and the β vectors. Details are found in the appendix.

The applicability will be demonstrated in the next section (4) and its performance will be discussed in Section 6.

4. EXAMPLES FOR THE APPLICATION OF THE MATRIX-EXPONENTIAL METHOD

In this section we give two examples that make use of the largely analytical character of the method presented above. The third example will demonstrate its numerical performance.

4.1. On the appearance of negative elements in the transmission and reflection matrices for spherical configurations.

It was found numerically that in the application of the matrix-exponential method to spherical problems negative elements in the transmission and reflection matrices can occur if the spherical extension is large and the corresponding optical depth is small. Since this problem seems to be present in all methods in which the transfer equation is integrated along fixed μ values (compare Peraiah, 1984) and its relation to the μ discretisation has not yet been discussed in literature, we will derive here the conditions the matrix $\mathbf{A}_\tau$ (Eq. 2.6) has to fulfill so that negative elements are avoided.

For this purpose we express the matrix exponential function by means of the reflection and transmission matrices. From Eqs. (3.10) and (3.14) we find

$$\exp\left\{\mathbf{A}\,\tau\right\} = \begin{bmatrix} \mathbf{u}_\tau & \mathbf{v}_\tau \\ \mathbf{w}_\tau & \mathbf{x}_\tau \end{bmatrix} = \begin{bmatrix} (\mathbf{t}_\tau^+)^{-1} & -(\mathbf{t}_\tau^+)^{-1}\,\mathbf{r}_\tau^+ \\ \mathbf{r}_\tau^-\,(\mathbf{t}_\tau^+)^{-1} & \mathbf{t}_\tau^- - \mathbf{r}_\tau^-\,(\mathbf{t}_\tau^+)^{-1}\,\mathbf{r}_\tau^+ \end{bmatrix} \tag{4.1}$$

For small optical depths (when negative elements are most likely to occur) we may approximate

$$\exp\left\{\mathbf{A}\,\tau\right\} = \mathbf{E} + \mathbf{A}\,\tau = \begin{bmatrix} \mathbf{E} + \boldsymbol{\alpha}\,\tau & \boldsymbol{\beta}\,\tau \\ \boldsymbol{\gamma}\,\tau & \mathbf{E} + \boldsymbol{\delta}\,\tau \end{bmatrix} \tag{4.2}$$

with

$$\mathbf{A} = \begin{bmatrix} \boldsymbol{\alpha} & \boldsymbol{\beta} \\ \boldsymbol{\gamma} & \boldsymbol{\delta} \end{bmatrix} \tag{4.3}$$

By comparison with Eq. (4.1) we obtain

$$\mathbf{t}_\tau^+ = (\mathbf{E} + \boldsymbol{\alpha}\,\tau)^{-1} = \mathbf{E} - \boldsymbol{\alpha}\,\tau \tag{4.4}$$

$$\mathbf{r}_\tau^+ = -(\mathbf{E} + \boldsymbol{\alpha}\,\tau)^{-1}\,\boldsymbol{\beta}\,\tau = -\boldsymbol{\beta}\,\tau \tag{4.5}$$

$$\mathbf{r}_\tau^- = \boldsymbol{\gamma}\,\tau\,(\mathbf{E} + \boldsymbol{\alpha}\,\tau)^{-1} = \boldsymbol{\gamma}\,\tau \tag{4.6}$$

$$\mathbf{t}_\tau^- = \mathbf{E} + \boldsymbol{\delta}\,\tau - \boldsymbol{\gamma}\,\tau\,(\mathbf{E} + \boldsymbol{\alpha}\,\tau)^{-1}\,\boldsymbol{\beta}\,\tau = \mathbf{E} + \boldsymbol{\delta}\,\tau \tag{4.7}$$

The second formulae are derived by using the approximation

$$(\mathbf{E} + \boldsymbol{\alpha}\,\tau)^{-1} = \mathbf{E} - \boldsymbol{\alpha}\,\tau \tag{4.8}$$

and by neglecting quadratic terms in τ.

Eq. (4.4) implies that if all elements of $\mathbf{t}_\tau^+$ are larger than zero, the elements of matrix $\boldsymbol{\alpha}$ must satisfy

$$\alpha_{ij} \begin{cases} < 1/\tau & \text{if } i = j \\ < 0 & \text{if } i \neq j \end{cases} \tag{4.9}$$

and vice versa. In a similar way the conditions for the elements for $\boldsymbol{\beta}$, $\boldsymbol{\gamma}$ and $\boldsymbol{\delta}$ are found so that the matrix $\mathbf{A}$ has to have a structure

$$\mathbf{A} = \left[\begin{array}{c|c} \begin{matrix} & & < 0 \\ & < \frac{1}{\tau} & \\ < 0 & & \end{matrix} & < 0 \\ \hline > 0 & \begin{matrix} & & > 0 \\ & > -\frac{1}{\tau} & \\ > 0 & & \end{matrix} \end{array}\right] \tag{4.10}$$

if all elements of the $\mathbf{t}_\tau$ and $\mathbf{r}_\tau$ matrices are to be larger than zero.

It has been shown (Schmidt, 1985) that these requirements can be met by an appropriate angle discretisation. However, if the additional constraints (as e.g. constant flux for a conservative medium) are taken into account, it turns out that the resulting integration roots and weights may lead to large errors in the scattering term of the transfer equation.

4.2. Perturbation theory

Many complex physical problems can only be approached by a perturbation theory, because e.g. mathematical methods for a solution of the full problem are not available. But even if an exact solution is possible a perturbation theory may offer significant advantages, e.g.

(i) the solution (up to some accuracy) may be obtained faster and/or more easily

(ii) the physics of the system (e.g. interactions between different points of the configuration space, instabilities) may become visible.

(iii) the intrinsic similarity of quite different physical phenomena may become evident.

In this section we want to develop a perturbation theory on the basis of the matrix exponential that clearly demonstrates (ii) and (iii) and in addition provides information on the convergence of iteration schemes.

Let us assume that the matrix $\mathbf{A}_\tau$ (Eq. 2.6) is composed of two terms

$$\mathbf{A}_\tau = \mathbf{A}^0 + \mathbf{A}_\tau^1 \tag{4.11}$$

where the first one is independent of the optical depth and has pairwise distinct eigenvalues (i.e. the conditions for matrix $\mathbf{A}$ of Section 3 are fulfilled) and $\mathbf{A}_\tau^1$ is bounded and integrable. The intensity vector is assumed to be given by a sum

$$\mathbf{I}_\tau = \sum_{i=0}^{\infty} \mathbf{I}_\tau^i \tag{4.12}$$

If we assume in addition that terms $\mathbf{A}_\tau^j\ \mathbf{I}_\tau^{k+1}$ (j=0,1; k=0,1,..) can be neglected when compared with $\mathbf{A}_\tau^j\ \mathbf{I}^k$, we obtain from Eq. (2.6) the following hierarchy

$$\frac{dI_\tau^0}{d\tau} = \mathbf{A}^0\ \mathbf{I}_\tau^0 + \mathbf{B}_\tau \tag{4.13}$$

$$\frac{dI_\tau^l}{d\tau} = \mathbf{A}^0\ \mathbf{I}_\tau^l + \mathbf{A}_\tau^l\ \mathbf{I}_\tau^{l-1} \qquad (l=1,2,..) \tag{4.14}$$

According to Eq. (3.5) they have the solution

$$\mathbf{I}_\tau^0 = \exp(\mathbf{A}^0\ \tau)\ (\mathbf{I}_0 + \int_0^\tau \exp(-\mathbf{A}^0\ s)\ \mathbf{B}_s\ ds \tag{4.15}$$

and

$$\mathbf{I}_\tau^l = \exp(\mathbf{A}^0\ \tau) \int_0^\tau \exp(-\mathbf{A}^l\ s)\ \mathbf{A}_s^l\ \mathbf{I}_s^{l-1}\ ds \quad . \tag{4.16}$$

Note that the boundary condition is satisfied by the zeroth order equation. The resolution of the recurrence formula (4.16) gives

$$\begin{aligned} \mathbf{I}_\tau = \exp\left\{ \mathbf{A}^0\ \tau \right\} \Bigg[\Bigg\{ \mathbf{E} + \int_0^\tau \mathbf{\Psi}_{\tau'} \Big[\mathbf{E} + \int_0^{\tau'} \mathbf{\Psi}_{\tau''} \Big[\mathbf{E} + \int_0^{\tau''} \mathbf{\Psi}_{\tau'''} \Big[\mathbf{E} + \\ \ldots \Big]\Big]\Big] \ldots\ldots d\tau'''\ d\tau''\ d\tau' \Bigg\}\ \mathbf{I}_0^0 \\ + \mathbf{\Omega}_\tau + \int_0^\tau \mathbf{\Psi}_{\tau'}\ \mathbf{\Omega}_{\tau'}\ d\tau' + \int_0^\tau \mathbf{\Psi}_{\tau'} \int_0^{\tau'} \mathbf{\Psi}_{\tau''}\ \mathbf{\Omega}_{\tau''}\ d\tau''\ d\tau' \\ + \int_0^\tau \mathbf{\Psi}_{\tau'} \int_0^{\tau'} \mathbf{\Psi}_{\tau''} \int_0^{\tau''} \mathbf{\Psi}_{\tau'''}\ \mathbf{\Omega}_{\tau'''}\ d\tau'''\ d\tau''\ d\tau' \\ + \ldots\ldots \Bigg] \end{aligned} \tag{4.17}$$

with

$$\Psi_s = \exp(-\mathbf{A}^0 s)\, \mathbf{A}_s^1 \exp(\mathbf{A}^0 s) \tag{4.18}$$

and

$$\Omega_s = \int_0^s \exp(-\mathbf{A}^0 s')\, \mathbf{B}_{s'}\, ds'. \tag{4.19}$$

Following Dyson (1949a, b; see also Schweber, 1961) the first term of Eq. (4.17) may be written as

$$\mathbf{U}_\tau = \exp\left\{ \mathbf{A}^0 \tau \right\} P \exp\left[+ \int_0^\tau \Psi_{\tau'}\, d\tau' \right] \mathbf{I}_0^0 \tag{4.20}$$

where P is the "optical depth ordering operator" i.e. it rearranges products of matrices labeled by the optical depth (e.g. Ψ_s) in the same order as the optical depth sequence of their label, the largest optical depth occuring first in the product.

In a similar way we may write for the second term of Eq. (4.17)

$$\mathbf{V}_\tau = \exp\left\{ \mathbf{A}^0 \tau \right\} \int_0^\tau P \exp\left[\int_0^{\tau'} \Psi_{\tau''}\, d\tau'' \right] \exp(-\mathbf{A}_0 \tau')\, \mathrm{B}_{\tau'}\, d\tau' \tag{4.21}$$

5. GENERAL SOLUTION OF THE TRANSFER EQUATION

In order to derive a general solution in the canonical form (3.13) we subdivide the matrix **A** and the transition matrix **Φ** into blocks referring to the intensities in the positive and negative directions

$$\mathbf{A}_\tau = \begin{bmatrix} \alpha_\tau & \beta_\tau \\ \gamma_\tau & \delta_\tau \end{bmatrix} \tag{5.1}$$

$$\boldsymbol{\Phi}_{\tau,s} = \begin{bmatrix} \mathbf{A}_{\tau,s} & \mathbf{B}_{\tau,s} \\ \mathbf{C}_{\tau,s} & \mathbf{D}_{\tau,s} \end{bmatrix} . \tag{5.2}$$

The inverse of Φ is denoted by

$$\boldsymbol{\Phi}^{-1}_{\tau,s} = \begin{bmatrix} \mathbf{a}_{\tau,s} & \mathbf{b}_{\tau,s} \\ \mathbf{c}_{\tau,s} & \mathbf{d}_{\tau,s} \end{bmatrix} \tag{5.3}$$

Since the transition matrix has the properties

$$\boldsymbol{\Phi}_{t,\tau} \cdot \boldsymbol{\Phi}_{\tau,t_0} = \boldsymbol{\Phi}_{t,t_0} \tag{5.4}$$

$$\boldsymbol{\Phi}^{-1}_{t,t_0} = \boldsymbol{\Phi}_{t_0,t} \tag{5.5}$$

(note that Eqs. (5.4), (5.5) and (2.10) imply that the transition matrices form a group) we can write Eq. (2.8)

$$\begin{bmatrix} \mathbf{I}^+_\tau \\ \mathbf{I}^-_\tau \end{bmatrix} = \begin{bmatrix} \mathbf{A}_{\tau,0} & \mathbf{B}_{\tau,0} \\ \mathbf{C}_{\tau,0} & \mathbf{D}_{\tau,0} \end{bmatrix} \left\{ \begin{bmatrix} \mathbf{I}^+_0 \\ \mathbf{I}^-_0 \end{bmatrix} + \int_0^\tau \begin{bmatrix} \mathbf{a}_{s,0} & \mathbf{b}_{s,0} \\ \mathbf{c}_{s,0} & \mathbf{d}_{s,0} \end{bmatrix} \begin{bmatrix} \mathbf{B}^+_s \\ \mathbf{B}^-_s \end{bmatrix} ds \right\} \tag{5.6}$$

Since now all second arguments of the transition (sub-) matrices are zero we will skip them subsequently. A straightforward rearrangement of the terms gives

$$\begin{bmatrix} \mathbf{I}^+_0 \\ \mathbf{I}^-_\tau \end{bmatrix} = \begin{bmatrix} \mathbf{A}^{-1}_\tau & -\mathbf{A}^{-1}_\tau \mathbf{B}_\tau \\ \mathbf{C}_\tau \mathbf{A}^{-1}_\tau & \mathbf{D}_\tau - \mathbf{C}\mathbf{A}^{-1}_\tau \mathbf{B}_\tau \end{bmatrix} \begin{bmatrix} \mathbf{I}^+_\tau \\ \mathbf{I}^-_0 \end{bmatrix} + \begin{bmatrix} -\mathbf{A}^{-1}_\tau & \mathbf{0} \\ -\mathbf{C}_\tau \mathbf{A}^{-1}_\tau & \mathbf{E} \end{bmatrix} \begin{bmatrix} \boldsymbol{\Gamma}^+ \\ \boldsymbol{\Gamma}^- \end{bmatrix} \tag{5.7}$$

with

$$\begin{bmatrix} \Gamma^+ \\ \Gamma^- \end{bmatrix} = \begin{bmatrix} \int_0^\tau \{\mathbf{A}_\tau\ (\mathbf{a}_s-\mathbf{b}_s) + \mathbf{B}_\tau\ (\mathbf{c}_s-\mathbf{d}_s)\}\ \boldsymbol{\zeta}_s\ \mathbf{Q}_s\ ds \\ \int_0^\tau \{\mathbf{C}_\tau\ (\mathbf{a}_s-\mathbf{b}_s) + \mathbf{D}_\tau\ (\mathbf{c}_s-\mathbf{d}_s)\}\ \boldsymbol{\zeta}_s\ \mathbf{Q}_s\ ds \end{bmatrix} \tag{5.8}$$

and

$$\begin{bmatrix} \mathbf{B}_s^+ \\ \mathbf{B}_s^- \end{bmatrix} = \begin{bmatrix} \boldsymbol{\zeta}_\tau & \mathbf{O} \\ \mathbf{O} & -\boldsymbol{\zeta}_\tau \end{bmatrix} \begin{bmatrix} \mathbf{Q}_\tau \\ \mathbf{Q}_\tau \end{bmatrix} \tag{5.9}$$

By comparison with the canonical equation (Eq. 3.13) we find

$$\begin{aligned} \mathbf{t}^+ &= \mathbf{A}_\tau^{-1} & \mathbf{r}^+ &= -\mathbf{A}_\tau^{-1}\ \mathbf{B}_\tau \\ \mathbf{r}^- &= \mathbf{C}_\tau\ \mathbf{A}_\tau^{-1} & \mathbf{t}^- &= \mathbf{D}_\tau - \mathbf{C}_\tau\ \mathbf{A}_\tau^{-1}\ \mathbf{B}_\tau \end{aligned} \tag{5.10}$$

and

$$\boldsymbol{\beta}^+ = -\mathbf{A}_\tau^{-1}\ \Gamma^+ \qquad \boldsymbol{\beta}^- = -\mathbf{C}_\tau\ \mathbf{A}_\tau^{-1}\ \Gamma^+ + \Gamma^- \tag{5.11}$$

These equations are completely analogous to those of Eq. (3.14). Note however, that right hand sides of Eqs. (3.14) are known explicitly from Eq. (3.10) whereas in this general case they are determined by the differential equation (2.9). We transform them with Eqs. (5.1), (5.2), and (2.9) into differential equations for the transmission and reflection matrices

$$\frac{d}{d\tau}\ \mathbf{t}_\tau^- = \boldsymbol{\delta}_\tau\ \mathbf{t}_\tau^- - \mathbf{r}_\tau^-\ \boldsymbol{\beta}_\tau\ \mathbf{t}_\tau^- \tag{5.12}$$

$$\frac{d}{d\tau}\ \mathbf{r}_\tau^+ = -\ \mathbf{t}_\tau^+\ \boldsymbol{\beta}_\tau\ \mathbf{t}_\tau^- \tag{5.13}$$

$$\frac{d}{d\tau}\ \mathbf{r}_\tau^- = \boldsymbol{\gamma}_\tau+\boldsymbol{\delta}_\tau\ \mathbf{r}_\tau^- - \mathbf{r}_\tau^-\ \boldsymbol{\beta}_\tau\ \mathbf{r}_\tau^- - \mathbf{r}_\tau^-\ \boldsymbol{\alpha}_\tau \tag{5.14}$$

$$\frac{d}{d\tau}\ \mathbf{t}_\tau^+ = -\mathbf{t}_\tau^+\ \boldsymbol{\alpha}_\tau - \mathbf{t}_\tau^+\ \boldsymbol{\beta}_\tau\ \mathbf{r}_\tau^- \tag{5.15}$$

with the initial conditions at $\tau = 0$

$$\mathbf{t}_0^+ = \mathbf{t}_0^- = \mathbf{E} \tag{5.16}$$

$$\mathbf{r}_0^+ = \mathbf{r}_0^- = \mathbf{0} \tag{5.17}$$

The source terms are now given by

$$\begin{aligned}\boldsymbol{\beta}_\tau^+ &= \int_0^\tau ((\mathbf{a}_s - \mathbf{b}_s) - \mathbf{r}_\tau^+(\mathbf{d}_s - \mathbf{c}_s))\, \boldsymbol{\zeta}_s\, \mathbf{Q}_s\, ds \\ &= \int_0^\tau ((\mathbf{r}_s^+ - \mathbf{r}_\tau^+)(\mathbf{t}_s^-)^{-1}(\mathbf{r}_s^- + \mathbf{E}) - \mathbf{t}_s^+)\boldsymbol{\zeta}\, \mathbf{Q}_s\, ds\end{aligned} \tag{5.18}$$

$$\begin{aligned}\boldsymbol{\beta}_\tau^- &= \int_0^\tau \mathbf{t}_\tau^-(\mathbf{c}_s - \mathbf{d}_s)\, \boldsymbol{\zeta}_s\, \mathbf{Q}_s\, ds \\ &= -\int_0^\tau \mathbf{t}_\tau^-(\mathbf{t}_s^-)^{-1}(\mathbf{r}_s^- + \mathbf{E})\, \boldsymbol{\zeta}_s\, \mathbf{Q}_s\, ds\end{aligned} \tag{5.19}$$

whereas the Eqs. (5.12) - (5.15) are already in a form suitable for numerical integration (see e.g. example of the next section), Eqs. (5.18) and (5.19) cannot be evaluated immediately since the norm of $(\mathbf{t}_s^-)^{-1}$ tends to infinity for $s \to \infty$ and since the cancellations in the term $(\mathbf{r}_s^+ - \mathbf{r}_\tau^+)$ can lead to unacceptable errors. In addition, Eqs. (5.18) and (5.19) require that $\mathbf{t}_s^\pm$ and $\mathbf{r}_s^\pm$ be stored at many intermediate s points.

Therefore we define

$$\mathbf{G}_s = \mathbf{t}_\tau^-(\mathbf{t}_s^-)^{-1} \tag{5.20}$$

$$\mathbf{H}_s = (\mathbf{r}_s^+ - \mathbf{r}_\tau^+)\ (\mathbf{t}_s^-)^{-1} \tag{5.21}$$

for which from Eqs.(5.12) to (5.15) the differential equations are found

$$\frac{d}{ds}\mathbf{G}_s = -\mathbf{G}_s\ (\boldsymbol{\delta}_s - \mathbf{r}_s^-\ \boldsymbol{\beta}_s\ ) \tag{5.22}$$

$$\frac{d}{ds}\mathbf{H}_s = -\mathbf{t}_s^+\ \boldsymbol{\beta}_s - \mathbf{H}_s\ (\boldsymbol{\delta}_s - \mathbf{r}_s^-\ \boldsymbol{\beta}_s) \tag{5.23}$$

The initial conditions are

$$\mathbf{G}_\tau = \mathbf{E} \qquad \mathbf{H}_0 = -\mathbf{r}_\tau^- \tag{5.24}$$

Note that the matrices $\mathbf{G}_s$ and $\mathbf{H}_s$ are well behaved: The norm is always between zero and one so that the integration does not give rise to problems. We derive now the source terms from the differential equations

$$\frac{d}{ds}\boldsymbol{\beta}_s^+ = (\mathbf{H}_s\ (\mathbf{E} + \mathbf{r}_s^-) - \mathbf{t}_s^+\ )\ \boldsymbol{\zeta}_s\ \mathbf{Q}_s \tag{5.25}$$

(initial condition $\boldsymbol{\beta}_\tau^+ = \mathbf{0}$) and

$$\frac{d}{ds}\boldsymbol{\beta}_s^- = -\ \mathbf{G}_s\ (\ \mathbf{r}_s^- + \mathbf{E}\ )\ \boldsymbol{\zeta}_s\ \mathbf{Q}_s \tag{5.26}$$

(initial condition $\boldsymbol{\beta}_0^- = \mathbf{0}$)

In the actual numerical calculations (Section 6) we have first integrate Eqs. (5.12) to (5.15) simultaneously. Subsequently, we evaluate Eqs. (5.23) and (5.26) and recalculate $\mathbf{r}_s^-$ and $\mathbf{t}_s^+$ (in order to avoid storing these matrices at many depth points). Finally, $\boldsymbol{\beta}_s^+$ and $\mathbf{G}_s$ are determined by integrating in the opposite direction ($\mathbf{r}_s^-$ is again recalculated for the same reason as above).

We will discuss the details of the numerical performance of this method in Section 7.

6. TWO EXAMPLES FOR THE APPLICATION OF THE GENERAL METHOD

6.1. The outer boundary temperature of a slab in radiative equilibrium

In this example we want to study for a grey atmosphere with optical depth T (finite) and flux F = const. the dependence of the outer boundary temperature on the center-to-limb-variation of the radiation impinging at the inner boundary. This is of relevance for configurations where an optically thin (to moderately thick) layer with no energy sources or sinks is on top of a region whose structure is not determined by the radiation field; such a situation may e.g. be found in cool white dwarfs and accretion discs.

Since in this case the Planck function is replaced by the mean intensity because of energy conservation the transfer equation (2.1) has to be solved with

$$\alpha = 1 \qquad \beta = 0$$
$$R = \frac{1}{2}\,\delta\,(\nu - \nu') \tag{6.1}$$

and subject to the boundary condition

$$I_0^- = 0 \qquad I_T^+ = I_{bc}^+ \text{ (given)} \tag{6.2}$$

The temperature Θ is then given by the condition that the energy absorbed (per unit volume and second)

$$\kappa\ J = \kappa\ \boldsymbol{\varphi}\ (\ I^+ + I^-\ ) \tag{6.3}$$

(κ = absorption coefficient, J = mean intensity, $\boldsymbol{\varphi}$ = row vector containing the weights for the angle integration) equals the emitted energy

$$\kappa B = \kappa \ \sigma \ \theta^4 \tag{6.4}$$

(B = Planck function in this Section). Therefore we have with Eq. (6.2)

$$\sigma \ \theta_T^4 = \boldsymbol{\Phi} \ (\mathbf{E} + \mathbf{r}_T^+) \ \mathbf{I}_{bc}^+ \tag{6.5}$$

and

$$\sigma \ \theta_0^4 = \boldsymbol{\Phi} \ \mathbf{t}_T^+ \ \mathbf{I}_{bc}^+ \tag{6.6}$$

We employ two angles per half-sphere (μ_1 = 2.11324866e-1 and $\mu_2 = 1 - \mu_1$ according to a Gaussian division, Abramowitz and Stegun, 1965) and denote by Z the ratio $I_{bc,1}^+/I_{bc,2}^+$.

Since the weights for both angles are equal we find

$$\rho =: \left[\frac{\theta_0^4}{\theta_T^4}\right] = \frac{(t_{11}^+ + t_{21}^+) \ Z + (t_{12}^+ + t_{22}^+)}{(1 + r_{11}^+ + r_{21}^+) \ Z + (1 + r_{12}^+ + r_{22}^+)} \tag{6.7}$$

Note that for an infinite atmosphere this ratio is

$$\left[\frac{\theta_0^4}{\theta_T^4}\right]_{\text{infinite}} = \frac{2/3}{T + 2/3} \tag{6.8}$$

The numerical calculations consist essentially of the integration of 16 coupled differential equations (eqs. 5.12 to 5.15) since we did not take advantage of symmetry relations. We use a Runge-Kutta-Fehlberg algorithm (order O(6)) with automatic step adjustment (Engeln-Müllges and Reutter, 1985) assuring the absolute and relative errors to be less than 1.e-4. On an ATARI 520 personal computer the typical CPU time

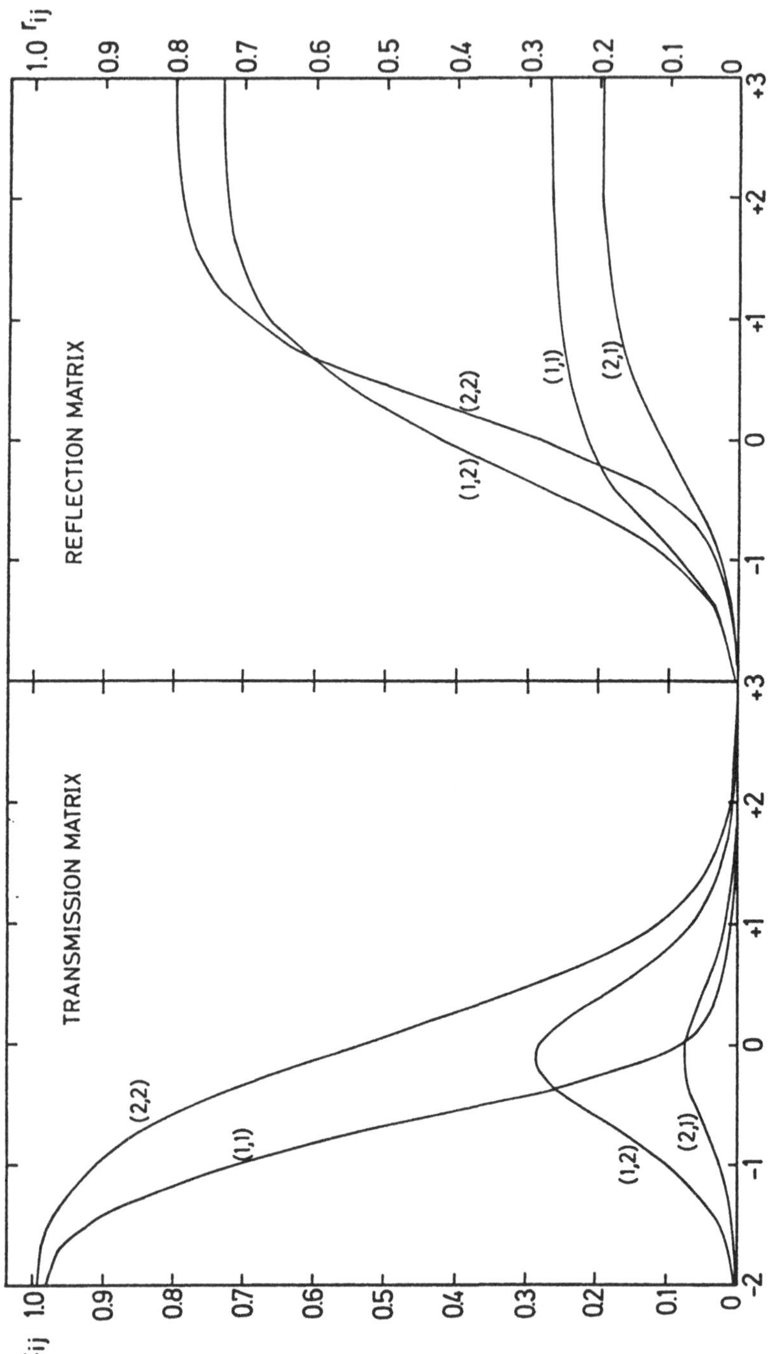

Fig. 1: The depth dependence of the elements of the transmission and reflection matrices for the problem discussed in Section 6.1. Qualitatively, the same behaviour is found in all transfer cases.

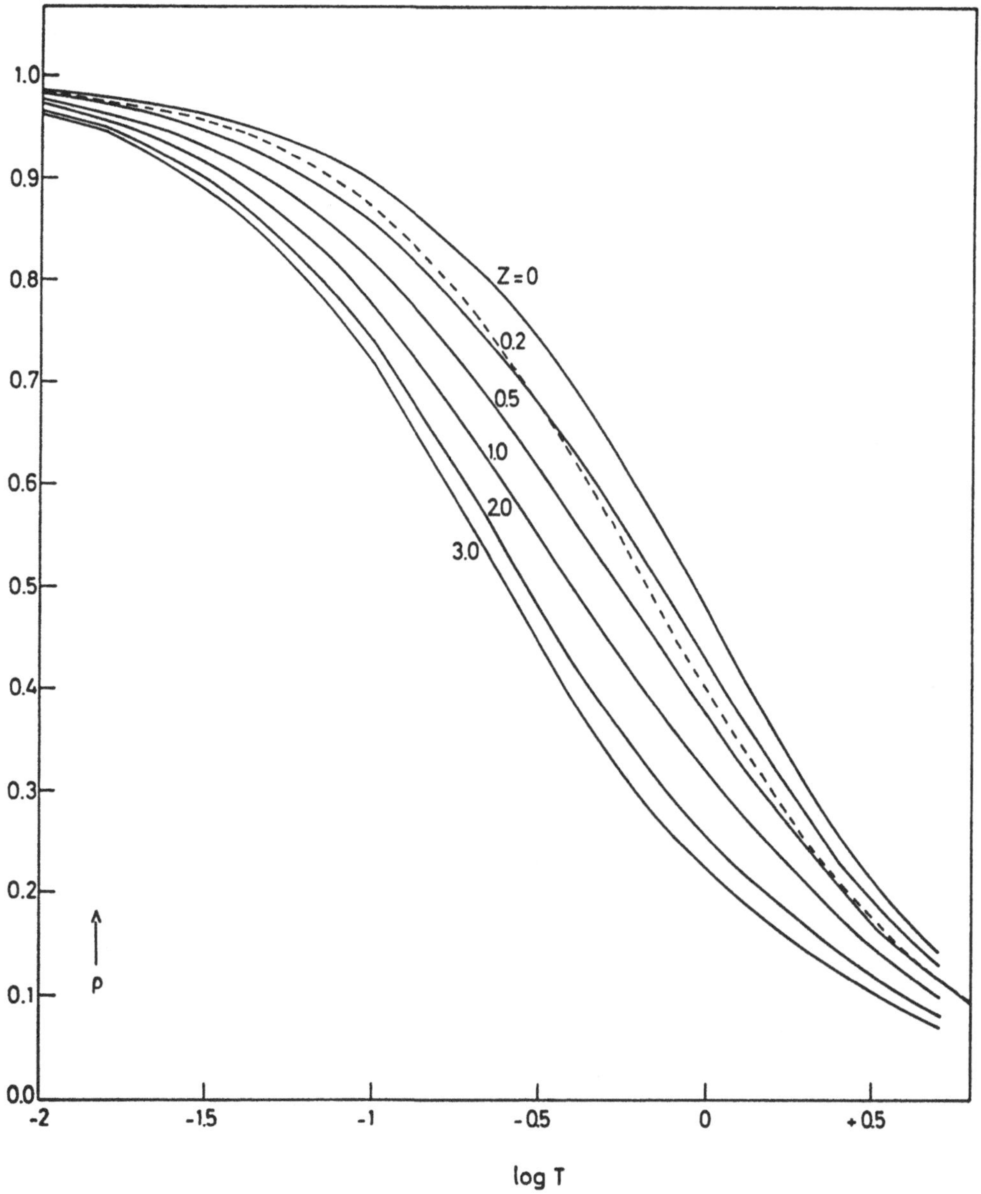

Fig. 2: The ratio ρ of the fourth powers of the temperature at optical depths $\tau = 0$ and $\tau = T$ for a finite grey atmosphere which is illuminated from the bottom by the intensities $T^{+}_{bc,1}$ and $T^{+}_{bc,2} = I^{+}_{bc,1}/Z$ as a function of T and Z (full lines). For comparison, the broken line gives the corresponding ratio for an infinite atmosphere (cf. Eq. 6.8).

amounts to a few minutes for a T value.

Fig. 1 shows the run of the elements of the transmission and reflection matrices with the total thickness T. Note that

$$\mathbf{t}_T^+ = \mathbf{t}_T^- \qquad \text{and} \qquad \mathbf{r}_T^+ = \mathbf{r}_T^-$$

since $\tau = T/2$ is a symmetry plane. It is seen that the diagonal elements of **t** decrease monotonically and reach very small values for moderate T. The off-diagonal terms t_{12}^+ and t_{21}^+ reach maxima and then fall off considerably more slowly than t_{11}^+ and t_{22}^+. All elements of the reflection matrix increase monotonically and tend for large optical depths to finite asymptotic values. Qualitatively, such a behaviour is found for transmission and reflection matrices in all transfer problems.

In Fig. 2 we plot $\rho(T)$ for various values of Z. It is seen that for small and large optical depths ($T \lesssim 0.03$ and $T \gtrsim 3$,resp.) all curves are rather close and therefore Z hardly affects the temperature ratio. In the intermediate region, however, the curves split considerably which means that temperature structure and possibly the stability of the radiative zone are significantly influenced by the angle distribution of the radiation entering at the bottom.,

6.2. Polarized radiation from a hot and dense plasma with a strong magnetic field

In this example we consider a plane parallel medium of high temperature (several KeV) and density (some 1.e15 cm^{-1}) which is under the influence of a strong magnetic field ($B \approx$ 1.e7 G) in the normal direction. Such conditions are found e.g. in the accretion columns of AM Her type objects (Wickramasinghe and Meggitt, 1985). The opacity is then essentially given by cyclotron and free-free absorption and electron scattering. Note that cyclotron absorption is extremely anisotropic. For the calculation of the radiation field we use a normal mode description (Gnedin and Pavlov, 1974, see also Kaminker, Pavlov, and Shibanov, 1982) in which the transfer equation can be written

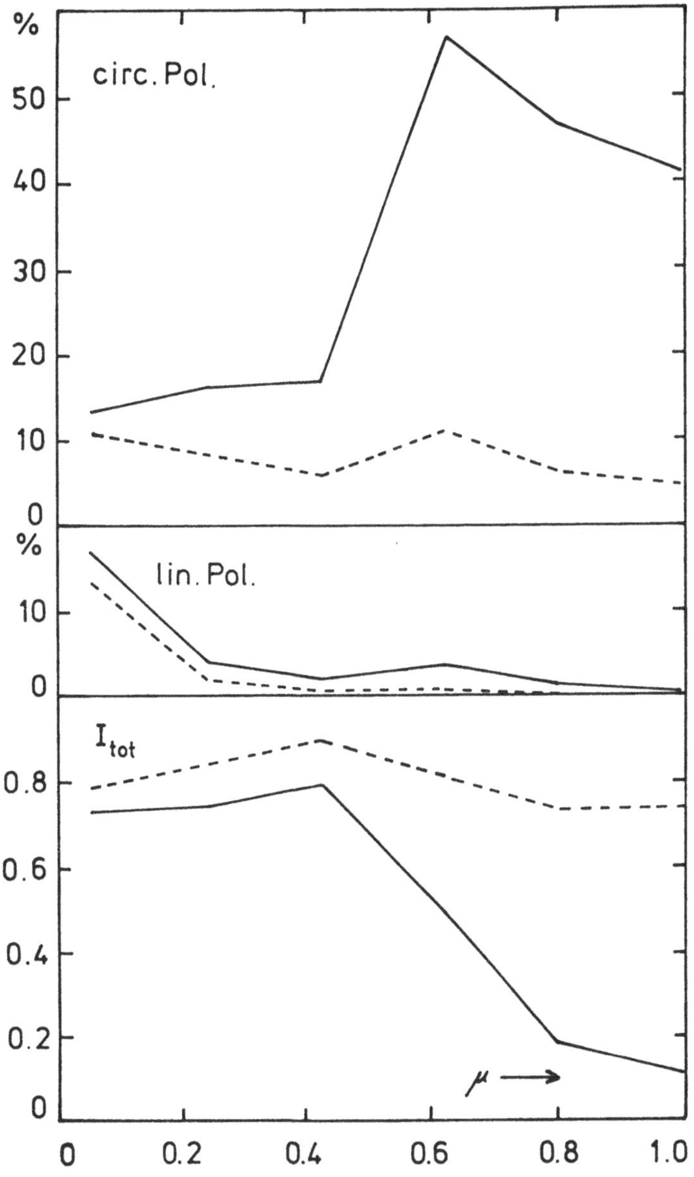

Fig. 3: Two examples for the angle dependence of the circular polarisation (upper panel), the linear polarsation (central panel), and the total specific intensitiy (in arbitrary units; lower panel) for plane parallel magnitized slabs of thicknesses 1.e8 cm (full lines) and 6.e9 cm (broken lines). Both cases are calculated for an harmonic number ω/ω_c = 8 (ω_c = cyclotron frequency), a temperature of 5 KeV, an electron density N_e = 1.e15, and a magnetic field B = 3.e7 G, typical values for accretion columns in AM Her systems.

$$\mu_k \frac{d}{dz} I_{j,k} = - (\kappa_j + \sigma_j) I_{j,k}$$

$$+ \sum_i \sum_l \sigma_{j,i,k,l} I_{i,l} \varphi_l + \kappa_j B \quad (6.9)$$

with the first index of the intensity I referring to the propagation mode (i.e. j=1 corresponds to the extraordinary ray and j=2 to the ordinary one) and the second to the angle. κ_j and σ_j are the total absortion and scattering coefficients for mode j. $\sigma_{j,i,k,l}$ describes the scattering of radiation from mode i and direction l to mode j and direction k. The coefficients are calculated following Kaminker et al. (1982) and Wickramasinghe and Meggitt (1985). Eq. (6.9) is of the form (2.6) and therefore can be solved by the formalism presented above.

In Fig. 3 we show the resulting angle dependence of the circular and linear polarisation and the total intensity for two cases of different depths. With 12 angle points (full sphere) the computer times using a non-optimized program on an IBM 3090 without vector processor ranged between 40 sec and 15 min for relative and absolute precisions of 1.e-8. The calculations are in very good agreement with those of Wickramasinghe, Meggitt, and Wehrse (1987) and confirm their result that for such conditions electron scattering may have strong effects on the emerging intensities and the degrees of polarisation and must be considered explicitely.

7. DISCUSSION

Since the matrix exponential method has a performance quite different from that of the general formalism, let us discuss this special case first: As can be seen from the two examples of Chapter 4 it is particularly suited for analytical calculations and this can give insight into the physics of line and continuum formation in various situations. In addition, it shows clearly the relation of radiative transfer to quantum electro dynamics, theory of molecular vibrations, solid state physics (cf. Abrikosov et al., 1974) etc. where one has to deal with similar systems of equations.

If a particular problem meets the necessary conditions for the matrix $\mathbf{A}_\tau$ (see Chapter 2) the matrix exponential method can also be used with advantage in the numerical solution : (i) The storage requirements are quite moderate; (ii) the computing time is independent of the optical depth; (iii) since very fast and accurate algorithms for the determination of eigenvalues and -vectors have been developped (cf. Engeln-Müllges and Reutter, 1985) the method is fast; (iv) since no iterations nor operations with ill-conditioned matrices are involved, a very high accuracy can be achieved essentially without any expense.

The limitations of this method are given by the restrictions that have to be imposed on matrix $\mathbf{A}_\tau$. It seems therefore not possible to treat polarisation or pure scattering problems since they imply two identical eigenvalues so that $\mathbf{A}_\tau$ cannot be transformed to diagonal form (i.e. the normal form contains Jordan boxes). Difficulties are also encountered if D, α, or R (eq. 2.1) depend on the optical depth. In these cases, however, the medium may be subdivided into a number of layers for which the depth dependences of D, α and R can be approximated by a suitable average value. The radiation field can then be determined from the $\mathbf{A}$ and $\mathbf{r}$ matrices and source terms by requiring continuity of the specific intensities at the boundaries of the layers (see e.g. Peraiah, 1984).

From this discussion it follows that the main fields of application for this method seem to be analytical investigations (as e.g. the study of the influence of the redistribution functions, stochastic radiative transfer, cf. Gierens, Traving and Wehrse, 1987; etc.) and numerical problems, for which the depth structures are relatively simple, but very high accuracies are required.

The main advantage of the general method is that it does not suffer from the limitations of identical eigenvalues or non-vanishing commutators: absorption coefficients, profile functions, de-excitation parameters etc. may depend on depth in an arbitrary way. Even the frequency and angle discretisation may change with τ so that negative elements in the reflection and transmission matrices (cf. Section 4.1) can be avoided. In addition, overdeterminations within the equations (that occurs e.g. in the Stokes description of polarized radiation) do not require any special precautions.

On scalar machines the computing time requirements for this general approach scale with the cube of the number of frquency-angle points n_{af}. On vector or parallel processors the time is essentially determined by the number of remaining scalar operations, which is rather small and is about proportional to n_{af}. In addition, the time is proportional to the total optical depth, which must be regarded as the main disadvantage for optically very thick configurations, in particular since the stepsize of the integration (Eqs. 5.12 - 5.15, 5.22, 5.23) for prescribed absolute and relative errors is determined by the component that just gets optically thick. However, in actual calculations with the highly efficient Runge-Kutta-Fehlberg algorithm it did not turn out to be a severe problem, even though the optical depths for different angle-frequencies differed by a factor larger than 1.e4.

Since in this method the solution of the transfer equation is obtained without the use of particular properties of the absorption coefficients, redistribution function etc., it is expected that it will mainly be employed in radiation problems with many couplings between the components and complex depth dependences, and in cases, for which a given prescribed error must not be exceeded and for which the time consumption is not critical

ACKNOWLEDGEMENTS

We are grateful to B. Baschek, K. Gierens, and G. Traving for very helpful discussions. R.W. thanks the Centre for Mathematical Analysis at the Australian National University for the hospitality during a 2 month visit during which the numerical properties of the general method were investigated. Illuminating discussions with D.T. Wickramasinghe on polarization are particularly acknowlegded.
We are indebted to B. Hoffmann and R. Plate for their help in the preparation of the manuscript. This work was supported by the Deutsche Forschungsgemeinschaft (SFB 132).

REFERENCES

Abrikosov, A.A., Gorkov, L.P., Dzyaloshinski, I.E. 1974, Methods of Quantum Field Theory in Staistical Physics, Dover, New York.

Bronson, R. 1970, Matrix Methods, Academic Press, New York.

Cannon, C.J. 1984, The Transfer of Spectral Line Radiation, Cambridge University Press, Cambridge.

Chandrasekhar, S. 1950, Radiative Transfer, Dover, New York.

Dyson,J.F. 1949a, Phys. Rev. **75**, 486.

Dyson, J.F. 1949b, Phys. Rev. **75**, 1736.

Engeln-Müllges, G., Reutter, F. 1985, Formelsammlung zur Numerischen Mathematik, Bibliographisches Institut, Mannheim.

Gierens, K, Traving, R., Wehrse, R. 1987, J. Q. S.R.T., in press.

Gnedin, Yu. N., Pavlov, G.G. 1974, Sov. Phys. JETP **38**, 903.

Gradshteyn, I.S., Ryzhik, I,M. 1980, Tables of Integrals, Series and Products, Academic Press, New York.

Kaminker, A.D., Pavolov, G.G, Shibanov, Yu.A. 1982, Astrophys. Sp. Sc. **86**, 249.

Kourganoff, V. 1963, Basic Methods in Transfer Problems, Dover, New York.

Mihalas, D. 1978, Stellar Atmospheres, Freeman, San Francisco.

Peraiah, A 1984, in: Methods in Radiative Transfer, Kalkofen, W., ed., Cambridge University Press, Cambridge.

Schmidt, M. 1985, Diploma Thesis, Heidelberg University

Schweber, S.S. 1961, An Introduction to Realtivistic Quantum Field Theory, Harper and Row, New York.

Wehrse, R. 1985, in: Progress in Stellar Spectral Line Formation Theory, Beckman, J.E., Crivellari, L., eds., Reidel, Dordrecht.

Wehrse, R., Kalkofen, W. 1985, Astron. Astrophys. **147**, 71.

Wickramasinghe, D.T., Meggitt, S.M.A. 1985, Mon. Not. Roy. Astr. Soc. **214**, 605.

Wickramasinghe, D.T., Meggitt, S.M.A., Wehrse, R. 1987, Astrophys. Sp. Sc , **131**, 571.

Zurmühl, R. 1964, Matrizen, Springer, Berlin.

APPENDIX

CALCULATION OF THE TRANSMISSION AND REFLECTION MATRICES FOR THE MATRIX-EXPONENTIAL METHOD

In terms of the submatrices **A**, **B**, ...**c**, **d** (see Eqs. 3.9 and 3.10) the transmission and reflection matrices (Eq. 3.14) can be written

$$\mathbf{t}_\tau^+ = (\mathbf{A}\,\lambda_\tau^+\,\mathbf{a} + \mathbf{B}\,\lambda_\tau^-\,\mathbf{c})^{-1}$$

$$\mathbf{r}_\tau^+ = -(\mathbf{A}\,\lambda_\tau^+\,\mathbf{a} + \mathbf{B}\,\lambda_\tau^-\,\mathbf{c})^{-1}\,(\mathbf{A}\,\lambda_\tau^+\,\mathbf{b} + \mathbf{B}\,\lambda_\tau^-\,\mathbf{d})$$

$$\mathbf{t}_\tau^- = \mathbf{C}\,\lambda_\tau^+\,\mathbf{b} + \mathbf{D}\,\lambda_\tau^-\,\mathbf{d} - (\mathbf{C}\,\lambda_\tau^+\,\mathbf{a} + \mathbf{D}\,\lambda_\tau^+\,\mathbf{c})$$

$$(\mathbf{A}\,\lambda_\tau^+\,\mathbf{a} + \mathbf{B}\,\lambda_\tau^-\,\mathbf{c})^{-1}\,(\mathbf{A}\,\lambda_\tau^+\,\mathbf{b} + \mathbf{B}\,\lambda_\tau^-\,\mathbf{d})$$

$$\mathbf{r}_\tau^- = (\mathbf{C}\,\lambda_\tau^+\,\mathbf{a} + \mathbf{D}\,\lambda_\tau^-\,\mathbf{c})\,(\mathbf{A}\,\lambda_\tau^+\,\mathbf{a} + \mathbf{B}\,\lambda_\tau^-\,\mathbf{c})^{-1} \tag{A.1}$$

In order to obtain numerically benign expressions we replace the λ^+ matrix by its inverse Γ_τ and cast the resulting equations in a form that for $\tau \to \infty$ singular terms are avoided, e.g.

$$\mathbf{t}_\tau^+ = (\mathbf{A}\,\lambda_\tau^+\,\mathbf{a} + \mathbf{B}\,\lambda_\tau^-\,\mathbf{c})^{-1}$$

$$= (\mathbf{A}\,\lambda_\tau^+\,(\mathbf{a} + \Gamma_\tau\,\mathbf{A}^{-1}\,\mathbf{B}\,\lambda_\tau^-\,\mathbf{c}))^{-1}$$

$$= (\mathbf{a} + \Gamma_\tau\,\mathbf{A}^{-1}\,\mathbf{B}\,\lambda_\tau^-\,\mathbf{c})^{-1}\,(\mathbf{A}\,\lambda_\tau^+)^{-1}$$

$$= (\mathbf{a} + \Gamma_\tau\,\mathbf{A}^{-1}\,\mathbf{B}\,\lambda_\tau^-\,\mathbf{c})^{-1}\,\Gamma_\tau\,\mathbf{A}^{-1} \tag{A.2}$$

After similar transformations we find for the remaining matrices

$$\mathbf{r}_\tau^+ = -\,(\mathbf{a} + \Gamma_\tau\,\mathbf{A}^{-1}\,\mathbf{B}\,\lambda_\tau^-\,\mathbf{c})^{-1}\,(\mathbf{b} + \Gamma_\tau\,\mathbf{A}^{-1}\,\mathbf{B}\,\lambda_\tau^-\,\mathbf{d}) \tag{A.3}$$

$$\mathbf{r}_\tau^- = (\mathbf{C} + \mathbf{D}\,\boldsymbol{\lambda}_\tau^-\,\mathbf{c}\,\mathbf{a}^{-1}\,\Gamma_\tau)\,(\mathbf{A} + \mathbf{B}\,\boldsymbol{\lambda}_\tau^-\,\mathbf{c}\,\mathbf{a}^{-1}\,\Gamma_\tau) \tag{A.4}$$

$$\mathbf{t}_\tau^- = \mathbf{C}\,(\mathbf{A} + \mathbf{B}\,\boldsymbol{\lambda}_\tau^-\,\mathbf{c}\,\mathbf{a}^{-1}\,\Gamma_\tau)^{-1}\,\mathbf{B}\,\boldsymbol{\lambda}_\tau^-\,(\mathbf{c}\,\mathbf{a}^{-1}\,\mathbf{b} - \mathbf{d})$$

$$+ \mathbf{D}\,\boldsymbol{\lambda}_\tau^-\,(\mathbf{d} - \mathbf{c}\,(\mathbf{a} + \Gamma_\tau\,\mathbf{A}^{-1}\,\mathbf{B}\,\boldsymbol{\lambda}_\tau^-\,\mathbf{c})^{-1}\,(\mathbf{b} + \Gamma_\tau\,\mathbf{A}^{-1}\,\mathbf{B}\,\boldsymbol{\lambda}_\tau^-\,\mathbf{d})) \tag{A.5}$$

For a polynomial depth dependence of the Planck function (cf. eq. 3.16) numerically benign expressions for the source vectors $\boldsymbol{\beta}_\tau^+$ and $\boldsymbol{\beta}_\tau^-$ (eq. 3.13) can be obtained in the same way:

$$\begin{aligned}\boldsymbol{\beta}_\tau^+ = -(\mathbf{a} + \Gamma_\tau\,\mathbf{A}^{-1}\,\mathbf{B}\,\boldsymbol{\lambda}_\tau^-\,\mathbf{c})^{-1}\,((\Gamma_\tau\,\mathbf{Q}_\tau^- - \mathbf{Q}_0^-)\,(\mathbf{a}\,\mathbf{p} + \mathbf{b}\,\mathbf{q})\\ + \Gamma_\tau\,\mathbf{A}^{-1}\,\mathbf{B}\,(\mathbf{Q}_\tau^+ - \boldsymbol{\lambda}_\tau^-\,\mathbf{Q}_0^+)\quad(\mathbf{c}\,\mathbf{p} + \mathbf{d}\,\mathbf{q}))\end{aligned} \tag{A.6}$$

$$\begin{aligned}\boldsymbol{\beta}_\tau^- = (\mathbf{C}\,\mathbf{A}^{-1}\,\mathbf{B} - \mathbf{D})\,\boldsymbol{\lambda}_\tau^-\,\mathbf{c}\,\mathbf{a}^{-1}\,(\Gamma_\tau\,(\mathbf{E} + \mathbf{A}^{-1}\,\mathbf{B}\,\boldsymbol{\lambda}_\tau^-\,\mathbf{c}\,\mathbf{a}^{-1}\,\Gamma_\tau)^{-1}\\ \mathbf{Q}_\tau^- - (\mathbf{E} + \Gamma_\tau\,\mathbf{A}^{-1}\,\mathbf{B}\,\boldsymbol{\lambda}_\tau^-\,\mathbf{c}\,\mathbf{a}^{-1})^{-1}\,\mathbf{Q}_0^-)\,(\mathbf{a}\,\mathbf{p} + \mathbf{b}\,\mathbf{q})\\ + (\mathbf{c} - (\mathbf{C} + \mathbf{D}\,\boldsymbol{\lambda}_\tau^-\,\mathbf{c}\,\mathbf{a}^{-1}\,\Gamma_\tau)\,(\mathbf{A} + \mathbf{B}\,\boldsymbol{\lambda}_\tau^-\,\mathbf{c}\,\mathbf{a}^{-1}\,\Gamma_\tau)^{-1}\,\mathbf{B})\\ (\mathbf{Q}_\tau^+ - \boldsymbol{\lambda}_\tau^-\,\mathbf{Q}_0^+)\,(\mathbf{c}\,\mathbf{p} + \mathbf{d}\,\mathbf{q})\end{aligned} \tag{A.7}$$

Index